# MICROLIVESTOCK

## Little-Known Small Animals with a Promising Economic Future



Board on Science and Technology for International Development

National Research Council



NATIONAL ACADEMY PRESS
Washington, D.C. 1991

NOTICE: The project that is the subject of this report was approved by the Governing Board of the National Research Council, whose members are drawn from the councils of the National Academy of Sciences, the National Academy of Engineering, and the Institute of Medicine. The members of the committee responsible for the report were chosen for their special competence and with regard for appropriate balance.

This report has been reviewed by a group other than the authors according to procedures approved by a Report Review Committee consisting of members of the National Academy of Sciences, the National Academy of Engineering, and the Institute of Medicine.

The National Academy of Sciences is a private, nonprofit, self-perpetuating society of distinguished scholars engaged in scientific and engineering research, dedicated to the furtherance of science and technology and to their use for the general welfare. Upon the authority of the charter granted to it by the Congress in 1863, the Academy has a mandate that requires it to advise the federal government on scientific and technical matters. Dr. Frank Press is president of the National Academy of Sciences.

The National Academy of Engineering was established in 1964, under the charter of the National Academy of Sciences, as a parallel organization of outstanding engineers. It is autonomous in its administration and in the selection of its members, sharing with the National Academy of Sciences the responsibility for advising the federal government. The National Academy of Engineering also sponsors engineering programs aimed at meeting national needs, encourages education and research, and recognizes the superior achievements of engineers. Dr. Robert M. White is president of the National Academy of Engineering.

The Institute of Medicine was established in 1970 by the National Academy of Sciences to secure the services of eminent members of appropriate professions in the examination of policy matters pertaining to the health of the public. The Institute acts under the responsibility given to the National Academy of Sciences by its congressional charter to be an adviser to the federal government and, upon its own initiative, to identify issues of medical care, research, and education. Dr. Samuel O. Thier is president of the Institute of Medicine.

The National Research Council was organized by the National Academy of Sciences in 1916 to associate the broad community of science and technology with the Academy's purposes of furthering knowledge and advising the federal government. Functioning in accordance with general policies determined by the Academy, the Council has become the principal operating agency of both the National Academy of Sciences and the National Academy of Engineering in providing services to the government, the public, and the scientific and engineering communities. The Council is administered jointly by both Academies and the Institute of Medicine. Dr. Frank Press and Dr. Robert M. White are chairman and vice chairman, respectively, of the National Research Council.

The Board on Science and Technology for International Development (BOSTID) of the Office of International Affairs addresses a range of issues arising from the ways in which science and technology in developing countries can stimulate and complement the complex processes of social and economic development. It oversees a broad program of bilateral workshops with scientific organizations in developing countries and conducts special studies.

This report was prepared by an ad hoc advisory panel of the Advisory Committee on Technology Innovation, Board on Science and Technology for International Development, Office of International Affairs, National Research Council. Staff support was funded by the Office of the Science Advisor, Agency for International Development, under Grant No. DAN-5538-G-SS-1023-00.

Library of Congress Catalog Card Number: 90-63998
**ISBN 0-309-04295-X**
**S-184**

## PANEL ON MICROLIVESTOCK

RALPH W. PHILLIPS, Deputy Director General (Retired), Food and Agriculture Organization of the United Nations, *Chairman*

EDWARD S. AYENSU, Senior Advisor to the President, African Development Bank, Abidjan, Ivory Coast.

BONNIE V. BEAVER, Professor of Veterinary Medicine, Department of Small Animal Medicine and Surgery, Texas A&M University, College Station, Texas, USA

KURT BENIRSCHKE, Professor of Pathology and Reproductive Medicine, University of California-San Diego, San Diego, California, USA

ROY D. CRAWFORD, Professor of Animal and Poultry Genetics, Department of Animal and Poultry Science, University of Saskatchewan, Saskatoon, Saskatchewan, Canada

TONY J. CUNHA, Distinguished Service Professor Emeritus, University of Florida, Gainesville, and Dean Emeritus, School of Agriculture, California Polytechnic University, Pomona, California, USA

DAVID E. DEPPNER, Director, New Forest Project, International Center, Washington, D.C., USA

ELIZABETH L. HENSON, Executive Director, American Minor Breeds Conservancy, Pittsboro, North Carolina, USA

DONALD L. HUSS, Menard, Texas, USA (Formerly Regional Animal Production Officer, FAO, Santiago, Chile)

DAVID R. LINCICOME, Guest Scientist, U.S. Department of Agriculture, Beltsville, Maryland, USA

THOMAS E. LOVEJOY, Assistant Secretary for External Affairs, Smithsonian Institution, Washington, D.C., USA

ARNE W. NORDSKOG, Professor Emeritus, Department of Animal Science, Iowa State University, Ames, Iowa, USA

LINDA M. PANEPINTO, Director, Miniature Swine Laboratory, Colorado State University, Fort Collins, Colorado, USA

KURT J. PETERS, Professor of Animal Breeding and Husbandry in the Tropics and Subtropics, University of Göttingen, Göttingen, West Germany, and Director of Research, International Livestock Centre for Africa, Addis Ababa, Ethiopia

JOHN A. PINO, Senior Fellow, National Research Council, Washington, D.C., USA

HUGH POPENOE, Director, International Program in Agriculture, University of Florida, Gainesville, Florida, USA

MICHAEL H. ROBINSON, Director, National Zoological Park, Washington, D.C., USA

KNUT SCHMIDT-NIELSON, James B. Duke Professor of Physiology, Department of Zoology, Duke University, Durham, North Carolina, USA

Albert E. Sollod, Associate Professor and Head, Section of International Veterinary Medicine, Tufts University, North Grafton, Massachusetts, USA

Lee M. Talbot, Visiting Fellow, World Resources Institute, Washington, D.C., USA

Clair E. Terrill, Sheep and Goat Scientist, U.S. Department of Agriculture, Beltsville, Maryland, USA

Christen M. Wemmer, Assistant Director for Conservation, National Zoological Park, Front Royal, Virginia, USA

Danny C. Wharton, Associate Curator Animal Departments, New York Zoological Park, Bronx Zoo, The Bronx, New York, USA

Charles A. Woods, Professor and Curator, Florida State Museum, University of Florida, Gainesville, Florida, USA

Thomas M. Yuill, Associate Dean for Research and Graduate Training, School of Veterinary Medicine, University of Wisconsin-Madison, Madison, Wisconsin, USA

* * *

Noel D. Vietmeyer, Board on Science and Technology for International Development (BOSTID), National Research Council, Washington, D.C., *Microlivestock Study Director and Scientific Editor*

## NATIONAL RESEARCH COUNCIL STAFF

F.R. Ruskin, *BOSTID Editor*
Mark Dafforn, *Technical Writer*
Mary Jane Engquist, *Staff Associate*
Elizabeth Mouzon, *Senior Secretary*
John Vreyens, *MUCIA Intern*

## CONTRIBUTORS

*The following individuals have made general contributions to the development of this book. All of the persons listed as research contacts in Appendix B also contributed—usually on one or two species that are their scientific specialty.*

ASHIQ AHMAD, Wildlife Management Specialist, Pakistan Forest Institute, Peshawar, Pakistan
ANGEL C. ALCALA, Division Research, Extension and Development, Silliman University, Dumaguete City, Philippines
HARTI AMMANN, Basel, Switzerland
PATRICK ANDAU, Forest Department, Sandakan, Sabah, Malaysia
S.P. ARORA, National Dairy Research Institute, Karnal, India
S. AYYAPPAN, CIFRI, Kausabyaganga, Bhubaneswar Orissa, India
WALTER BAKHUIS, Caribbean Marine Biological Institute, Willemstad, Curaçao, Netherlands Antilles
JAMES R. BARBORAK, CATIE, Turrialba, Costa Rica
PUSHKAR NATH BHAT, Indian Veterinary Research Institute, Izatnagar, Uttar Pradesh, India
STEVE BENNETT, Curepe, Trinidad, West Indies
K.P. BLAND, Department of Physiology, Royal (Dick) School of Veterinary Studies, University of Edinburgh, Edinburgh, Scotland
MELVIN BOLTON, Yeppoon, Queensland, Australia
JOSEPH BONNEMAIRE, Ecole Nationale Supérieure des Sciences Agronomiques Appliquées, Dijon, France
R.D.S. BRANCKAERT, Faculté des Sciences Agronomiques, Université du Burundi, Bujumbura, Burundi
PETER BRAZAITIS, Herpetology, New York Zoological Society, The Bronx, New York, USA
L. DE LA BRETONNE, JR., Louisiana Cooperative Extension Service, Louisiana State University, Baton Rouge, Louisiana, USA
P. BRINCK, Department of Animal Ecology, University of Lund, Lund, Sweden
LESLIE BROWNRIGG, CIAT, Cali, Colombia
D. HOMER BUCK, Illinois Natural History Survey, Kinmundy, Illinois, USA
GERARDO BUDOWSKI, Natural Renewable Resources Programme, CATIE, Turrialba, Costa Rica
DAVID BUTCHER, Taronga Zoo, Mosman, New South Wales, Australia
JULIAN O. CALDECOTT, World Wildlife Fund Malaysia, Kuching, Sarawak, Malaysia
GARY CALLIS, Texline, Texas, USA
J.K. CAMOENS, Asian Development Bank, Manila, Philippines
A. CHRISTOPHER CARMICHAEL, The Museum, Michigan State University, East Lansing, Michigan, USA
ROBERT H. CHABRECK, School of Forestry and Wildlife Management, Louisiana State University, Baton Rouge, Louisiana, USA
A.M. CHAGULA, Research, Ministry of Agriculture, Dar-es-Salaam, Tanzania

Charan Chantalakhana, Department of Animal Science, Kasetsart University, Bangkok, Thailand
Peter R. Cheeke, Rabbit Research Center, Oregon State University, Corvallis, Oregon, USA
G.S. Child, Forest Resources Division, FAO, Rome, Italy
A.S. Chopra, Ministry of Agriculture, Department of Agriculture and Cooperation, New Delhi, India
W. Ross Cockrill, 29 Downs Park West, Bristol, England, BS6 7QH
Crisostomo Cortes, Dairy Promotion and Extension Section, Dairy Development Division, Manila, Philippines
Wyland Cripe, College of Veterinary Medicine, University of Florida, Gainesville, Florida, USA
A. Ben David, Holon, Israel
C. Devendra, International Development Research Centre, Singapore
Rodney Dillinger, International Agency for Apiculture Development, Rockford, Illinois, USA
Director, Natal Parks Board, Pietermaritzburg, South Africa
Rollande Dumont, Ecole Nationale Supérieure des Sciences Agronomiques Appliquées, Dijon, France
N.G. Ehiobu, Department of Agricultural Sciences, College of Education, Agbor, Nigeria
Donald Farner, Department of Zoology, University of Washington, Seattle, Washington, USA
John A. Ferguson, Overseas Development Administration, Eland House, London, England
Abelardo Ferrer D., Quinta Nueva Exparta, San Bernardino, Caracas, Venezuela
Lynwood A. Fiedler, Section of International Programs, U.S. Fish and Wildlife Service, Denver Wildlife Research Center, Denver, Colorado, USA
H. Fischer, Tropical Science Centre, Division of Tropical Veterinary Medicine, Justus-Liebig University, Giessen, West Germany
J. Furtado, Commonwealth Science Council, London, England
Frank Golley, Institute of Ecology, University of Georgia, Athens, Georgia, USA
E. Gonzales J., Instituto de Producción Animal, Universidad Central de Venezuela, El Limon-Maracay, Venezuela
Graham Goudie, Mainland Holdings, Lae, Papua New Guinea
Alistair Graham, Tanglewood, Crowborough, East Sussex, England
Gordon Grigg, Zoology, University of Sydney, Sydney, New South Wales, Australia
M.R. De Guman, Jr., Food and Fertilizer Technology Center, Taipei, Taiwan
Colin P. Groves, Department of Prehistory and Anthropology, The Australian National University, Canberra, ACT, Australia
J. Hardouin, Institut de Médicine Tropicale "Prince Leopold," Antwerp, Belgium
Geoffrey Hawtin, International Development Research Centre, University of British Columbia, Vancouver, B.C., Canada

GORDON HAVORD, Technical Advisory Division, UNDP, New York, New York, USA
TIN HLA, Veterinary Department, Director General's Office, Rangoon, Burma
JAMES HENTGES, Department of Animal Science, University of Florida, Gainesville, Florida, USA
W.F. HOLLANDER, Department of Genetics, Iowa State University, Ames, Iowa, USA
RENE E. HONEGGER, Herpetology, Zurich Zoo, Zurich, Switzerland
JACK HOWARTH, School of Veterinary Medicine, University of California, Davis, California, USA
HUANG CHU-CHIEN, Institute of Zoology, Academia Sinica, Beijing, China
ANGUS HUTTON, Gympie, Queensland, Australia
H.A. JASIOROWSKI, Animal Production and Health Division, FAO, Rome, Italy
MUHAMMAD YAQUB JAVAID, Directorate of Fisheries, Government of the Punjab, Punjab, India
J. MANGALARAJ JOHNSON, Nudumalai Sanctuary, Vannarpet, Udagamandalani, India
MATI KAAL, Tallinn Zoo, ESSR Tallin, USSR
STELLAN KARLSSON, Simontorp Aquaculture AB, Blentarp, Sweden
JACKSON A. KATEGILE, International Development Research Centre, Nairobi, Kenya
ROBERT E. KENWARD, Institute of Terrestrial Ecology, Furzebrook Research Station, Wareham, Dorset, England
JAMES M. KEARNEY, Miami, Florida, USA
F. WAYNE KING, Florida State Museum, Gainesville, Florida, USA
H.-G. KLOS, Zoologischer Garten Berlin, Berlin, West Germany
NELS M. KONNERUP, Bova International, Stanwood, Washington, USA
NAVU KWAPENA, Office of Environment and Conservation, Boroko, Papua New Guinea
THOMAS E. LACHER, Huxley College of Environmental Studies, Western Washington University, Bellingham, Washington, USA
JOHN K. LOOSLI, Gainesville, Florida, USA
PETER LUTZ, Rosenstiel School of Marine and Atmospheric Science, University of Miami, Miami, Florida, USA
CRAIG MACFARLAND, CATIE, Turrialba, Costa Rica
CONSTANCE M. MCCORKLE, Department of Rural Sociology, University of Missouri, Columbia, Missouri, USA
ROBERT E. MCDOWELL, Department of Animal Science, Cornell University, Ithaca, New York, USA
JEFFREY A. MCNEELY, International Union for Conservation of Nature and Natural Resources, Gland, Switzerland
ADRIAN G. MARSHALL, Institute of South-East Asian Biology, University of Aberdeen, Aberdeen, Scotland
RICHARD R. MARSHALL, Veterinary Medicine, Sutter Hospitals Medical Research Foundation, Sacramento, California, USA
G.H.G. MARTIN, Department of Zoology, Kenyatta University College, Nairobi, Kenya

IAN L. MASON, Edinburgh, Scotland
JOHN C. MASON, Pacific Biological Station, Nanaimo, B.C., Canada
J. MAYO MARTIN, Fish Farming Experimental Station, Stuttgart, Arkansas, USA
YOLANDA MATAMOROS, Escuela de Medicina Veterinaria, Universidad Nacional, Heredia, Costa Rica
ROBIN MCKERGON, Livestock Development Corporation, Lae, Papua New Guinea
M. MGHENI, Faculty of Agriculture, Sokoine University of Agriculture, Morogoro, Tanzania
P. MONGIN, Station de Recherches Avicoles, INRA-Centre de Tours, Nouzilly, Monnaie, France
JOSE ROBERTO DE ALENCAR MOREIRA, Agricultural Research Center of the Humid Tropics, Belém, Pará, Brazil
W.L.R. OLIVER, Jersey Wildlife Preservation Trust, Jersey, Channel Islands, United Kingdom
WERNER PAUWELS, Basel, Switzerland
W.J.A. PAYNE, Worcester, England
IAN PLAYER, Wilderness Leadership School, Bellair, Natal, South Africa
JAMES H. POWELL, JR., Plainview, Texas, USA
WILLIAM R. PRITCHARD, School of Veterinary Medicine, University of California, Davis, California, USA
HECTOR HUGO LI PUN, International Development Research Centre, Bogotá, Colombia
VICENTE T. QUIRANTE, Small Ruminant Collaborative Research Project, Bureau of Animal Industry, Manila, Philippines
DAN RATTNER, The Institute of Animal Research, Kibbutz Lahav, D.N. Negev, Israel
C.V. REDDY, Faculty of Veterinary Science, Andhra Pradesh Agricultural University, Rajendranagar, Hyderabad, India
RHOEHEIT, Institute of Applied Science and Technology, University Campus, Turkeyen, Guyana
CHARLES T. ROBBINS, Department of Zoology, Washington State University, Pullman, Washington, USA
CARMEN MA. ROJAS G., CATIE, Turrialba, Costa Rica
D.H.L. ROLLINSON, Sardinia, Italy
JULIO E. SANCHEZ P., Museo Nacional, San José, Costa Rica
JEFF SAYER, World Conservation Centre, Gland, Switzerland
G. SEIFERT, Tropical Cattle Research Centre, CSIRO, Rockhampton, Queensland, Australia
ANDRES ELOY SEIJAS, Servicio Nacional de Fauna Silvestre, Maracay, Venezuela
S.K. SHAH, Institute of Animal Sciences, National Institute of Health, Islamabad, Pakistan
SIEH CHENXIA, Department of Animal Science, Nanjing Agricultural College, Nanjing, People's Republic of China
B.P. SINGH, College of Veterinary and Animal Science, Chandra Sekhar Azad University of Agriculture and Technology, Mathura, Uttar Pradesh, India
C. CATIBOG SINHA, Forest Research Institute, College, Laguna, Philippines

A.J. SMITH, Tropical Animal Health, Royal (Dick) School of Veterinary Studies, University of Edinburgh, Edinburgh, Scotland, Great Britain
A. MITHAT EFENDI, Ankara, Turkey
HENRY STODDARD, Shamrock Veterinary Clinic and Fisheries, Cross City, Florida, USA
SUKUT SULARSASA, Faculty of Animal Husbandry, Gadjalu Mada University, Yogyakarta, Indonesia
D.L. SUTTON, Agricultural Research and Education Center, Fort Lauderdale, Florida, USA
NICHOLAS SMYTHE, Smithsonian Tropical Research Institute, Panama
J. SZUMIEC, Polish Academy of Sciences, Experimental Fish Culture Station, Chybie, Poland
N. TABUNAKAWAI, Ministry of Primary Industries, Suva, Fiji
FRANK M. THOMPSON, Wild Animal Brokers, Bradenton, Florida, USA
ALLEN D. TILLMAN, Stillwater, Oklahoma, USA
DON TULLOCH, Winnellie, Northern Territory, Australia
CONRADO A. VALDEZ, Dairy Development Division, Bureau of Animal Industry, Manila, Philippines
LUIS VARONA, Havana, Cuba
PRAN VOHRA, Department of Avian Sciences, College of Agricultural and Environmental Sciences, University of California, Davis, California, USA
ANTOON DE VOS, Whitford, Auckland, New Zealand
GRAHAME WEBB, Conservation Commission of the Northern Territory, Winnellie, Northern Territory, Australia
DAGMAR WERNER, Fundación Pro Iguana Verde, Heredia, Costa Rica
GARY WETTERBERG, Department of the Interior, Washington, D.C., USA
CHARLES H. WHARTON, Clayton, Georgia, USA
F.W. BERT WHEELER, College Station, Texas, USA
ROMULUS WHITAKER, Madras Crocodile Bank, Perur, Tamil Nadu, India
WILDLIFE CONSERVATION INTERNATIONAL, New York Zoological Society, The Bronx, New York, USA
R.R. YEO, USDA-ARS, University of California, Davis, California, USA
BRUCE A. YOUNG, Department of Animal Science, University of Alberta, Edmonton, Alberta, Canada
CHAROON YOUNGPRAPAKORN, The Samutprakan Crocodile Farm and Zoo, Samutprakan, Thailand
THOMAS M. YUILL, School of Veterinary Medicine, University of Wisconsin, Madison, Madison, Wisconsin, USA
W. ZEILLER, Miami Seaquarium, Miami, Florida, USA

# Preface

The purpose of this report is to raise awareness of the potential of small livestock species and to stimulate their introduction into animal research and economic development programs. It is geared particularly towards benefiting developing nations.

"Microlivestock" is a term we have coined for species that are inherently small, such as rabbits and poultry, as well as for breeds of cattle, sheep, goats, and pigs that are less than about half the size of the most common breeds. These miniature animals are seldom considered in the broad picture of livestock development, but they seem to have a promising future. Wherever land is scarce it seems reasonable to assume that, things being equal, small animals would be more attractive than large ones. And land for livestock is becoming increasingly scarce.

In this report we have emphasized multipurpose species with promise for smallholders. In some species, the promise is immediate; in others, it is long term, and much research must be undertaken before that promise can be realized or even understood.

We have included wild species that seem to have potential as future livestock. Some are threatened with extinction but are described here because their economic merits may be the key to acquiring support for their protection. Also, we have highlighted rare breeds of domesticated species because the current tendency has been to concentrate on a small number of large breeds, and many potentially valuable breeds are becoming extinct through neglect.

The book was prepared after an intensive survey of more than 300 animal scientists in 80 countries. They suggested more than 150 species for inclusion. The staff then drafted chapters on about 40 species and these drafts were reviewed by more than 400 researchers worldwide. The thousands of resulting comments, corrections, and additions were integrated into the drafts. The panel then met to review the product, to select the most promising species, and to rework the chapters based on their own experiences and joint conclusions. The result is the current 35 chapters. Most of the case studies and accounts of innovations highlighted in the various sidebars were developed by the staff study director.

Collectively, this study covers many species, but it by no means exhausts all the microlivestock possibilities. Lack of space and time precludes discussion of creatures such as edible insects, snails, worms, turtles, and bats, which in some regions are highly regarded foods. Similarly, we have not included aquatic life. These decisions were arbitrary; perhaps invertebrates and aquatic species can be included in future volumes.

This report is addressed to government administrators, technical-assistance personnel, and researchers in agriculture, nutrition, and related disciplines who are concerned with helping developing countries achieve a more efficient and balanced exploitation of their biological resources. Hence, we deal with the animals in a general way and do not cover details of biology, husbandry, or economics. A selection of readings that contains such technical information is cited in Appendix A.

A further goal of this project has been to explore the common ground between the disparate arms of animal science: to show that specialists in wildlife, zoology, and livestock science have much to learn from one another's field of expertise; to show that "fanciers" of pigeons, pheasants, chinchillas, iguanas, and other species may have much to offer livestock breeders—including germplasm; and that those who raise "obsolete" breeds are not only playing a vital role in the protection of rare genes but can offer the benefit of their experience to commercial livestock producers.

Throughout this report, the scientific names of mammals follow those in: *Mammal Species of the World: A Taxonomic and Geographic Reference*. 1982. J.H. Honacki, K.E. Kinman, and J.W. Koeppl, editors. Published by Allen Press, Inc.; and the Association of Systematics Collections, Lawrence, Kansas, USA. All dollar figures are in U.S. dollars; all ton figures are in metric tons.

This report has been produced under the auspices of the Advisory Committee on Technology Innovation (ACTI) of the Board on Science and Technology for International Development, National Research Council. ACTI was mandated to assess innovative scientific and technological advances, with particular emphasis on those appropriate for developing countries. In this spirit, therefore, the current report includes some extremely unusual species. Whether these will eventually prove practical for widespread use is uncertain, but we present them here for researchers and others who look forward to challenges and enjoy the satisfaction of successful pioneering. The domestication of new poultry, as well as the management of rodents, iguanas, and small deer and antelope, should be viewed in this spirit.

Current titles in the ACTI series on managing tropical animal resources are:

- *The Water Buffalo: New Prospects for an Underutilized Animal*
- *Little-Known Asian Animals with a Promising Economic Future*
- *Crocodiles as a Resource for the Tropics*
- *Butterfly Farming in Papua New Guinea.*

The production of these books has been supported largely by the Office of the Science Advisor of the U.S. Agency for International Development (AID), which also made this report possible.

**WARNING**

**If misunderstood, this book is potentially dangerous. Because of the severity of the food crisis, the panel has selected some animals—mainly in the rodent section—that are highly adaptable and grow quickly. These seem appropriate for raising only in areas where they already exist, which are clearly identified in those chapters. Such potentially invasive animals should not be introduced to other environments because they could become serious pests. In any trials, local species should always be given priority.**

How to cite this report:
National Research Council. 1991. *Microlivestock: Little-Known Small Animals with a Promising Economic Future*. National Academy Press, Washington, D.C.

# Contents

# Art Credits

| Page | |
|---|---|
| 14 | *Small Farmer's Journal* |
| 33 | Balai Penelitian Ternak, Bogor, Indonesia |
| 46 | Tom Phillips, The Anstendig Institute, San Francisco |
| 62 | Brenda Spears |
| 90 | CAB International, Wallingford, United Kingdom |
| 114 | Drawing from *Lewis Wright's Poultry* by J. Batty, reproduced by permission Nimrod Bood Services, Liss, United Kingdom |
| 124 | Drawing by Charles W. Schwartz, reproduced by permission from *Wildlife of Mexico: The Game Birds and Mammals*, by A. Starker Leopold, courtesy University of California Press |
| 156 | *Small Farmer's Journal* |
| 166 | Courtesy, Department of Library Services, American Museum of Natural History |
| 206 | David W. Macdonald |

| | |
|---|---|
| 216 | Charles A. Woods |
| 232 | Reproduced from *The Rodents of West Africa*, © Trustees of the British Museum (Natural History). |
| 250 | Charles A. Woods |
| 270 | FAO, Santiago, Chile |
| 290 | Reprinted from *Animals of Southern Asia* by M. Tweedie, courtesy Paul Hamlyn Publishing, part of Reed International Book. |
| 320 | Huang Chu-Chien |
| 346 | Horacio Rivera |
| 362 | Drawing by Sarah Landry. Reprinted by permission of Harvard University Press from *The Insect Societies*, by Edward O. Wilson, p. 97. Cambridge, Mass.: The Belknap Press of Harvard University Press, copyright © 1971 by the President and Fellows of Harvard College. |

Drawings on pages 198, 240, and 256 are reproduced by permission from *The Random House Encyclopedia*, copyright © 1983 by Random House, Inc.

Silhouettes on pages 284, 297, 313, 319, 334, are reproduced from *Hoofed Mammals of the World* by Ugo Mochi and T. Donald Carter, reproduced with the permission of Charles Scribner's Sons. Copyright © 1953 by Ugo Mochi and T. Donald Carter, copyright renewed 1981 by Edna Mochi. All rights reserved.

The maps on pages 203, 210, 219, 227, 236, 243, 253, 260, 265, 274, 292, 293, 309, and the drawings on pages 276 and 314 are adapted from *Grzimek's Animal Life Encyclopaedia* and are reproduced by permission of Coron Verlag, Lachen am Zürichsee, Switzerland.

Drawings on pages 178, 224, 262, and 306, are reprinted by kind permission of Andromeda Oxford Ltd. and first published in the *Encyclopedia of Mammals* by David W. Macdonald, Facts on File (New York).

Drawings on pages 326 and 336 are by Clare Abbott and are reprinted from *The Mammals of the Southern African Subregion*, courtesy the University of Pretoria.

Cover Design by David Bennett

# MICROLIVESTOCK

*In the developing countries, there are over 100 million farms of less than five hectares, supporting about 700 million people, who represent about 17 percent of the world population. Even more significant is the fact that about 50 million farms have less than one hectare of land.*

C. Devendra and Marcia Burns
*Goat Production in the Tropics*

*We may now be in the wind down stage of bigger is better animal selection trend and it has certainly been a wild ride. . . . the lesson now being learned is that the bigger breeding animals . . . cost more to maintain, are often slower to reproduce, and may even have a shorter lifespan.*

Kelly Klober
*Small Farmer's Journal*

# Introduction

Like computers, livestock for use in developing countries should be getting smaller and becoming more "personal." Conventional "mainframes," such as cattle, are too large for the world's poorest people; they require too much space and expense. "Miniframes," such as the conventional breeds of sheep and goats, have an increasingly important role to play. But tiny, "user-friendly" species for home use are the ones highlighted in this report. We have called them "microlivestock."[1]

There are two types of microlivestock. One consists of extremely small forms of conventional livestock—such as cattle, sheep, goats, and pigs. The other consists of species that are inherently small—poultry, rabbits, and rodents, for instance.

Microlivestock are important because the developing world's animal production is only a fraction of what it should be. Throughout Latin America, Asia, and Africa, the poor eat almost no meat, milk, or eggs—the most nutritious foods. It is estimated, for example, that in Mexico 25 million campesinos cannot afford meat. In poor countries, even the middle class eats less meat in a year than the populations of North America and Europe eat in a month. Malnutrition is common and its effects, especially on children, can be debilitating. It is one of mankind's most serious imbalances—and most pressing problems.

Rural families in the Third World usually subsist mainly on the products from their homes or farms. Thus, if we are to help their livestock production, more attention must be given to animals that are sized for their situations. Examples discussed in the report are summarized here.

**Microbreeds** Small breeds of cattle, sheep, goats, and pigs are common in the developing world. Because they are often raised for

[1] For purposes of this study we have coined terms such as "microlivestock," "microbreeds," "microcattle." These words emphasize the commonality of smallness among species as diverse as sheep, chickens, and iguanas. Although words such as "dwarf" and "miniature" could have been used, they already have connotations within animal science, and they lack the spark of newness and future promise that we are attempting to foster.

subsistence rather than for commerce, the national and global contribution that they make is often overlooked. These small, hardy animals deserve much greater recognition.

Later chapters highlight dozens of promising microbreeds. All are less than half average size; some are far smaller than that. The "mini-Brahman" cow of Mexico is only 60 cm tall and weighs 140 kg; the southern Sudan dwarf sheep of eastern Africa can weigh as little as 11 kg; the Terai goat of Nepal weighs less than 12 kg; and the cuino pig of Mexico weighs merely 10 kg.

**Poultry** The widespread use of poultry in Third World villages demonstrates the importance of small, easily managed, household livestock. Small size, the ability to forage for themselves, and a natural desire to stay around the house put chickens, ducks, guinea fowl, and other birds among the most vital resources of rural Asia, Africa, and Latin America. Scratching a living out of the dirt, dust, ditches, and debris, these often-scrawny creatures are a resource to be taken seriously. For the most poverty stricken, a bony bird may be the only source of meat during much of a lifetime.

Among poultry, there are many underrated, but highly promising, species, including:

- *Pigeon*. These birds forage widely but return home, thereby providing the farmers with squab, one of the tastiest of all meats.
- *Quail*. Small and efficient, they, too, are suited to home rearing, and in Japan and a few other countries, large numbers are raised commercially in very small space.
- *Muscovy*. A native of South American rainforests, this bird is a major poultry resource of France, Taiwan, and a few other countries. Tame, tolerant, and tough, it deserves greater recognition everywhere.
- *Guinea Fowl*. One of the most self-reliant of all domestic birds, this native of Africa is raised in huge numbers in Europe—notably France. Its potential for increased production elsewhere is exceptional.
- *Turkey*. The traditional turkey of Mexico still exists as a scavenger bird in villages and household backyards. Unlike the highly selected modern breeds, it is self-reliant, robust, and disease-resistant.

**Rabbits** Like chickens, rabbits exemplify the vast possibilities that microlivestock offer for increasing meat production in the most poverty-stricken parts of the world. Captive rabbits have been popular as food

Opposite: Microlivestock—pigeons, turkeys, chickens, and goats—being fed in a village in Togo. (FAO)

at least since the time of the Romans. Rabbit rearing has been well established in Europe and China, and now national rabbit projects have begun in many developing nations.

**Rodents** Some 7,000 years ago, guinea pigs were domesticated as a source of food for the high Andes; even today in the uplands of Bolivia, Peru, and Ecuador, most Indians raise them inside their homes and regard them as an essential part of life. For many Indians, these indoor livestock are the main source of meat. Prolific, tractable, and easy to feed, house, and handle, guinea pigs are even kept in downtown apartment buildings—often in boxes under the bed.

Other rodents might also be suited to domestication; for instance, the potentially tamable, clean-living species of South American fields and woodlands—agouti, capybara, hutia, mara, coypu, paca, and vizcacha. Two remarkable domestication programs have been started in Africa: the grasscutter in West Africa and the giant rat in Nigeria. Both animals provide popular "bushmeat" and researchers are now learning to raise them in captivity. (Because of its tangy taste, Ghanaians actually pay three times more for grasscutter meat than for beef!)

**Antelope** Another wild African mammal with potential for "household animal husbandry," the blue duiker, is a rabbit-sized antelope. In some areas of Central and Southern Africa the demand is so great that its population is declining at an alarming rate. Duiker rearing, if it can be made successful, might provide both food and an economic alternative to slaughtering the wild populations. It is reported that duikers are easy to maintain and they reproduce well in captivity.

The meat of several other tiny antelope species is also much sought in many African countries, and these animals are also suitably sized to feed the average family at one meal.

**Deer** Several species of tiny deer—smaller than many dogs—might make useful microlivestock, although much research is needed before their true potential can be judged. Normal-sized deer were once considered too easily frightened to be reared as domestic livestock, but several species are now raised on thousands of deer farms in New Zealand as well as in at least a dozen other countries.

Mouse deer and musk deer (which, strictly speaking, are not true deer at all) are of microlivestock size and are also possible future livestock. The musk deer produces one of the most valuable materials in the animal kingdom—more valuable, in fact, than gold. The musk from the male's glands is used in oriental medicines as well as in European perfumes.

**Iguanas** Over much of the Caribbean and Latin America, iguanas are a traditional source of food. Indeed, the meat of these large herbivorous lizards is so delicious they are being hunted to extinction throughout their wide range. Their eggs are much enjoyed also. Programs in Panama, Costa Rica, El Salvador, Curaçao, and Argentina have developed simple methods to hatch and rear three iguana species.

**Bees** Honey bees are present almost everywhere, and honey and wax are high-value products that demand little processing and can be stored and transported easily. Innovations in equipment and technique have made beekeeping successful in the tropics without requiring sophisticated hives or elaborate training. Raising bees can also benefit the many crops that require pollination.

## THE MICROLIVESTOCK ADVANTAGES

Although animal science has traditionally emphasized bigness, smallness has its advantages. Some of these are summarized below.

### Economic

Microlivestock lend themselves to economic niches that are not easily filled by large livestock. Much of their potential is for subsistence production. They are promising for the many peasants who, being outside the cash economy, are now unable to purchase meat, milk, cheese, or eggs. These people can afford only animals that can be raised within the home or backyard under ambient climatic conditions and on feeds that are cheap and easily available.

A subsistence farmer is likely to benefit more from small species than from large because of several factors:

- The animals are less expensive to buy.
- They are less of a financial risk to maintain. (A farmer with several small animals is less vulnerable to loss than a farmer with a single large animal, a feature that is particularly important in subsistence farming where success determines whether the family will survive.)
- They give a faster return on investment. (Small size generally signifies high reproductive capacity and a fast turnover.)
- They provide flexibility. (Farmers can more easily change the size of their herd or flock to match the amount of feed available at a given time. Also, they can sell animals according to the family's fluctuating needs for cash or food.)

## RATIONALE FOR LIVESTOCK PRODUCTION

Many have argued that livestock raising should be discouraged, that it is a primary cause of desertification through overgrazing, and that it is an inefficient converter of basic material and energy into human food. "Grow more pulses, grow more grains," these people cry. Their arguments can be valid where the land has high potential for permanent cultivation. Much of the world's surface, however, does not fit into that category, and it is in these areas and for those people who have no access to arable lands that a convincing case for livestock can most easily be made. As W.J.A. Payne has written:*

- *Livestock, particularly ruminants, can process forage and waste crop materials inedible by man into nutritionally desirable food products, many of high protein, mineral and vitamin content and including some of high caloric value.*

- *Approximately 40 percent of total land available in developing countries can be used only for some form of forage production and a further 30 percent is classified as forest with some potential for the production of forage. Some 12 percent of the world's total population live in areas where people depend almost entirely on the products obtained from ruminant livestock.*

- *Livestock provide a range of extremely valuable by-products. Dung is not only a fertilizer and soil-stabilizer but also a fuel of often considerably greater value than the fodder consumed in its production. Other by-products, especially hides and wool, form the bases of rural enterprises that may provide significant incomes to the poorest members of society.*

- *Animal, plant and human life are ecologically interdependent. The establishment of agricultural systems in which livestock are integrated with crops, forestry and aquaculture is essential for the improvement of overall productivity.*

Livestock produce food that adds to the nutritional quality and variety of human diets. Although it is possible for humans to exist without them, these foods are relished and sought after by the majority of humanity. These foods include meat, eggs and processed products such as *biltong* and cheese.

* W.J.A. Payne, "The desirability and implications of encouraging intensive animal production enterprises in developing countries," from *Intensive Animal Production in Developing Countries*, A.J. Smith and R.J. Gunn, eds. Occasional Publication No. 4, British Society for Animal Production, 1981.

- They provide a steadier source of income.
- They increase the chances of successful breeding because greater numbers are usually kept. (This also means that breeding stock is more likely to be retained in times of scarcity.)
- They are more easily transported. (Who hasn't traveled in a Third World bus or train without chickens, ducks, or guinea pigs as fellow passengers?)
- In some cases they are more efficient converters of food energy.

There are also a number of other benefits to small species.

- Reduced Spoilage. A portion of meat that comes in a "package" of a size that can be readily consumed by a family is important in areas where refrigeration is unavailable or uneconomic. A family can eat the meat produced by most microlivestock in one meal or in one day to minimize the risk of spoilage.
- Efficient Use of Space. The space required for handling and feeding microlivestock is proportionately less than that required for large animals. Low space requirements make many microlivestock (such as guinea pigs, rabbits, pigeons, and quail) available to landless rural inhabitants who have no room for a cow. This is particularly important with respect to feed production.
- Cheaper Facilities. Facilities and equipment required for microlivestock are, by and large, smaller and simpler than those required for large animals. They often can be made from local products or scrap material or both.
- Ease of Management. Farmers and villagers can manage small animals more easily than large, which is an advantage in the many places where women and children are the main keepers of livestock.
- Increased Productivity. Small animals tend to fit well into existing farming systems, thereby expanding the resource base and recycling nutrients. Some—for example bees, ducks, and geese—can feed themselves by scavenging.
- By-Products. Many species have fur, feathers, skins, and other by-products that are often more valuable than their meat, milk, or eggs. Examples include the feet and tails of rabbits, musk from musk deer, and pelts from rodents such as coypu. Processing such by-products creates diversification for the farmer and perhaps jobs for the village.

## Feed

In general, small species tend to expand the food base by using a wider array of resources than do major livestock such as cattle. Many can be raised on feeds that people discard: fibrous residues, industry by-products, or kitchen wastes. Some collect minute feeds that otherwise go unused. For example, chickens and pigeons gather scattered seeds, turkeys gobble up insects, geese graze water weeds, iguanas feed in the tops of trees, and bees collect nectar and pollen from flowers that may be miles away.

Even some grazing microlivestock prefer different forages from those preferred by cattle. Antelope and deer, for instance, browse tree leaves; capybara and grasscutters eat reeds. Combining microlivestock with conventional livestock results in a more complete utilization of forage resources and greater animal production per hectare.

Under conditions of abundance, small size may be of no advantage in mammals, but if feed is limited, it is of great help. A small animal (or its keeper) needs to cover less area to fulfill its daily requirements, so that microlivestock may grow fat in areas where the forage is too sparse to support a larger animal. This is particularly vital when there are seasonal bottlenecks. For example, feed may be plentiful enough for most of the year to supply many large animals; however, the dry season may greatly restrict the numbers that can be kept.

Although small animals generally require proportionately higher inputs of feed, they also grow proportionately faster (see sidebar opposite). In addition, species such as rabbits, guinea pigs, and grasscutters digest fibrous matter with surprising efficiency, even though they are not true ruminants like cattle, sheep, and goats.

## Reproduction

Many small animals have high reproductive capacity with short gestation periods, large numbers of offspring, and rapid juvenile growth. They tend to reach sexual maturity at a younger age than large animals, and the interval between the generations can be very short. Thus, meat or other products can be produced more rapidly and more evenly throughout the year.

Cows, for example, produce a maximum of one calf per year. A pig, on the other hand, may produce 7 or more young; a rabbit, 30 or more; a chicken, more than 100.

## Adaptability and Hardiness

The survival rates and manageability of many small breeds and species can be outstanding. Smallness is often an adaptation to harsh environment. Indeed, a major promise for microlivestock is in special environmental niches. Where cold, heat, temperature fluctuations,

## FOOD UTILIZATION VERSUS BODY SIZE

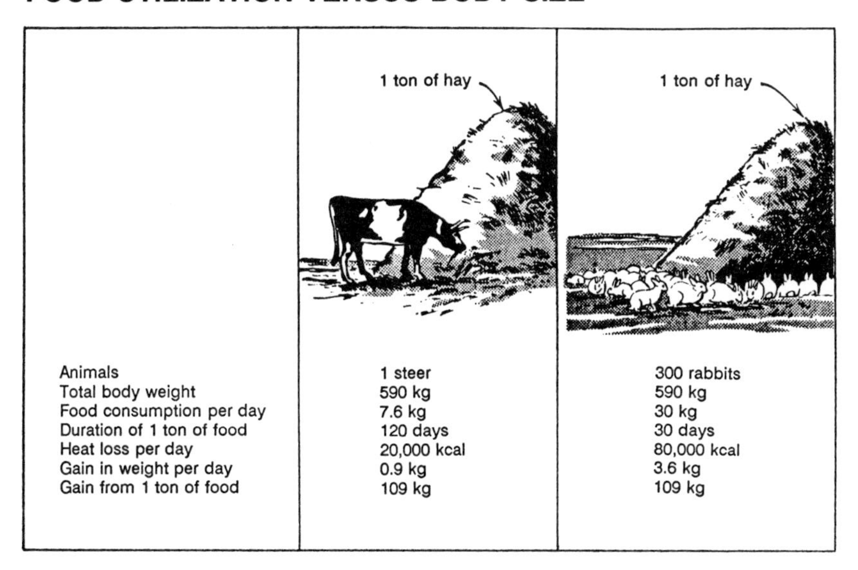


| | | |
|---|---|---|
| Animals | 1 steer | 300 rabbits |
| Total body weight | 590 kg | 590 kg |
| Food consumption per day | 7.6 kg | 30 kg |
| Duration of 1 ton of food | 120 days | 30 days |
| Heat loss per day | 20,000 kcal | 80,000 kcal |
| Gain in weight per day | 0.9 kg | 3.6 kg |
| Gain from 1 ton of food | 109 kg | 109 kg |

Compared to large animals, smaller animals have a higher metabolic rate per unit weight and waste more energy per kg body weight per day as heat. This would seem at first sight to make them less efficient than large animals. This is not necessarily the case, however. The figure shows how rabbits can utilize feed just as efficiently as a steer. Their greater rate of heat loss per unit weight is compensated by a correspondingly greater rate of growth per unit weight.

With rabbits, 10 20-kg animals can be raised on the same feed as one 200-kg cow, producing the same amount of meat and doing it faster.

* Figure adapted from *Fire of Life: An Introduction to Animal Energetics*, by Max Kleiber

aridity, or humidity are extreme, microlivestock are likely to show their greatest advantage. Chickens, guinea fowl, goats, and many other small species already live in villages, homes, and backyards in harsh and disease-prone climates, and are usually given no care and sometimes no food: they have to scavenge for their sustenance and survive as best they can. Such selection pressures result in animals of remarkable adaptability, tolerance, and robustness.

Some microlivestock can produce under conditions where conventional species die. The capybara, for instance, is at home in the Latin American lowlands, where the climate is hot and humid and floods cause seasonal inundations. Cattle, by contrast, die because of malnutrition, foot rot, or drowning. Other microlivestock species with a wide tolerance to ecological extremes include the turkey, pigeon, and bee. And some dwarf breeds of cattle, sheep, goats, and pigs show surprising tolerance to trypanosomes, the parasites that make conventional breeds impossible to maintain throughout much of Africa.

Some small species can be raised in cities, where poverty and malnutrition are often worse than in rural areas. It is estimated, for instance, that one million livestock exist in Cairo, not counting the pigeons that are raised on countless rooftops. Goats and cattle are common in urban India, and many Third World cities have far more chickens than people.

## MICROLIVESTOCK LIMITATIONS

Raising microlivestock is not a panacea for the Third World's food problems. Efforts to develop them will not be without difficulties. Some likely problems are noted here.

### High Energy Requirements

Smaller animals tend to have a higher feed requirement per unit of body weight than large animals. Anatomical and physiological constraints prevent them from meeting their relatively high energy requirements simply by increasing the rate of food ingestion. Therefore, for optimum production, some small animals, particularly nonruminants, require feed that is higher in protein and lower in fiber than large animals. This is particularly true when the small animals are compared with ruminants such as cattle, sheep, and goats.

### Increased Labor Requirements

The advantages of low investment, fast return on capital, flexibility, and efficient resource utilization are offset by higher demands for labor. Keeping small animals often requires considerable effort, and its economic viability may depend on the availability of cheap and willing labor. Many small animals are raised at home by family members, such as children, the elderly, and the handicapped, who have time available and whose labor costs are nominally zero.

### Diseases

Some potential microlivestock are undomesticated, and resistance to diseases and parasites is one justification for their consideration.

However, the general healthiness of a species when it is free-ranging can be a misleading guide for its husbandry. Confining any animal in high density invariably increases the potential for the spread of infectious diseases and parasites. Moreover, mismanagement can foster respiratory and gastrointestinal diseases (such as salmonella or coccidiosis) that are rare among scattered populations.

Some microlivestock are potential reservoirs for diseases that affect not only local animals but people as well. This may limit their successful development in some areas. Although the dangers are often exaggerated, controls may be needed, particularly of rinderpest, tickborne diseases, and those diseases communicable to humans.

## Predation

Small size makes microlivestock easy prey.

## Lack of Research

Techniques to manage some microlivestock species are not yet well established. The development of appropriate husbandry techniques, as well as a better understanding of the animals' particular biological and behavioral characteristics, will be needed before major progress can be made. These species (for instance rodents, deer, and iguanas) may require collection of different genotypes, as well as studies of diseases, nutrition, and management.

## Complex Logistics

It is complicated and expensive to reach millions of widely scattered peasants, each having only a handful of small animals. Even though total production may far exceed that of commercial farms raising large animals, the smallholdings are often dispersed, their animals are often used for subsistence rather than commerce, and their managers are often ill-trained and illiterate.

## Legislative Restrictions

The use of some microlivestock species may be restricted by legislation. For instance, some countries have meat and veterinary laws that work against the development of species other than cattle.

Others have laws to protect wildlife, which could be important in the case of species such as antelope, deer, paca, and iguana.

### Lack of Markets

Microlivestock need not be just for home or local consumption; they can also be raised for market. But some commercial programs, including some with rabbits and guinea pigs, have failed because no public demand was developed.

### Resistance to New Species

People have close associations with livestock, and in most cultures they do not easily accept animals or animal foods that are radically different from their traditional ones. In general, the ties between certain ethnic groups and a particular species or breed is very strong (one reason, for example, why European colonists introduced their own large breeds of cattle and sheep to Africa and Asia, paying little attention to small indigenous breeds). Moreover, people who are used to bringing the animal to the feed rather than the feed to the animal may resist a small animal that has to be penned up and fed by hand.

### Opposition to Smallness

Finally, there seems to be an innate human trait that considers bigger to be better, especially among the common livestock. For example, because many tropical cattle are small, there is strong inclination on the part of those responsible for livestock improvement to dismiss them or to increase their size by crossing them with large breeds.

## FUTURE OF MICROLIVESTOCK

Small animals are likely to become increasingly important. As human populations increase, the space available for growing forage decreases, and this phenomenon favors small animals. Many villagers already have little or no pastureland. Some live in areas (the rice-growing areas of Southeast Asia, for instance) where crops are grown on almost every square meter almost every month of the year. Microlivestock are potentially important for urban areas of developing countries as well. There, too, land is at a premium and is usually inadequate for raising conventional livestock.

So far, however, microlivestock have been largely ignored. Compared with cattle, they have been accorded little scientific effort. In the drive towards larger animals, stimulated by experience in the temperate zone, the virtually unstudied gene pool of small species and breeds has been mostly bypassed. There have been few attempts to assess or improve their farm productivity.

This is unfortunate, and it is perhaps due to the fact that small animals may be less efficient at digesting certain foods and therefore technically less attractive than large, "modern" breeds. But to Third World peasants, an animal's efficiency is far less important than its survivability and manageability. If an animal cannot be raised under village conditions, its feed-use efficiency or milk yield is irrelevant.

Microlivestock production should be integrated into most rural-development projects. Small animals offer a way to improve the lives of people who are hard to reach by other methods. Only by expanding research on the husbandry, hygiene, nutrition, reproduction, physiology, and breeding can the promise of animals sized for small farms and villages be fulfilled. Moreover, the costs will be small compared with those of programs for large animals.[2]

Specifically, experiment stations should produce and promote methods and materials for use in rearing microlivestock. Donors and development institutions, planners, and policymakers should note the potentials of microlivestock and the benefits that can be derived from them. Teaching manuals and materials are needed, and classes in microlivestock husbandry should be included in rural school curricula.

Raising personal livestock on weeds and table scraps in cages beside the house or boxes under the bed will, in many instances, get quality protein to the most poverty stricken more effectively than raising large livestock on pastures.

Although small size confers many advantages, the question is not whether the large or small animal is "best," but rather how well each can meet a person's varying requirements. In a given situation, livestock can be too small or too large. But the fact remains that not everyone who wants meat or money has the resources to acquire, keep, manage, or utilize a large animal.

The key is balance. Both microlivestock and traditional livestock deserve serious attention. Indeed, it seems likely that the two will seldom compete. Most microlivestock complement traditional livestock because of unique physical, physiological, behavioral, or economic characteristics. They increase the range of options for the millions of poor for whom the choice may not even be between large and small livestock, but between microlivestock and no livestock at all.

[2] For example, in El Salvador a highly successful, nationwide, rabbit-development project costs less than the price of a single stud bull.

*It's an unfortunate fact that small animals don't have the prestige among Third World farmers that large animals do (perhaps this arose because children can look after goats and sheep but it takes men to look after cattle). Even sheep and goats are not accorded the same stature as cattle.*

Hugh Popenoe

*Breeds and varieties were created from mutant genes and thus have become living reservoirs of these genes, holding them for use in future generations of mankind.*

Anonymous

# Part I

# Microbreeds

Cattle, goats, sheep, and pigs supply millions of people around the world with the bulk of their cash and animal products. Yet scores of breeds—especially in the tropics—are left out of livestock development projects merely because they are considered too small. These "microbreeds"[1] have sometimes been considered genetic dead ends because they appear undersized and puny. Many of these traditional animals—some in local use for thousands of years—are disappearing, and even the small ancestors of large modern breeds are becoming extinct.

These small breeds deserve to be studied and developed in their own right. Throughout Africa, Asia, and Latin America, these usually hardy animals are especially adapted to traditional husbandry practices and harsh local conditions. Some have remarkable qualities and are well adapted to resist hostile weather, ravaging pestilence, and poor diets. In remote places and in areas of extreme climate, they are often vitally important for basic subsistence.

Indeed, because of stress or disease, or insufficient forage, land, or money, microbreeds may be the only practical livestock in many settings. Their individual output may be low, but it can be efficient considering the lack of care and poor feeds they are given. Their availability and the growing number of small-sized farms in the developing world make them increasingly worthy of consideration.

The following chapters in this section describe microcattle, microgoats, microsheep, and micropigs.

[1] A term used in this report to characterize cattle, goats, sheep, and pigs that are less than about half the size of selectively bred types.

# 1

# Microcattle

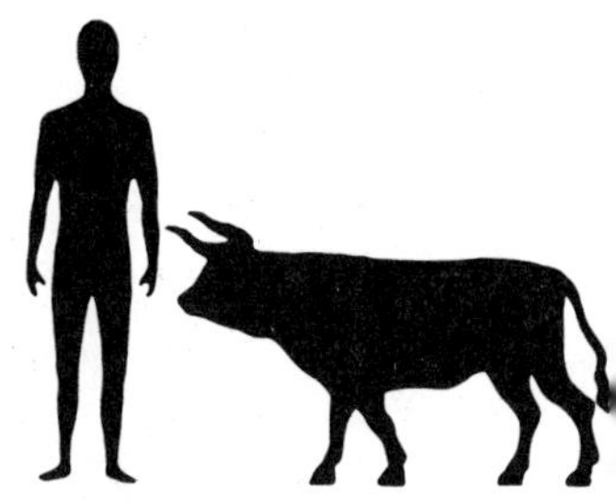

For the purposes of this report, "microcattle"[1] are considered to be small breeds of cattle (*Bos taurus* and *Bos indicus*) with a mature weight of about 300 kg or less. In many areas of the developing world, these are actually the animals most widely held by farmers and pastoralists. They are often treasured because of their resilience and simple requirements. Many survive and produce under harsh conditions, grow rapidly, calve easily, show good maternal ability, yield lean meat, or have other advantages.

Microcattle have generally been ignored in the push towards larger animals, but they seem inherently suitable for traditional and small-farm husbandry. As rural people in developing countries improve their own productivity, as they become more aware of nutritional needs, and as they depend more upon cash economies, microcattle could become vital means for improving personal, dietary, and economic status.

## AREA OF POTENTIAL USE

Worldwide.

## APPEARANCE AND SIZE

Cattle have been classified in many ways, but they are generally designated as humped or humpless types. However, clear distinctions among them are sometimes difficult or impossible to make because they have intermingled for thousands of years. Representative microcattle types are listed at the end of the chapter.[2]

[1] Cattle are so common and sought after in so many countries that their inclusion in this report seems appropriate, even though the smallest cow is obviously larger than other species described later.

[2] A recent book by John P. Maule describes some 300 indigenous and new breeds of tropical cattle. (See Selected Readings.)

## DISTRIBUTION

More than two-thirds of the world's 1.3 billion cattle are found in the developing world; one-third is in the tropics. As noted, a considerable number of these could be called "microcattle."

## STATUS

Many strains of microcattle are threatened with extinction because of replacement or crossbreeding with larger types. This is in some respects shortsighted because promoting just a few breeds contributes to narrowing of the genetic base, and valuable traits may be lost when selection is done to conform to any preconceived standard, including large size.

## HABITAT AND ENVIRONMENT

Microcattle are adapted to a wide variety of habitats. Many types thrive—even with little or no attention—in climates that are hot, humid, arid, or beset by diseases and parasites.

## BIOLOGY

Cattle are ruminants and digest fiber well, although they are selective foragers and prefer tender grasses and low-growing legumes.

As with other tropical cattle, microbreeds generally reach physical and sexual maturity in 2 or 3 years. Many can breed year-round when conditions are favorable (gestation lasts about 9 months). Cows may remain fertile 10 years or more, and can live more than 20 years.

## BEHAVIOR

Cattle usually graze from as few as four hours to as many as eight hours a day. If feed is of poor quality, they must forage (and ruminate) longer to receive adequate nutrition.

Opposite: "Daniel," the prize bull of the 1981 Warren County Fair, Pennsylvania, USA, exemplifies the universal trend to enlarge livestock. He stands more than 1.8 m tall and weighs over 1,360 kg. Cattle have received far more modern development than other livestock species, much of it aimed at increasing size. The result: today's best-developed breeds are far too big for the major needs of the Third World. (G. Lester)

Microcattle are commonly docile and undemanding animals, and many small breeds are surprisingly responsive to humans.

## USES

Like conventional breeds, microcattle produce the same well-known products: meat, milk, manure, hides, horn, blood, and bone. They are also used for traction.

Small cattle often produce only modest amounts of milk and meat per animal. However, given higher stocking rates, a herd of microcattle is often able to outyield larger, genetically improved animals on a per-hectare basis, especially under stressful conditions. When their ability to survive adversity and poor management is taken into account, they may often be far and away the most efficient cattle for traditional husbandry.

Surprisingly, there is a place even for small draft animals. They tend to be active, thrifty (efficient), and more maneuverable in tight spaces, and so are adapted for use in the small fields, terraces, and paddies that are becoming increasingly common. The small hill cattle of Nepal, for instance, are valued because they can negotiate steep slopes and narrow terraces on Himalayan mountainsides.

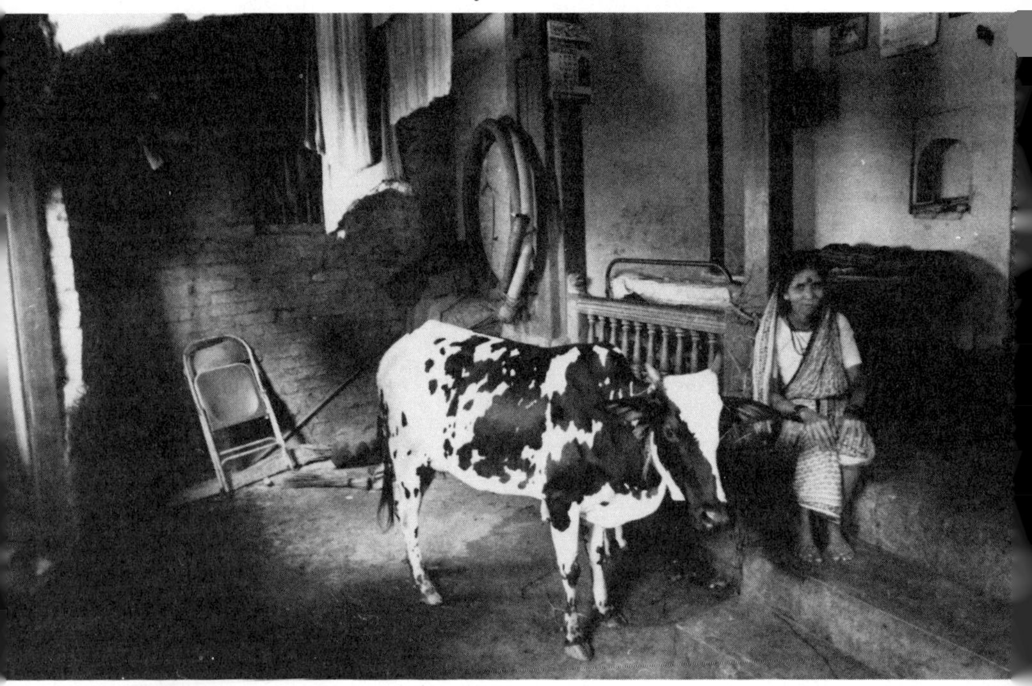

Microcow maintained inside a home in Bombay, India. In many parts of the world indoor animals are common; smallness and docility, therefore, are vital traits. (Carol Guzy, *The Washington Post*)

## HUSBANDRY

Microcattle are handled like their larger counterparts, but herding, tethering, fencing, and hobbling are generally easier.

## ADVANTAGES

Cattle are familiar animals that are accepted in nearly all cultures; their meat, milk, manure, and leather are in demand almost everywhere. In many societies, beef is preferred over other meats, even by those who can rarely afford it.

In most areas, organized breeding, production, and marketing associations are already in place. Microcattle can also integrate well into traditional forms of husbandry, whether in pastoral herds of hundreds or as solitary backyard milk cows.

Under humid and hot conditions, microcattle probably suffer less than larger breeds because their greater ratio of skin area to body mass enhances their ability to shed heat.[3]

The number of cattle that can be kept on a given parcel of land may be increased, sometimes even doubled, with smaller animals. Microcattle can also be penned and fed cut-and-carry forage more easily than can larger cattle, and more of them can be maintained on the same amount of feed. This permits more continuous production and less financial hardship when an animal perishes.

Small cattle may require less labor because they are generally easier to handle, herd, confine, and transport. They usually have few problems with calving, and as a rule require little or no assistance.

Some microcattle have unusual tolerances to disease. In Africa, for instance, there are breeds that tolerate or resist trypanosomiasis, a parasitic disease that makes large areas of that continent uninhabitable for most other cattle breeds. Others seem more tolerant of internal or external parasites, theileriosis (east coast fever), rinderpest, or other afflictions.[4]

## LIMITATIONS

Microcattle often lack the prestige of larger breeds.

When given quality forage and supplemental feeding, small unimproved cattle may not match the overall productivity of the large,

[3] However, high temperatures are not the only cause of heat buildup. The heat of fermentation can raise rumen temperatures above 40°C—especially when the animals are feeding on hard-to-digest roughage, which is often the bulk of the diet in regions with poor grazing.

[4] International Livestock Centre for Africa, 1979.

## BONSAI BRAHMAN

In Mexico, researchers are deliberately creating microcattle. Since 1970, Juan Manuel Berruecos Villalobos, former director of the Veterinary Medicine school at the National Autonomous University, has directed this enterprise. He and his colleagues have miniaturized cows by selecting the smallest specimens out of a herd of normal-sized Brahman cattle and breeding them with one another. After five generations, adult females

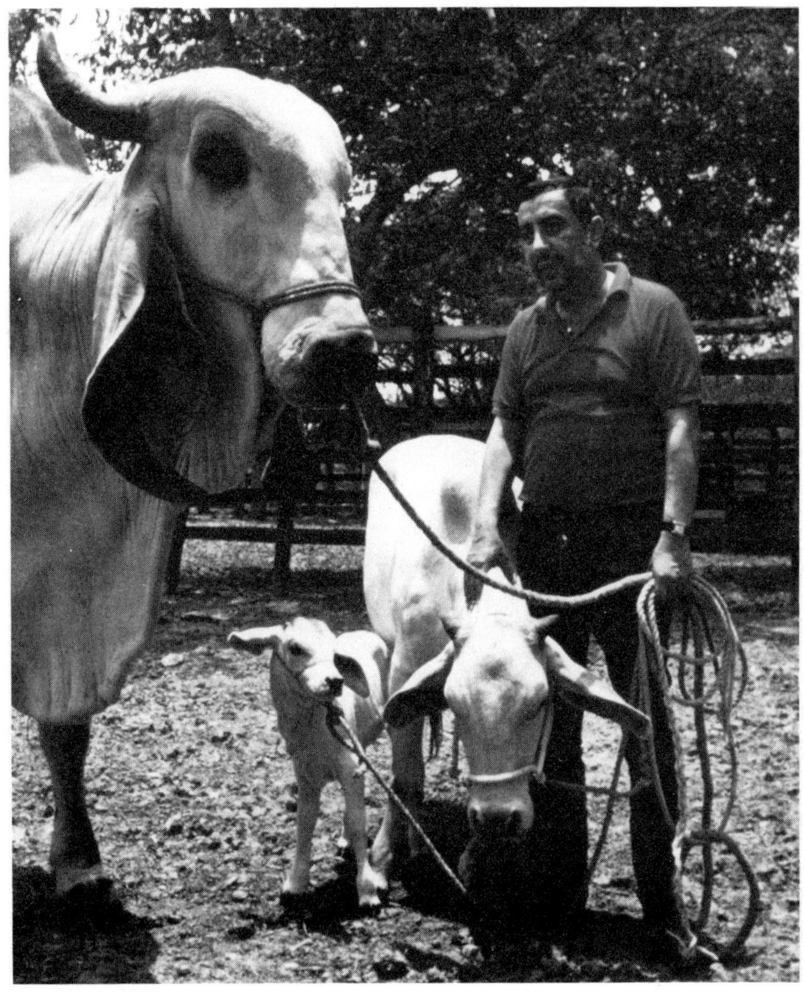

A miniature Brahman cow and her calf stand next to a Brahman of normal size. Although only one-fifth the weight, a miniature Brahman provides two-thirds as much milk as a normal cow. (Sergio Dorantes, Sygma)

average 150–180 kg; adult males 200–220 kg. A few of the smallest cows are now only 60 cm tall and 140 kg in weight. Merely one-fifth of normal weight, they are shorter than the turkeys that share the barnyard with them. Indeed, they even get lost in the grassy pastures so that the farmers cannot see them.

This program seems to have yielded a productive animal that can be cheaply and easily maintained in a small space. Berruecos has demonstrated that the tiny cows can be stocked on one-third the area needed to support one normal-sized cow. He reports that they are giving remarkable amounts of milk: up to four liters a day, compared with six liters from their full-sized counterparts. On a feed-intake to weight-gain basis, the tiny cattle are no less efficient than their normal-sized counterparts.

Although 17 years have gone into the selection of what Berruecos calls his "bonsai cattle," the process is not yet finished. Future goals include testing embryo transplants to see if one normal-sized cow can support multiple "microfetuses" (possibly as many as eight). This would help to rapidly increase the numbers of the miniature form, which weigh merely 4–5 kg at birth.

All in all, the Mexican researchers see miniaturization as a new option for governments and farmers increasingly squeezed by shrinking farm land and rising production costs. Small livestock, they say, are a way to produce more food on less land faster. For example, a campesino with almost no land can have one or two bonsais, but could never maintain a standard-sized cow.

---

highly developed breeds. Their greatest potential may prove to be for traditional husbandry and for grazing marginal areas where survival is more important than feed efficiency.

## RESEARCH AND CONSERVATION NEEDS

Their adaptability and robustness make microcattle worthy of preservation, study, and greater use, and they should be incorporated into many ongoing programs.

Selective breeding, although infrequently attempted, can probably improve productivity significantly. Records of breed history should be established, and unusual or special characteristics noted and the information disseminated.

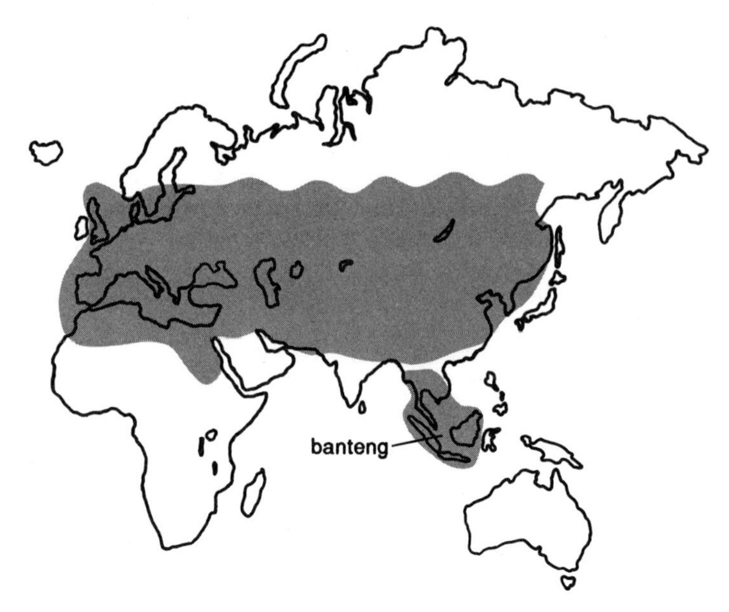


Original distribution of wild cattle and banteng. (Based on Mason, 1984)

In areas where small, indigenous breeds are being replaced, representative populations should be maintained and studied to increase understanding of their adaptive diversity and to retain a genetic storehouse for the future.

## REPRESENTATIVE EXAMPLES OF MICROCATTLE[5]

### Dwarf West African Shorthorn

West African coastal forests, and inland. Female 125 kg; male 150 kg. Adaptation to harsh, humid climates and good resistance to trypanosomiasis and other diseases allow these small animals to exist where other cattle die. They are perhaps the smallest cattle of all (often weighing less than 100 kg). In the areas of worst disease and highest rainfall, this hardy animal is often found thriving, but half-wild.

[5] Weights are sometimes difficult to obtain and verify, and many (perhaps most) breeds have never been scientifically examined. When weight is listed by gender, they are representative; when a range is given, it includes both males and females.

**Muturu** Nigeria. Female 160 kg; male 210 kg. This notable subtype is slightly larger. It is the most trypanotolerant of all cattle, showing no symptoms or loss of vitality. It is widely kept, mostly as a village scavenger and often as a pet, and yields a high percentage of meat.

## N'Dama

West Africa. 200–400 kg. These active, stocky animals utilize low-quality forage, produce good beef, and are used as light draft oxen. Milk production, though poor, improves with feeding level. N'Dama mature early and are exceptionally fertile, and they have already become important in breeding programs. They are resistant to trypanosomiasis, and can exist where temperatures average 30° C with 1,500 mm annual rainfall. In the least hospitable areas, N'Damas ranging down to 200 kg are often the only cattle that can remain productive.

## Rodope

Southeastern Europe. Female 200 kg; male 350 kg. A humpless multipurpose breed—draft, milk, and beef—that is exceptionally hardy. The milk is high in butterfat. Possibly adaptable to the subtropics. It is rapidly being lost to crossbreeding.

## Zebu

Zebus are among the most important tropical domestic animals.[6] However, the dwarfs are not well known, although in many areas they are preferred, especially as draft animals. Zebus use less water, even though their sweat glands are larger and more numerous than those of most other cattle. All have a low basal metabolism and resist heat well. In general, they also have high resistance to ticks and other parasites.

**Taiwan Black** Taiwan. Female 250 kg; male 250 kg. Well adapted to poor tropical conditions, these work animals are also used for meat.

[6] Although many questions above bovine evolution remain unresolved, all humped cattle are usually considered to have some zebu blood. However, not all of them are classified as zebus.

**Kedah-Kelantan** Malaysia. Female 200 kg; male 250 kg. Hardy, well-adapted cattle with exceptional fertility on a poor diet, both sexes are used as draft animals as well as sources of meat and cash.

**Sinhala (Dwarf Zebu)** Sri Lanka. Female 200 kg; male 250 kg. An ancient type of zebu, preferred for its handiness in cultivating small paddies and terraced fields.

**Nuba Dwarf** Sudan. 180–220 kg. These work animals are well proportioned but are not slaughtered for meat, and milk production is low. Although tolerant to trypanosomiasis, their numbers have dwindled because of crossbreeding.

**Small Zebu** Somalia. 160–230 kg. These small native cattle are used for beef, milk, and for work. They are well adapted to poor feed in a desolate environment.

**Abyssinian Shorthorn Zebu (Showa)** Central highlands of Ethiopia. Female 225 kg; male 305 kg. These widespread, small-humped cattle are very hardy. They produce beef and are generally milked, with surplus production about 2–4 kg daily. Resistant to many parasites, they also have a gentle disposition and make good work animals.

**Dwarf Zebu (Mongalla)** Tanzania, Uganda, and Kenya. Female 150 kg; male 250 kg. A highly variable, long-entrenched, small, East-African zebu with some nonzebu blood. Pastoralists favor it because of its hardiness. Although slow-maturing, it is well-fleshed, can yield excellent beef, and some types are milked.

**Mashona** Zimbabwe. Female 200 kg; male 250 kg. This hardy zebu-sanga type (see below) is widespread in drier areas and has a high resistance to disease and parasites. Since the 1940s, it has been bred for beef production and selected animals now weigh more than 500 kg.

**Mini-Brahman** Mexico. 135 kg. Downsized from 450-kg Brazilian zebus through selective breeding by Mexican researchers, these gentle animals are reported to yield two-thirds as much milk (3–4 liters daily) as the parent stock. Because of much higher stocking rates on grass, production per hectare is reportedly greater than with full-sized animals (see sidebar, page 22).

## Criollo

Central and South America. Descendants of Spanish and Portuguese cattle imported over 400 years ago, "criollo" cattle have adapted to a wide range of harsh climates. Many varieties are small: mature females often weigh 200–300 kg or less. They sometimes produce little beef or milk under traditional conditions and management, but they are extremely hardy and survive when other cattle perish. Through importation and crossbreeding, many local types have been lost or are threatened.

**Chinampo** Baja California, Mexico. 200–350 kg. Extremely tolerant of wild desert conditions, these docile criollo cattle exist largely on scrub and cactus. They get most of their water from succulent plants, have a low metabolic rate and body temperature, and are mostly active at night.

**Florida Scrub** Florida, USA. 225–300 kg. Genetically isolated for more than 300 years, the Florida Scrub is very hardy in harsh, subtropical conditions. It has good resistance to ticks and screwworm, and can subsist on forage with a high roughage content.

## Sanga

This type—an ancient cross between longhorns or shorthorns and humped animals—is found throughout eastern and southern Africa. It weighs from 150 to 500 kg or more. Some types have been selectively bred or crossed with European cattle and are quite productive.

**Bavenda** Transvaal, South Africa. 240–290 kg. This hardy and disease-tolerant tropical variety is small and prolific. It is generally used for draft, barter, and beef. However, it has been crossbred with larger animals so frequently that the smaller types are almost extinct; most "Bavendas" now weigh more than 300 kg.

**Ovambo**[7] Northeastern Namibia. Female 160 kg; male 225 kg. A calm and docile animal with a small hump, it is used by seasonal pastoralists for beef and milk.

[7] This is a small type of the Kaokoveld sanga. Its small size may perhaps be due to a mineral deficiency rather than its genes.

## BANTENG: THE CUTEST COW

The banteng (*Bos javanicus*) is a small Southeast Asian bovine with a promising future.* It is a different species from cattle. The two will interbreed, but the hybrid offspring are normally sterile.

Although almost entirely neglected by the animal science community, the banteng is remarkable for an ability to thrive under hot, humid, and disease-ridden conditions where cattle often grow poorly. The sexes are easily distinguished: males are jet black; females are golden brown. Both have bright white socks and rumps as if they had been freshly whitewashed.

Kluang, Malaysia. Banteng in an experimental herd. (N.D. Vietmeyer)

Wild banteng are found in remote areas of countries from Burma to Indonesia. But only Indonesia has used it as a farm animal so far. It has more than 1.5 million domesticated banteng—some 20 percent of the country's total "cattle" population. Indonesian farmers value the animal's agility, which allows them to cultivate fields too narrow for cattle to turn the plow. In addition, gourmets consider banteng meat the tastiest of all. Indonesia appreciates the banteng so much that it has established a genetic sanctuary on the island of Bali—banning cattle in order to maintain the banteng's genetic purity.

Outside Indonesia, only a few scientists have studied this animal, but it seems clear that it is particularly useful under tropical conditions. In heat and humidity, it thrives; even when cattle are starving, one rarely sees a skinny banteng. And demand for its meat is never ending.

* Often called "Bali cattle," this animal is better called "banteng" because it is neither a cow nor exclusively from Bali. Under good conditions, it can be 50 percent bigger than the microcattle considered in this chapter, but in practice most banteng are within the 300-kg upper limit we have chosen to define microcattle.

** More details can be found in the companion volume *Little-Known Asian Animals with a Promising Economic Future*.

**Nilotic** Sudan. 180–300 kg. These cattle of southern Sudan show great variation in size, partly due to environmental factors. They are generally resistant to local parasites and worms, have good potential for increased beef production, and their milk is very important locally.

## Chadian "Native" and Dwarf Black Cattle

Chad. Female 225 kg; male 275 kg. These two types are small, humped meat animals that graze the sparse savanna and are very drought resistant. Little scientific information exists about them.

## "Arab Cattle"

Middle East. Small types (female 225 kg; male 300 kg) are used for meat and some milk, especially in Lebanon. There are many local

Ethiopia: Microcattle in market. (World Bank)

forms with variable appearance, but all have small humps. Well adapted to grazing sparse vegetation on rough land, they are becoming rare due to crossbreeding.

## Hill Cattle

Nepal. Female 160 kg; male 200 kg. A widespread type often re-crossed with Indian zebu animals, they are bred to be small. They are thrifty creatures that maintain themselves well on poor forage. Bulls make sure-footed draft animals on rough ground and slopes, and the cows are milked.

## Tibetan Dwarf

Tibet. Less than 250 kg. These humpless cattle are used as pack animals and can tolerate poor forage and high altitudes.

## Yellow Cattle

Southwest and south China. Female 220 kg; male 380 kg. In the subtropics and tropics, small multipurpose types of Yellow Cattle withstand high temperature and humidity. They are used mainly for work and meat, and seem well adapted to poor feed, harsh conditions, and rugged terrain. The Chowpei (190–380 kg) is a hardy working breed of more temperate areas in Hubei Province.

## Cheju Hanwoo

Korea. Female 230 kg; male 280 kg. A yellowish-brown Cheju Island native that has almost no calving difficulty, it is well adapted to poor grazing conditions in harsh environments and is docile and obedient.

## Madura

Indonesia. Female 220 kg; male 300 kg. An ancient cross between humped cattle and the banteng (see sidebar), these heat- and disease-resistant hybrids also have good grazing and mothering ability, and are kept in the most extreme humid tropical environments. Breeding for fighting and racing has given them a poor disposition.[8]

## Dexter Cattle

Ireland and North America. 220–360 kg. This breed can be traced back to eighteenth-century Ireland and is believed to have been developed by peasant farmers living on rough land. It is exceptionally hardy and produces both milk and meat. In North America, it has become popular among city folk who acquire country property, as this microbreed is particularly well suited to their usually tiny farms.

[8] These interesting animals are described in the companion volume *Little-Known Asian Animals with a Promising Economic Future*.

# 2

# Microgoats

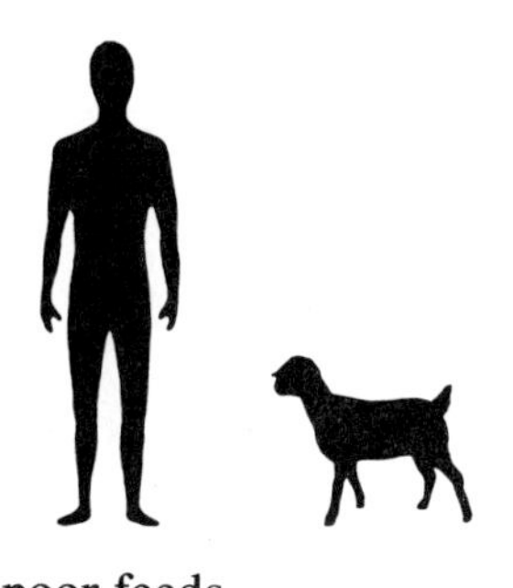

More than 90 percent of the world's nearly half-billion goats (*Capra hircus*) are found in developing countries; many weigh less than 35 kg fully grown.[1] Such "microgoats" are noted for their high reproductive rates, rapid growth, early maturity, tasty meat, and rich milk, as well as for their robust constitution, ease of handling, and tolerance of climatic stress and poor feeds.

To many people—especially where pigs and poultry are not common—meat and milk from microgoats are the primary animal proteins consumed during a lifetime. Perhaps the world's best foragers, goats eat practically anything made of cellulose, and are not dependent on grass. Because of their unselective feeding behavior, they are capable of living where the feeds—tree leaves, shrubs, and weeds—are too poor to support other types of livestock.

Such microgoats deserve wider recognition, for they are often the poor person's only source of milk, meat, and cash income. They are cheap to acquire and easy to maintain, even by people with little property and scarce resources.

## AREA OF POTENTIAL USE

Worldwide, especially in arid and semiarid climates.

## APPEARANCE AND SIZE

Goats generally have a long snout and an upright tail, by which they can be distinguished from most sheep. The mouth is unusual in having

[1] Small goats are found not only in developing countries, however. One of the smallest, the American pygmy, was developed in the United States. The most productive European milk goats are derived from once-small Norwegian and Swiss animals.

a mobile upper lip and a grasping tongue, which permits the animal to nibble even tiny leaves on spiny species.

Common commercial goat breeds generally weigh between 60 and 100 kg, with some weighing more than 200 kg. Microgoats may weigh less than 15 kg. Representative examples are listed at the end of the chapter.

## DISTRIBUTION

Worldwide, with half in Asia and one-third in Africa.

## STATUS

The FAO projects that world numbers may nearly double by the turn of the century. Goats are thus not endangered, but in some areas select populations of feral goats are being deliberately eradicated, with the consequent loss of potentially valuable genes. Some small breeds are also threatened by excessive crossbreeding with larger types.

## HABITAT AND ENVIRONMENT

One of the most adaptable of all livestock, goats can persist in conditions from arid to humid, and from sea level to high altitude. They are especially well adapted to hot, semiarid climates and to rocky, barren terrain.

## BIOLOGY

These ruminants can subsist on many feedstuffs that would otherwise be left to waste. Although selective browsers, they often prefer coarse leaves (including palm fronds) and shrubbery to palatable forage grass.

Most microgoats mature quickly, and in the tropics they can generally breed year-round. Their reproductive potential has often been underestimated; kidding is rarely difficult, and many types produce twins and sometimes even triplets or quadruplets.

In hot, dry areas, goats require less attention than other livestock, and smaller goats have the added advantage of better heat dissipation. Some microgoats may also show disease resistance. For example, tolerance to trypanosomiasis makes them an important livestock in many regions of Africa.

Mali. Small goats can survive in the most desolate conditions and live off the most unpalatable feeds. (IDRC)

## BEHAVIOR

Goats are generally gentle, but can be easily frightened. They may become stubborn and aggressive when threatened or thwarted, and can prove hard to confine.

If their feed smells of other animals—particularly of other goats—they usually shun it unless nothing else is available.

## USES

Microgoats mainly produce meat and form an important part of the diet in southern Asia, the Middle East, Africa, and Latin America, especially the Caribbean. Goat is sometimes a preferred meat, and there are few social or religious prohibitions against eating it.

Some microgoats are good milkers, and under stressful conditions they may keep producing when other livestock are dry. Goat milk is a valuable dietary supplement: it is nutritious, easily digestible, and usually commands premium prices. It makes excellent cheese and yogurt and can be used by people allergic to cow's milk.

Microgoats produce some of the finest and most valuable fibers in the world. Angora and Cashmere goats often weigh less than 30 kg fully grown, for example.

Goats produce a fine-textured, durable leather that finds extensive uses both locally and internationally. Horns, hooves, blood, and bone meal also have commercial value. Manure is another important product, and comes in fairly dry pellets that are easy to collect, store, and distribute.

Goats perform important functions in land management. Seeds of many trees (*Acacia* and *Prosopis*, for example) are "scarified" by passing through the goats' digestive system, fostering germination and natural revegetation. With care, goats can also be used to clear land of weeds and brush.

## HUSBANDRY

Goats are often allowed to roam and scavenge for their own food. They form strong territorial attachments and can be trained to stay within a designated area. However, they cannot be kept from investigating—and quite probably devouring—anything within that territory. They are persistent browsers, so it is essential to prevent overstocking as well as raids on crops.

Variety of diet is important, and goats show much individuality in feed preferences. They are often raised on crop residue and kitchen refuse.

Goats can be run with other livestock without creating serious competition. The goats browse weedy shrubs, whereas the sheep and cattle graze more on grasses.

Although perhaps the hardiest of all livestock, most breeds benefit when they are provided shelter from rain and high-noon sun. Abrupt chilling and poor ventilation can cause severe respiratory problems. They are also susceptible to various maladies, such as internal parasites, especially when confined. The highest mortality, however, is caused when very young kids are not supplied with adequate feed and clean, dry shelter.

## ADVANTAGES

In most developing countries goats are already prominent in rural life. Common almost everywhere in Africa, Asia, and Latin America, they are dependable multi-use animals. They are particularly important in providing ready cash, such as for school fees, taxes, marriages, or funerals.

Goats integrate well in mixed agriculture, for example, by consuming leafy wastes, clearing land, and contributing fertilizer. In many places they are raised almost exclusively by women and children. If confined, goats require only simple, inexpensive shelters or pens, which makes them especially important as subsistence animals. In many situations, they may be the most efficient and economic producers for smallholders.

These animals have a relatively fast rate of growth and early reproductive age, even under harsh conditions. They can graze rougher terrain than cattle and most sheep, can go for longer periods without water, and forage well in wooded areas where grass is lacking. They can derive most or all of their diet from roughage unusable by humans; high-energy feeds, such as protein supplements or carbohydrate supplements, are usually not needed even to fatten them for slaughter.

Goats are generally healthy and are not affected by many of the parasites and diseases that ravage other livestock. Some resistance to mange, internal parasites, foot-and-mouth disease, and other livestock scourges has been reported.

## LIMITATIONS

In some places (notably, in industrialized nations) there is a strong prejudice against goats and goat meat.

Smallness makes microgoats targets for predators and thieves.

Many small goats are poor milkers, especially under hardship conditions; however, even small amounts of milk can often fulfill a child's daily nutritional requirement or reinforce a nursing mother's diet.

Goats are independent and may wander away if not watched, and they can be difficult to pen. They may also have an unpleasant odor when kept confined (males are particularly malodorous during rutting season).

## FRESH GENES

A rare wild animal with spectacular horns, the bezoar (*Capra aegagrus*) is the goat's wild ancestor. People domesticated it before 7000 B.C., probably in the mountains along the Iran/Iraq border. Until recent times, it remained widely scattered across the vast region between Greece and Pakistan, but it now exists only in pockets and is threatened with extinction.

This would be a tragedy because the bezoar is a resilient wild species that crosses readily with domestic goats, and it could pass on its genetic inheritance for heat, drought, and cold tolerance; disease resistance; and other survival qualities.

Fascinating science and valuable results probably await those willing to study this hardy, handsome creature and to explore the reharnessing of its genetic endowment. Today the bezoar is considered merely a trophy for hunters. The power of its genes to refresh—perhaps even revolutionize—the world's 500 million goats has been lost to sight.

Cretan bezoar. One of Europe's last remaining wild goats, the bezoar is down to a mere 100 specimens on two small Greek islands. Here these potentially valuable creatures (locally known as "agrimi") are just a misstep away from extinction. (Zoological Society of San Diego)

Distribution of the bezoar. The arrow indicates the area where it was probably first domesticated, resulting in the goat as we know it. (From Mason, 1984)

Goats are often disparaged for degrading land and destroying vegetation because they continue to survive on overutilized lands often laid waste by mismanagement of sheep or cattle.

## RESEARCH AND CONSERVATION NEEDS

The microgoat's potential has hardly been realized. More research on performance and husbandry is needed to preserve and restore small breeds. Selective breeding for prolificacy, viability, and rapid growth, as well as more selective on-site culling, could greatly improve both meat and milk yields and quality.

Management systems that exploit smallness, stabilize production, and preserve the environment should be introduced and publicized in appropriate goat-rearing areas. Careful assessments of indigenous management methods should be made, particularly emphasizing their desirable characteristics. Improving hygiene in the wet season and supplemental feeding in the dry season are also important, as are disease- and parasite-control measures.

The undomesticated ibex and markhor could possibly be major contributors in the development of new, useful breeds for tropical and arid regions (see sidebar, page 42).

## REPRESENTATIVE BREEDS OF MICROGOATS

### West African Dwarf (Djallon)

West and Central Africa. Female 20 kg; male 30 kg. Adapted to humid lowlands, this widespread goat is particularly valuable for meat and skin production. Generally, it is bred for meat, but milk is sometimes an important secondary product. Sexual maturity is very early (3–6 months), and quadruplets occasionally occur (most goat breeds normally produce only single births). Related types go by the names "Cameroon Dwarf," "Dirdi," and "Nigerian Dwarf."

### Nubian Dwarf

United States. 35–40 kg (often less). A stable miniature variety of the milking Nubian, this microgoat has been developed recently in the

Ghana. Half-grown West African Dwarf goats. Their average weight is 11.5 kg. (C. Devendra and M. Burns. *Goat Production in the Tropics*. Commonwealth Agricultural Bureaux, Farnham Royal, Buckinghamshire, UK, 1983.)

United States by crossing standard-sized Nubians with the West African Dwarf. It combines a good milk output with high levels of butterfat.

## American Pygmy

United States. 15–25 kg. Derived from the West African Dwarf, it is noted for its hardiness and good nature, good milk production, and adaptability to various climates. There are several varieties, some for milking, others for meat.

## Sudanese Nubian

Northern Sudan. 25–30 kg. Widespread milk goats of riverain and urban areas.

## Sudanese Dwarf

Southern Sudan. 11–25 kg. A very hardy desert goat similar to the West African Dwarf, it averages 15 kg, but some mature individuals may weigh as little as 11 kg. Used for meat and hides, it produces little milk.

## Small East African

Kenya, Uganda, Tanzania. 20–30 kg. A widely neglected meat and hide animal found over a wide range, it is fast growing (sexual maturity at four months) and extremely hardy.

## Mauritian

Mauritius. 25–30 kg. A prolific, year-round breeder raised for meat production, it is often confined in simple shelters from birth to slaughter. Perhaps because of this isolation, mortality is less than 10 percent, even with little or no veterinary care.

## Criollo

Latin America. "Criollo" is a name given to several breeds of ancient Iberian blood with local adaptations to many unfavorable conditions. They are often small and hardy.

## WILD RELATIVES

Several wild relatives will cross with the goat. Surprisingly, they have the same chromosome number (2n = 60), and the offspring are frequently fertile. Although essentially unknown to agricultural science, these hybrids may offer a new gene pool for creating new farm animals and for improving the world's goats. They seem to combine the self-reliance of wild species with the usefulness of domestic ones. Artificial insemination and other modern techniques could make them easier to produce today than ever before.

### Ibex*

A project in Israel has already produced a cross between the goat and the Nubian ibex (*Capra ibex*). The Sinai Desert goat, the breed that was used, ranks next to the camel in its ability to go without water—it often drinks only twice a week—but its meat has such a strong flavor that most people consider it dreadful. On the other hand, the ibex is compact and muscular and produces tender, mild meat that steak lovers find delicious. The product from crossbreeding the two is a creature seemingly able to endure extreme temperatures and drought, make use of poor pasture, and produce wonderful steaks.

A herd of several hundred of these hybrids (dubbed "ya-ez") has been created at Kibbutz Lahav in the northern Negev Desert area. Both sexes are fertile, and they can be bred with each other or with either parent. The meat is already in demand on the menus of elegant Tel Aviv hotels.

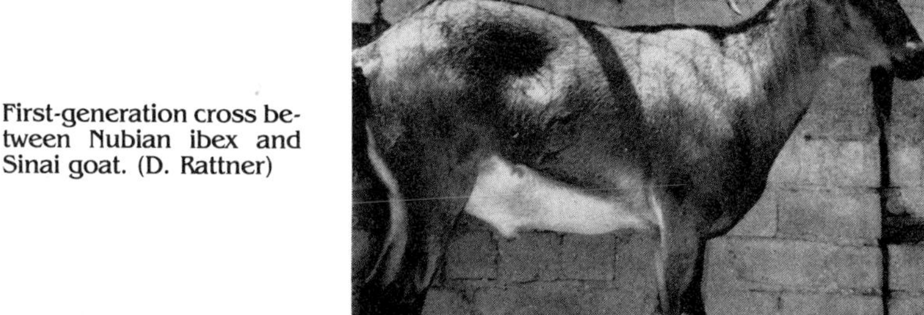

First-generation cross between Nubian ibex and Sinai goat. (D. Rattner)

Gilgit, Pakistan. A 2-year-old markhor/goat hybrid maintained as a stud male for improving a goat herd. Genes from such half-wild animals help enrich many domestic goats of this region. (G. Rasool)

**Markhor****

In Pakistan's northern uplands, it is not uncommon to find hybrids between domestic goats and the mountain goat known as "markhor" (*Capra falconeri*). Each year in Chitral and Gilgit, they can be found in the goat markets.

Markhors inhabit high elevations in rugged mountains and thrive on diets so meager as to be useless to goats. The hybrids are produced when markhor males—perhaps ousted by more dominant males—come in contact with feral domestic goats. However, some farmers raise young markhor and goats together (to overcome mutual resistance) and produce their own hybrids.

For a single hybrid animal, local goatherds pay up to 5,000 rupees, a princely sum in this impoverished region. Traditionally, villagers have kept them as stud animals. They appreciate the animal's genetic endowment. Markhors tolerate extremes of cold and snow, are nimble and skilled at escaping predators, and survive on scanty fodder. Moreover, they have a high reproduction potential because they generally produce twins. As a result, they also tend to give more milk, and it is rich in nutritive value. Instead of long body hairs, markhors possess insulating underfur—a soft and valuable raw material for the famous Kashmiri shawls.

Apparently, the hybrids can possess many of these qualities together with a calm disposition. Thus they could be useful in themselves and as conduits for passing such traits on to goats.

---

* Information from D. Rattner.
** Information from G. Rasool.

**Creole** Caribbean. Females 20 kg; males 25 kg. Robust meat goats of Spanish or West African origin that are kept throughout the Caribbean.

**Crioulo** Brazil. 30–35 kg. A meat and skin goat derived from Portuguese ancestors, it is hardy, prolific, undemanding, and adapted to harsh environments.

## Chapper

Pakistan. Female 20 kg; male 24 kg. Originating in dry regions, this meat and milk goat is a nonseasonal breeder with outstanding potential.

## Barbari

Pakistan, India. Females 20–25 kg; males 20–40 kg. A prolific, fast-growing "urban" goat with high twinning and low mortality. Often kept inside houses, they adapt well to confinement and are important for both milk and meat.

## Gaddi (White Himalaya)

Hill districts of northern India. 25–30 kg. Kept for meat and their long, lustrous white hair, they are pure-breeding and healthy.

## Changthangi (Ladakh)

Kashmir, India. Male 20 kg. A pashmina (cashmere) goat of India, it is adapted to a high altitude, high humidity climate with extremes of temperature.

## Terai

Nepal. 8–12 kg. A very small, hardy animal of the southern lowlands, it kids year-round (sometimes twice), and often produces twins.

## Southern Hill Goat

Nepal. 12–16 kg. A small, mid-altitude goat resembling the Terai.

## Black Bengal (Teddy, Bangladesh Dwarf)

Eastern India and Pakistan. Female 10 kg; male 14 kg. A widespread, humid-area, meat goat that is early maturing and very prolific. It kids twice a year, and produces 60 percent twins and 10 percent triplets. It produces a superior leather.

## Katjang

Southeast Asia, China, and Pacific Islands. In places, less than 20 kg. A widespread, highly variable, hardy goat adapted to humid conditions, it usually has twins or triplets. Used for meat and skins, with exceptional females being milked.

## Chinese Dwarf (Tibetan, Jining, Fuyang, or Chengdu Grey)

China. 20–40 kg. Well adapted to the humid tropics, it normally twins and is a good meat producer.

## Heuk Yumso

Korea. Female 25 kg; male 35 kg. A prolific cold-climate goat with a year-round breeding season. The meat is highly prized, and often sells at a premium due to its supposed health-giving effects.

## Hejazi

Middle East. Female 20 kg; male 20 kg. A meat goat, usually black, for harsh desert conditions.

## Sinai (Black Bedouin)

Sinai, Egypt and Negev Desert, Israel. Female 20 kg; male 50 kg. Native to dry, hot deserts, this milk and meat goat matures at 5–8 months and has a twinning rate over 50 percent. A most important characteristic is its drought tolerance. The female, for instance, can drink only once a day—at a pinch, once every other day—without losing appetite or reducing milk flow.

TOM PHILLIPS

# 3

# Microsheep

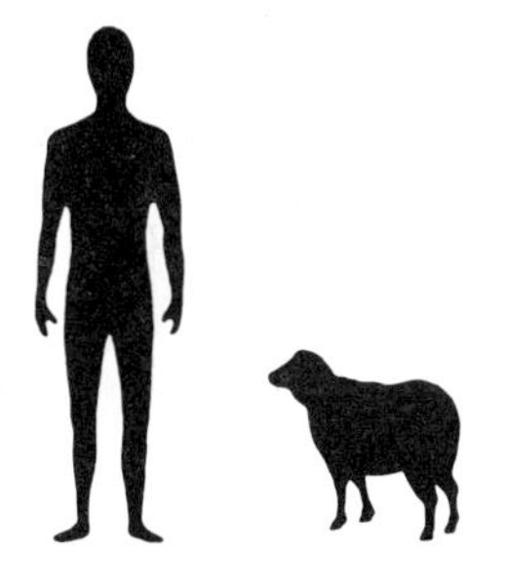

Among the hundreds of breeds of sheep (*Ovis aries*) in the world, those weighing less than 35 kg when mature have been largely ignored. Although these are common, the impression lingers that they are too small to be useful. Yet this virtually untapped gene pool is especially well adapted to traditional Third World animal husbandry. Given attention, these "microsheep" could boost meat, milk, skin, wool, and pelt production in many villages and small farms of Africa, Asia, and Latin America.

Many microsheep thrive in environments that tax the ability of larger breeds to survive. They are adapted to poor feeds and can be grazed in uncultivated wastelands unsuited to any other livestock except goats or camels. Because of their size, microsheep can fatten in areas where forage is so scattered and sparse that larger animals cannot cover enough ground to fill their bellies each day. In addition, their foraging complements that of other livestock. For example, sheep can graze rough grasses and weeds that cattle find unpalatable. Some survive even the stress of extreme aridity and for this reason are the predominant livestock in North Africa and the Middle East.

Many small breeds can be disease resistant. Some, for example, are widespread in the zones of Africa where trypanosomiasis is prevalent. They are generally less adversely affected by foot-and-mouth disease than are cattle, and some small native sheep seem to have fewer problems with insects and parasites than do most other livestock, including temperate-area sheep.

Giving more attention to the management and improvement of microsheep could pay back abundantly in the form of food, income, and improved land utilization in many parts of the developing world.

## AREA OF POTENTIAL USE

Worldwide, but notably in drier regions of the tropics.

## APPEARANCE AND SIZE

An average weight for temperate sheep breeds is about 70 kg,[1] but the smallest microsheep weigh less than 20 kg fully grown. Many tropical microsheep are "hairless," and have little or no wool. These are often difficult to distinguish from goats, but (like all sheep) they generally have blunter snouts, more fat, and hanging tails. Some have greatly enlarged rumps or tails that store fat. Unlike goats, sheep have no odor-producing glands.

Some representative microsheep are described at the end of this chapter.

## DISTRIBUTION

More than one billion sheep occur worldwide, and they occupy every climatic zone in which people live. At least half are in developing countries.

## STATUS

Although more than 1,000 breeds are recognized, only a handful dominate the world's sheep industries. Lesser-known breeds are rapidly becoming extinct (especially in developed countries, although scattered efforts are being made to preserve them). Elsewhere, genetic resources have not been properly evaluated, and potentially valuable stock is being lost before it is even understood.

## HABITAT AND ENVIRONMENT

Sheep are among the most adaptable animals. Various types are kept in areas of extreme heat, cold, altitude, aridity, humidity, and rainfall. They are especially widespread in hot, dry climates, but some breeds also thrive in humid areas.

[1] The largest, such as the New Lincoln breed of New Zealand, may weigh up to 250 kg.

Ecuador: In the Andes, as in many parts of the Third World, small sheep are a common resource. (E.G. Huffman, World Bank Photo)

## BIOLOGY

Sheep make efficient use of a wide variety of fodder: tree leaves, forbs, grasses, crop residues, and agricultural by-products, for instance. They often survive privation by calling on their reserves of body fat.

In the tropics, sheep reach sexual maturity in about a year. Many breeds lamb year-round, which allows for a continuous production of premium meat. Gestation takes about five months, and lambing is usually timed to occur when feed is most abundant and nutritious. Microsheep often bear two or more young and, under good management, may produce lambs annually for more than five years.

## BEHAVIOR

These shy animals flock together and, in general, are managed with little effort. They are easily panicked, however, and rams can become aggressive during rutting or when threatened.

## THE LITTLE SHEEP THAT COULD

*Dozens of the world's neglected breeds of tiny sheep should be preserved from extinction, for many will undoubtedly prove to have outstanding qualities. Current efforts to save the Navajo sheep in the United States exemplify what can be achieved.*

The Navajo is a microsheep, and is perhaps the oldest breed of sheep in the United States. It may have been introduced to North America in 1540 by the Spanish explorer Francisco Vázquez de Coronado, who was seeking the mythical Seven Golden Cities of Cíbola in the region that is now Arizona and New Mexico. Smaller than many dogs, a full-grown Navajo sheep may weigh only 30 kg, but it became a big part of the culture of the Southwest. Although the Navajos and other local Indians had never seen sheep before the 1500s, they soon became shepherds and weavers, and their rugs made from the unique wool of this wiry little animal remain famous even today.

Navajo sheep have white or brown wool hanging in ringlets around their bodies. The fleece is a double coat: long, coarse guard hairs on the outside and short wool on the inside. It yields warm, waterproof, and long-lasting products. Many of the sheep have four horns because the Indians believed that this trait was sacred, and they favored four-horned rams for breeding purposes.

The number of Navajo sheep was reduced sharply between 1930 and 1950 because of severe overgrazing and replacement by improved wool breeds. In recent times there has been so little commercial and scientific interest in this microsheep that by the 1970s only a handful of purebred specimens survived. Since the late 1970s, however, Lyle McNeal, a Utah State University professor, has been working to save it from extinction. By 1988 he had a burgeoning flock at the university and was learning that this supposedly obsolete dwarf is amazingly useful.

The breed originated in the arid south of Spain (where it is called the "churro"), and it thrives in the hot, dry climate. Unlike normal breeds, it can exist in the desert without supplementary food and with little water. As McNeal has pointed out, any sheep that can survive and raise a lamb in the aridity and searing heat of the American Southwest has to be superior. He has found that the ewes have a strong maternal instinct, which is vital for protecting lambs against the coyotes that are common in the region.

Ogden, Utah. Lyle McNeal and the Navajo sheep. (Lynn R. Johnson, *Salt Lake City Tribune*)

Thanks to the efforts of McNeal and his colleagues, Indians are beginning to use Navajo sheep again; by 1988 there were more than 400 on the Navajo reservation, with their wool fetching premium prices. This tough little sheep could prove valuable not only for American Indians but for poor people in many other dry regions as well.

## USES

Microsheep are mainly kept for meat production, but—especially in arid regions—for milk as well. Their meat is usually lean with little "muttony" taste.

Wool or hair is taken from many breeds, although the yield is often small. Skins from hair sheep, thinner than cowhide, are widely used and are in international demand. In some places, manure is considered an important product. In Nepal, thousands of small sheep are used as pack animals, especially to carry salt into mountain valleys.

## HUSBANDRY

Most sheep are maintained in free-ranging flocks. Many are grazed (often tethered) over a small area during the day and confined in a "fold" at night. Others are penned or kept as village scavengers. These are usually fed supplements of household scraps.

Sheep form an integral part of a mixed farming economy; for example, they may graze pastures during the wet season, and survive on crop residues and field weeds during the dry season. They have excellent foraging capabilities and are often kept alongside goats. This broadens the variety of forages utilized and often increases total production from a single piece of land, for sheep and goats have complementary feeding habits and male goats help protect the sheep from some predators.

In spite of the heavy toll that predators (such as feral dogs) can take on lambs and ewes, the largest proportion of sheep in the tropics are lost through lack of basic care. Modest supplemental feeding of lambs and inexpensive preventive medicines can do much to lower mortality and boost production.

## ADVANTAGES

Sheep are multipurpose animals, and almost everywhere they produce several products. The rich milk is often preferred to that of cows or goats, especially for making cheese and yogurt.

Lambs form an important part of the household economy for much of the rural world, and only rarely is social or religious stigma attached to keeping or eating them. Indeed, sheep are the traditional feast animals of several religions, and in some places sheep meat is preferred to beef and sells at a premium.

By and large, all sheep products can be processed, utilized, or marketed by the producer. In addition, sheep marketing and transportation systems exist in most countries, at least to some degree.

Sheep are efficient producers and can provide a quick turnover for food and cash. On the brush and coarse grasses of marginal lands, they may be more productive than cattle, and on grass they may outproduce goats. As long as they are not overstocked, sheep do not degrade vegetation; unless starving they will not debark trees. Small breeds cause little erosion, even on steep slopes, heavily traveled paths, or near water holes.[2] In South Asia, they have been continuously stocked on the same ground for thousands of years without causing apparent harm.

Because sheep have a natural tendency to accumulate fat, they "finish" well on grazing and usually do not require a high-energy finishing diet.

## LIMITATIONS

Despite their general healthiness, sheep are affected by many internal parasites and diseases, a few of which are communicable to man. They are especially susceptible to infectious conjunctivitis (pinkeye).

Predators and thieves can be greater threats than sickness. Labor inputs can be high because of the almost continual protection sheep need.

Some mutton has a strong taste that many find unappealing. However, the taste is carried mainly by the fat, and the generally lean microsheep are often commended for their fine-textured, sweet meat.

## RESEARCH AND CONSERVATION NEEDS

The numerous breeds of small sheep should be investigated. Assessments should be made for the animals' ability to thrive under adverse conditions and for resistance to particular diseases and parasites.

As noted, even minimal extension services and veterinary support for sheep could greatly decrease mortality, especially among lambs.

Improving microbreeds without increasing their size is one of the most interesting challenges facing sheep scientists today. While efforts should be made to conserve and select within types, research should also be conducted on hybrid vigor. Efforts to improve the pelt and fleece of microsheep should also be encouraged.

More studies on the interactions between sheep and cropping systems are needed. Sheep (and the manure they produce) could become

[2] The small Soay sheep, for instance, is used in Cornwall, England, to graze banks of highly erodible china-clay spoil too unstable to carry heavier animals.

## SMALL SHEEP IN THE FOREST

*Even in countries with long traditions of raising large sheep, there are opportunities for using small, agile, hardy breeds. The following is an example.*

Seeking safer methods for stopping brush from smothering newly planted trees, U.S. government foresters have turned from chemical defoliants to flocks of sheep. Court decisions in 1983 and 1984 barred the use of herbicides along Oregon's Pacific Coast. Various alternatives were tried, and the animals proved the most successful. Sheep are now the favored method for controlling unwanted vegetation. Indeed, they have changed the foresters' whole approach to managing reforestation.

Formerly, the U.S. Forest Service allowed the brush to grow on logged-over sites and then sprayed it down before planting tree seedlings. Now it plants grass to suppress brush and reduce erosion. The sites are later fertilized, tree seedlings are planted, and within a year sheep are brought in to graze.

Alsea, Oregon. A flock of sheep suppressing brush that would otherwise smother young fir trees. (Gerry Lewin/*New York Times* Pictures)

Today, in the district around Alsea, Oregon, sheep nimbly skirt old stumps to graze on the lush vegetation. Three times each summer since 1984, about 2,000 sheep have been guided across the replanted areas by a herder and a range conservationist. The sheep eat both the grass and the new buds on brush, but they leave most fir-tree seedlings untouched. The key, according to Rick Breckle, a forester, is to have enough sheep to graze an area evenly and to keep them moving so they don't resort to nibbling the young trees.

Previously, chemical brush treatments had annually cost $135–$353 per hectare. Now, sowing grass and grazing sheep costs about $300 per hectare. And there is a product to sell: the adult sheep don't fatten well, but the lambs bring a useful income at the end of the summer. What is more, Breckle reports that the trees seem to be growing faster—probably because of the manurings they receive.

This method seems likely to be effective elsewhere—at least with trees that are unpalatable or too tall for their growing points to be nibbled. Malaysia, for instance, doubled its sheep population between 1986 and 1989, in part because it has begun raising sheep between the trees in rubber plantations. With the use of agroforestry increasing worldwide, small sheep could find a whole new application.

---

important components of forestry (see sidebar), crop rotation, alley cropping, and other forms of sustainable agriculture. For instance, sheep are especially effective for weed control in plantation crops such as oil palm and rubber as well as in forests.

## REPRESENTATIVE EXAMPLES OF MICROSHEEP

### West African Dwarf

Senegal to Nigeria, and south to Angola. Female 25 kg; male 35 kg. Well adapted to warm, humid conditions. Prolific, and good disease resistance. Major meat producer in West Africa. Fast growing: by six months of age they approach adult weight.

### Landim (Small East African)

East and Central Africa. 23–40 kg. Prolific, adaptable, long fat-tailed type. Large litter size for a sheep. In one recent test, ewes averaged more than 1.4 lambs.[3]

[3] Wilson et al., 1989.

### Berber

Atlas Mountains. 25–41 kg. Needing little feed and remaining constantly outdoors, these extremely hardy sheep are exploited for meat and their coarse, hairy wool. They fatten easily when well fed.

### Arab

North Africa. 40–50 kg. This thin-tailed sheep is exceptionally robust, and is resistant to extremes of temperature, drought, and poor nutrition. Primarily a meat producer, its wool is used for coarse cloths and carpets.

### Southern Sudan Dwarf

One of the many small breeds of eastern and southern Africa, its weight ranges from 15 to 25 kg, but it may weigh as little as 11 kg. Yielding a fine, short fleece, this hardy, frugal sheep is often run with cattle to maximize grazing.

### Hejazi

Deserts of Arabia. 32 kg. A popular and ancient fat-tailed meat producer that is highly acclimatized to drought and privation.

### Zel (Iranian Thin-Tailed)

Caspian region of northern Iran. Female 30–32 kg. Well adapted to subtropical regions, they produce coarse wool, milk, and excellent meat that lacks the "mutton taste" and odor of some sheep meats.

### Greek Zackel

Mountain and island types. Female 30 kg; male 40 kg. These common sheep are active, hardy, and resistant to extremes of climate and disease. Primarily a milking sheep, their wool is used locally and lambs are slaughtered for special occasions.

## Sitia

Crete. Female 25 kg; male 30 kg. Another of the hardy, screw-horned "zackel" sheep common to the Balkans, they are adapted to poor pasturage and extensive herding. Quick maturing and highly fertile, they can be exploited for milk as well as for meat and coarse wool.

## Common Albanian

Female 25 kg; male 35 kg. Similar to the Greek Zackel, they are used as triple-purpose animals: meat, milk, and wool. They survive in low, marshy areas where parasites are common.

## Zeta Yellow

Yugoslavia. Female 25 kg; male 35 kg. A small, hardy sheep used for milk and some meat, its primary product is wool. Often unshorn for several years, the long fibers are woven into expensive carpets.

## Pag

Yugoslavia. Female 20–30 kg; male 25–35 kg. These wool, milk, and meat sheep are frugal and well adapted to scant vegetation and rocky terrain. Although they have a low birth rate and carcass yield, their milk and wool are commercially exploitable.

## Roccia (Steinschaf)

Northern Italy, Austria. Female 30 kg; Male 30–35 kg. These "stone sheep" resemble a goat in their ability to exploit the poor pastures of high, steep, rocky mountains. Although not highly productive, they are hardy and frugal and commonly produce twins.

## Corsican

Corsica (France). 25–30 kg. A hardy native breed that is well adapted to rather sparse feed conditions. Coarse wool, both white and black, is well suited for hand processing.

## WILD AND WOOLLY

*Considering that dozens of countries depend on the productivity of more than a billion domesticated sheep, it is remarkable that their wild ancestor is accorded no attention. This fast-declining animal is now little more than a trophy for hunters, a fact that should be of vital international concern.*

Sheep were domesticated in the Middle East and Central Asia in the Stone Age era between 8,000 and 11,000 years ago.* Their wild ancestor was almost certainly the mouflon (*Ovis orientalis*). However, domestication may have occurred in more than one place, and two other wild creatures, the urial (*Ovis vignei*) and the argali (*Ovis ammon*), also possibly provided genes to some sheep breeds.

The mouflon, urial, and argali still exist in the mountains of Central Asia, and a European subspecies of mouflon is also found in the Mediterranean, but only on Corsica, Cyprus, and Sardinia.** Because they live in remote, rugged, upland areas, these wild sheep are usually undisturbed, but the numbers are decreasing everywhere.

This may be a serious loss because these animals could be extremely valuable. They are capable of crossing with domestic sheep, and the offspring are viable and fully fertile.*** For developing new meat-producing breeds, their potential seems almost limitless.

During the thousands of years that sheep have been protected by humans, their wild ancestors have continued to face predators, parasites, disease, extreme cold, and seasonal starvation. Their genetic endowment, forged and tempered in unforgiving harshness, could be a benefit for all future sheep generations. These animals appear to resist various diseases. Their meat is reported to be of excellent quality, notably lacking the strong mutton flavor that many people find objectionable. They have relatively short, thin tails—a feature that might eliminate the need for docking (tail removal) in the domestic flock. Some (for instance, the Asian mouflon and the urial) have rates of effective reproduction up to 1.6 lambs per ewe, more than twice the average of most domestic types, especially under the conditions where these wild creatures live.

That mouflon and other wild sheep could have practical utility is suggested by research at Utah State University. Scientists there have mated mouflon with farm sheep to create sheep better able to defend themselves against coyotes and other natural dangers. Half-wild, half-tame sheep hybrids have existed on a ranch in southern Utah for the past decade. Also, in Cyprus similar mouflon x sheep hybrids have shown considerable promise.

At the very least, this wiry little mountain sheep could be a model for educating students and the public. It is a living reminder of the fantastic changes that can be induced in animals by selection for various traits. Also, it is a "map" to the history of sheep domestication. Studies of mouflon genes, blood, immunology, morphology, physiology, horn structure, skeleton, fleece, temperament, and a host of other features would help unravel the ancestry. These studies and various biochemical analyses would be a fascinating contribution to agriculture, science, history, and the public perception of the origins of our natural resources.

Genes from wild sheep are not likely to quickly benefit wool production. Lack of fleece is one reason why these creatures have been neglected, but throughout most of Asia and in North Africa, sheep are bred primarily for meat and milk, and there is a growing worldwide interest in the use of hair sheep. All of this brings new possibilities for the use of this old resource.

---

* The first evidence for a domesticated sheep comes from 11,000-year-old remains discovered at Zawi Chemi Shanidar in Iraq. Almost certainly, it happened when mouflons were attracted to the lush crops in farmers' fields.

** The specimens on these islands are probably feral domesticates, but they seem little different from truly wild mouflon. They are probably relics of the first domestic sheep taken to Europe by Neolithic farmers sometime between 6,000 and 7,000 B.C. On Corsica and Sardinia they escaped and have lived wild ever since.

Another relic is the Soay (see page 60), found on islands in the Outer Hebrides of Scotland. They, too, are primitive, but their fleece proves them to be rather more domesticated than the mouflon.

*** Mouflon is the only wild sheep with the same chromosome number as domestic sheep (2n=54). The urial (2n=58) and the argali (2n=56) have different chromosome numbers, but nevertheless, both will hybridize with domestic sheep, and the offspring are also fertile.

## Entre Douro e Minho

Portugal. Female 15–18 kg; Male 20–25 kg. These independent sheep yield a good wool in mountainous terrain that would otherwise be nonproductive.

## Churra do Campo

Portugal. Female 20 kg; male 30 kg. A coarse-wooled sheep extensively kept in Portugal's dry interior for milk and wool.

### Galician

Spain. Female 18 kg; male 25 kg. A milking breed that survives on poor pasture, it also produces a marketable wool.

### Soay

Scotland. Female 25 kg; male 30 kg. Adapted to wide temperature variations. Possibly the most primitive domesticated sheep of Europe, probably unchanged from Viking times. Immune to foot rot. A wool sheep with short brown fleece that is shed annually.

### North Ronaldsay (Orkney)

Northern Scotland. 27–32 kg. Surviving year-round on seaweed, this rare breed is adapted to high salt intake and the associated digestive problems. Yield 1–2 kg medium-coarse wool.

### Criollo

Latin America. Derived from "native" Spanish Churro and Merino sheep. Many are small and very hardy.

**Navajo-Churro** Southwestern United States. Female 45 kg; male 70 kg. Maternal, and very resistant to internal parasites and hoof rot. Although the Navajo subsists and reproduces on little feed and scarce water in desert regions, it was widely replaced by improved breeds earlier in this century. Because of its hardiness, however, and the use of its wool in traditional weaving, its numbers are rebounding (see sidebar, page 50).

**Florida Native** Southeastern United States. Females 35–45 kg; males 45–60 kg. This long-isolated and highly variable sheep is adapted to harsh subtropical climates and is known for its ability to forage. A medium-wool breed, it is very resistant to intestinal parasites. Verging on extinction due to neglect and uncontrolled crossbreeding.

### Virgin Islands White Hair (St. Croix)

Caribbean. Female 35–45 kg; male 45–55 kg. Hair sheep with some wool in young animals. Well adapted to warm humid conditions, it has fairly good disease and parasite resistance and produces good meat. Prolific, it breeds most of the year and commonly has twins.

## Magra (Chokhla)

Northwest India, Pakistan.[4] 20–25 kg. Adapted to hot, dry areas, the extremely white and shiny fleece is valued for carpet wool. Slow-maturing and low fertility (lambing at 45 percent) plus extensive cross-breeding have led to serious declines in population.

## Marwari

Northwest India. 25–30 kg. A widespread, white-fleeced sheep that has a high resistance to disease and worms, good fertility, and low mortality. They do well in large flocks.

## Mandya (Bandur)

Southwest India. Female 25 kg; male 35 kg. An outstanding meat breed with good mutton quality, it adapts well to mixed farming and has unusually low lamb mortality.

## Hu (Huyang, Lake Sheep)

China. Female 35 kg; male 45 kg. These fat-tailed sheep have a six-month lambing interval and are very prolific. They are used under intensive management to produce meat, wool, and a valuable lambskin.

## Javanese Thin-Tail

Indonesia. 25–40 kg. Widely held as a "bank account," these meat, manure, and skin sheep are well known for being prolific. Although single lambs are not uncommon, litters of six have also been recorded.

[4] South Asia has many fascinating breeds of microsheep. India has the Balangir (Orissa), Chotanagpuri (Bihar and West Bengal), Coimbatore (Tamil Nadu), Ganjam (Orissa), Hassan (Karnataka), Kilakarasal (Tamil Nadu), Madras Red (Tamil Nadu), Malpura (Rajasthan), Mecheri (Tamil Nadu), Nali (Haryana), Ramnad White (Tamil Nadu), Sonadi (Rajasthan and Gujerat), and Tiruchy Black (Tamil Nadu). Pakistan has the Baltistan (Baltistan), Buchi (Punjab), Kooka (Sind), and Kaghani, Michni, Tirahi, and Waziri (all North-West Frontier). Females of all these breeds weigh less than 25 kg at maturity.

# 4

# Micropigs

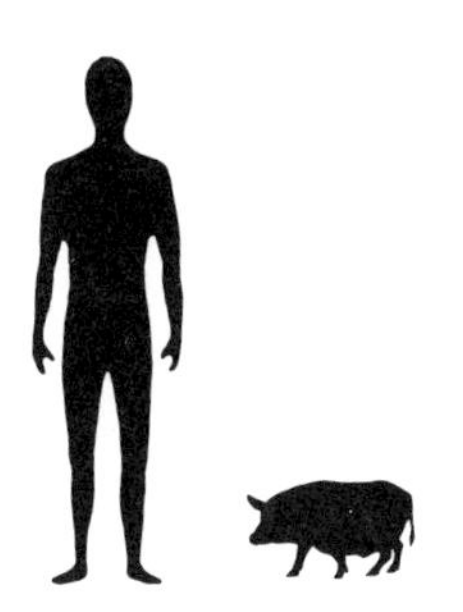

Most breeds of swine (*Sus scrofa*) are too large to be considered microlivestock, but there are some whose mature weight is less than 70 kg. These micropigs ®[1] are particularly common in West Africa, South Asia, the East Indies, Latin America, and oceanic islands around the world. At least one, the Mexican Cuino, may weigh a mere 12 kg full-grown.

Many miniature swine have been developed for use in medical research, but their agricultural potential has been largely ignored. This is unfortunate, for micropigs of all types—native, feral, and laboratory—deserve investigation. Swine provide more meat worldwide than any other animal, and micropigs are potentially important sources of food and income for poor people in many parts of the developing world.

Smallness makes for nimble and self-sufficient pigs, in contrast to large, lethargic breeds. Small breeds are easier to manage and cheaper to maintain; the threat of injury from angry or frightened animals is lessened; and the sows are less likely to crush newborn piglets, often a major cause of mortality in large breeds. Some micropigs—particularly those from hot regions or wild populations—also have a higher resistance to heat, thirst, starvation, and some diseases.

Pigs adapt to a wide variety of management conditions, from scavenging to total confinement; some are even kept indoors.[2] They gain weight quickly, mature rapidly, and help complement grazing livestock because they relish many otherwise unused wastes from kitchens, farms, and food industries, as well as other foods such as small roots, leafy trash, or bitter fruits that are not consumed by humans or ruminants.

[1] We use the word in its generic sense to represent all extremely small types. Charles River Laboratories, Inc. (see Research Contacts) has trademarked the term "micropig," particularly for their strain of the Yucatan miniature pig, which is becoming widely used in medical research.

[2] Small pigs are raised as indoor livestock in Vietnamese cities, for example.

For these reasons, micropigs could become useful household and village livestock in the developing world, and they deserve greater attention than they now receive. Although their growth may not be as rapid as that of improved breeds raised under intensive commercial production, with modest care and minimum investment, backyard micropigs can produce sizable yields of meat and other products, as well as improved income for rural and even urban populations.

## AREA OF POTENTIAL USE

Worldwide, especially in warm, humid areas.

## APPEARANCE AND SIZE

Like full-sized breeds, micropigs are stout-bodied, short-legged animals with small tails and flexible snouts ending in flat discs. Examples of some micropigs are listed at the end of the chapter.

## DISTRIBUTION

Domestic pigs are found all over the world, but their concentrations vary greatly. Africa has the fewest per capita, but in recent years they have gained increasing favor in the sub-Saharan regions. In Latin America, pigs have long been a major component of backyard agriculture. In the Middle East, an early center of domestication, pigs are not widely kept today because of religious dietary restrictions. In the Far East, they are the major meat source, and China has more pigs than any other country. And in the Pacific region, pigs and chickens are often the only meat available.

## STATUS

Pigs are becoming more popular: their worldwide numbers increased by about 20 percent in the 1970s. However, in most countries commercial pig production has focused on a mere handful of breeds, and much genetic diversity is unstudied or even threatened with extinction. Some microbreeds have already been lost, and others are dwindling in numbers.[3] Many European breeds have been completely lost. The Cuino and some other Latin American criollo types are threatened, as are most of Africa's traditional breeds. China, however, has made notable efforts to preserve its native types.

[3] Because pigs breed at an early age and can have many large litters, physical characteristics can be lost quickly. Even within a single litter, piglets can show great diversity.

## HABITAT AND ENVIRONMENT

Although, as previously noted, pigs are found all over the world, they are in general adapted to warm, humid climates where many other livestock species are more susceptible to diseases and environmental stresses. They are also raised at high altitudes, such as in the Andes and Tibet. Although there are few climatic limitations to pig production, only about 20 percent of the world's pigs are currently kept in the tropics.

## BIOLOGY

Pigs are omnivores, willing and able to eat almost anything.[4] Unlike most other livestock, they eat their fill and sleep as the food digests, allowing humans to establish a convenient eating and sleeping schedule.

Pigs are prolific; a few Chinese breeds routinely have litters of 20 or more. Micropigs are no exception; litters of 6–10 are common. Piglets gain weight rapidly and can be weaned after a few weeks. Sexual maturity is sometimes attained as early as 4–6 months, depending on breed and environment. Pigs are usually slaughtered at 6–7 months of age, allowing them to be produced on an annual cycle. They can live 10–20 years.

Because of their smaller size, micropigs have a relatively greater skin-to-weight ratio than today's commercial breeds, and therefore they probably shed heat more effectively. Certainly they seem to perform better in tropical heat and humidity, which normally keep the heavier types from reaching their maximum productivity. Studies have suggested that an optimal size for some tropical environments—because of metabolic and feed efficiency—may be less than 65 kg.[5]

## BEHAVIOR

Pigs are social animals; they enjoy companionship and ferociously defend their young and sometimes even the humans who care for them. They are employed as guard animals in some areas and have been used extensively in behavioral research.

Contrary to common belief, pigs are clean and tidy if provided adequate space. Larger breeds, however, wallow in mud to stay cool in hot weather and require a wallow or shade (except for some Latin American types, which seem less susceptible to heat). Some light-colored pigs sunburn easily.

Pigs will dig up earth with their mobile snouts; some breeds do it constantly.

[4] Although pigs have some ability to ferment cellulose in their intestines, they do not digest it as well as ruminants do, and cannot utilize large amounts of roughage.

[5] Williamson and Payne, 1965.

## USES

Fresh pork is the major pig product in tropical areas. It usually fetches premium prices, and in many places (such as the Pacific Islands and China) it is the most important red meat available to rural people. Nutritious and tasty, it is one of the easiest meats to preserve, needing only salt or melted fat. Processed products such as bacon and sausages can be important for both home consumption and cash sales.

Pig fat (lard) is a good source of food energy, and can substitute for cooking fats and oils. It is easily melted and clarified, is widely used to make soap, and is a valuable commercial product.

Pig skin, once degreased, is easily tanned into leathers that are popular for garments, shoes, and other products demanding soft, light, and flexible leathers.

Pig manure is a good fertilizer. Because the animals are often kept in confinement, it can be easily collected.

## HUSBANDRY

In many places, pigs are kept as free-roaming scavengers. They can be trained (by coaxing with feed, salt, or affection) to keep close to home, thereby helping to minimize destructive scavenging.

Herding is a higher level of management that requires more effort, but it allows pigs to be integrated into other types of agriculture while utilizing feeds that otherwise go to waste.

Because their exercise needs are minimal and dominance is quickly established within litters, pigs are the easiest hoofed livestock to raise in small enclosures (sties). However, fencing must be secure, and if sties are small, the animals must be moved frequently to prevent diseases and parasites from building up.

## ADVANTAGES

Pigs are well-known, often traditional, animals in many areas, and people usually do not have to be taught how to manage and use them. Efficient scavengers, they can live, grow, and reproduce with a minimum of investment or specialized care.

Opposite: Riverside, California. Recently, Vietnamese pot-bellied pigs have become popular pets in the United States. They sell at high prices and are kept indoors. Purportedly they "love baths, never have fleas, and do not shed hair. They are also easily housebroken and are a nonrooting variety of pig." Their real potential, however, is in Third World villages. (Steve Ellison/*People* Weekly © 1988 Time Inc.)

Pigs are highly efficient converters of feed to meat. They can provide the greatest return for the least investment of any hoofed livestock because of their fecundity, low management costs, broad food preferences, and rapid growth.

Pigs normally accumulate fat during adolescent growth (making "finishing" feeds less necessary). Some micropigs (especially those from feral ancestors) have the ability to quickly mobilize and store these body-fat reserves; in times of extreme scarcity, it aids their survival.[6]

Pigs work well in multiple-cropping schemes. They are often used to help clear small plots by uprooting weeds, shrubs, and even small trees. In Southeast Asia, they are frequently raised in conjunction with aquaculture, their manure providing food for the fish.

## LIMITATIONS

If improperly managed or maintained in filthy conditions, pigs may quickly succumb to disease and parasite epidemics. Most diseases are communicated only among pigs, but some can be transmitted to humans. For this reason, pork should always be fully cooked.

Some cultures never eat pork. Others do, but nonetheless accord pigs and their keepers low status.

Young pigs are vulnerable to many predators.

## RESEARCH AND CONSERVATION NEEDS

A major survey of small pig breeds is needed. They have the potential to be valuable producers in their own right, as well as to improve other pig breeds. For instance, they represent a little-known reservoir of disease resistance and climatic adaptation. Governments, research stations, universities, and individuals should make special efforts to preserve types that have outstanding or unusual qualities.

When it is necessary to eradicate feral pig populations (as is common on Pacific islands), representative stocks should be preserved. These rugged animals have been genetically isolated for decades or even centuries and are likely to carry valuable traits for survival under adversity.

Large breeds may be promising candidates for genetic "downsizing," which has already produced the many types of miniature pigs that are used in medical research.

[6] Information from I. L. Brisbin.

## THE LITTLEST PIG

*Although this chapter highlights the world's smallest breeds, there exists a pig that is even smaller. It is, however, an entirely different species and it is on the brink of extinction.*

The pigmy hog (*Sus salvanius*) is a shy and retiring wild creature of northeastern India. It is merely 60 cm long, with a shoulder height of 25 cm, and weighs less than 10 kg. It was once found widely along the southern foothills of the Himalayas. Today, however, it apparently occurs in only one area, the Manas National Park in Assam. Despite this protection and the fact that it is listed among the 12 most endangered species on earth, it still falls victim to hunters and to habitat destruction—especially illegal grass fires.

If saved from extinction, this minute species—barely reaching a person's calf—might become useful throughout the world. Its chromosome number is the same as that of the common pig, and its physiological processes are probably also similar. Therefore, were its numbers to be built up, it might become a valued and well-known resource for laboratories and small farms. Its daily food intake and its space requirements are only a fraction of a normal pig's. It probably has exceptional tolerance to heat, humidity, and disease.

This is not a domesticated species, and there is therefore much to learn before its usefulness can be clearly seen. Indeed, whether it can be reared in captivity is uncertain. Some attempts have ended in disaster, but this seems to have been the result of mismanagement.

Before there is any possibility of developing it, however, the pigmy hog must be preserved from ultimate loss. The last specimen could go into a villager's pot at any time now.

Adult male of the common pig (wild boar) and pigmy hog drawn to same scale. (W.L.R. Oliver)

## REPRESENTATIVE EXAMPLES OF MICROPIGS

### West African Dwarf (Nigerian Black, Ashanti)

West Africa. Mature weights of 25–45 kg are reported. In the humid lowland forests of West Africa this breed has long been kept by villagers, often as a scavenger. Indigenous to the hot, humid tsetse zones of West Africa, it seems resistant to trypanosomiasis.

### Chinese Dwarfs

China (and Southeast Asian countries such as Vietnam) has long had small pigs—often characterized by numerous teats and large litters—associated with traditional intensive agriculture as well as scavenging conditions. Some Chinese pigs weighing less than 70 kg are adapted to tropical and subtropical conditions, but the smallest (20–35 kg) live in the cold climates and high altitudes of Gansu, Sichuan, and Tibet. Small black Chinese pigs were crossed with European types in the early 1800s and produced the foundation stock of many modern Western breeds.

### Criollo

There are a number of "native" breeds throughout Latin America commonly known as "criollo." Many are quite small. Although, apparently, they are slow to mature and bear small litters, they adapt well to environmental extremes and are widely kept by rural inhabitants for food and income. Criollos are little studied and are being replaced by imported breeds before their possibly outstanding qualities can be quantified.

**Cuino** This micropig from the highlands of central Mexico may be descended from small Chinese types and is the smallest domestic pig, weighing as little as 10–12 kg fully grown. Hardy and an efficient scavenger, it can grow quickly when feed—especially corn—is abundant. A century ago the cuino was a widespread household animal and was used for a time for experimental work in central Mexico. It is now little known and could be threatened with extinction.

**Black Hairless (Pelon, Tubasqueno, Birish)** These small pigs of central and northern South America survive in hot, humid, adverse climates. They are adapted to bulkier feeds than most pigs and can thrive on fruit wastes. Many local types exist.

**Nilo (Macao, Tatu, Canastrinha)** This small, widespread, black, hairless pig of Brazil is often kept inside the house.

**Yucatan Miniature Swine** A subtype of the black hairless from Mexico's hot, arid Yucatan Peninsula, it was imported into the United States in 1960. It has been downsized for laboratory use in the United States and is known as the Yucatan Micropig®. Weight at sexual maturity has been lowered through selective breeding from 75 kg to, currently, between 30 and 50 kg, with an ultimate goal of 20–25 kg. There is no evidence of "dwarfism," stunting, or loss of reproductive performance, and it appears to hold notable promise as microlivestock for developing countries. The parent stock, used for meat and lard production in Yucatan, is renowned for gentleness, intelligence, resistance to disease, and relative lack of odor. Exceptional docility, even in older boars and sows with litters, makes them easy to handle without the need for specialized housing or equipment.[7]

## Other Laboratory Breeds

Other miniature laboratory pigs have potential for tropical use. These include the Goettingen, Hanford, Kangaroo Island, Ohmini, Pitman-Moore, and Sinclair (Hormel). In general they weigh 30–50 kg when ready for slaughter and mature at less than 70 kg.

## Ossabaw

United States. 20–30 kg. Feral on Ossabaw Island, South Carolina, for more than 300 years, this pig is well adapted to environmental extremes. Unlike most domestic animals, it can maintain itself in coastal salt marshes. It has perhaps the highest percentage of fat of any pig. The piglets are very precocious, self-reliant, and robust.[8]

## Kunekune (Pua'a, Poaka)

New Zealand. Female 40 kg; male 50 kg. Perhaps of Chinese origin, these black-and-white spotted pigs are docile, slow, and easy to contain. Although late maturing, they can fatten on grass alone. Like other native breeds throughout the Pacific region (for example, the Pau'a of Hawaii), stock is being lost through crossbreeding, displacement by other breeds, and eradication efforts.

[7] Information from L. Panepinto.

[8] Information from I. L. Brisbin. Small feral pigs are established on many islands worldwide, often left by seafarers as a future source of meat.

*To find certain disease-resistant genes in poultry it may be necessary to go looking in the backyard chicken flocks in Latin America, Africa, or Asia.*

Kelly Klober
*Small Farmer's Journal*

*. . . Policies are needed to encourage development of a labor-intensive small-scale livestock sector, which would increase employment and provide a major market for surplus cereals. This sector, however, is particularly restrained by poor technology, poor public support services, and poor marketing channels. Third World livestock production of this type could provide a natural focus for foreign assistance that earlier seemed inappropriate because of concerns about global food scarcity.*

John W. Mellor
International Food Policy Research Institute

*Successful development of agriculture often requires an intimate understanding of the society within which it is to take place—of its systems of values, of its customary restraints. . . . It has been necessary to understand what incentives the farmer needs to change, what practical difficulties he encounters introducing change, what his traditional pattern of land use is and how this pattern or system can be upset by thoughtless innovation.*

John de Wilde
*Experience with Agricultural Development in Tropical Africa*

# Part II

# Poultry

Chickens, ducks, muscovies, geese, guinea fowl, quail, pigeons, and turkeys epitomize the concept of microlivestock. Throughout Africa, Asia, and Latin America they are (collectively) the most common of all farm stock. In many—perhaps most—tropical countries, practically every family, settled or nomadic, owns some kind of poultry. In the countryside, in villages, even in cities, one or another species is seen almost everywhere; in some places, several may be seen together. Although raised in all levels of husbandry, these birds occur most often in scattered household flocks that scavenge for their food and survive with little care or management.

Their size bestows microlivestock advantages, including low capital cost, low food requirements, and little or no labor requirements. They are also "family sized": easily killed and dressed, with little waste or spoilage.

These poultry species help meet the protein needs of the poorest people in the world. Some are raised even in areas where domestic cattle cannot survive because of afflictions such as trypanosomiasis and foot-and-mouth disease. Some are maintained under conditions of intensive confinement—provided a source of feed is available—and can be produced in areas with insufficient land for other meat-producing animals.

In addition, these birds grow quickly and mature rapidly. (For instance, a chicken can, under proper conditions, reach maturity in 2–6 months.) They adapt readily to being fenced or penned much, or all, of the time. And, compared with the major farm livestock, their life cycles are short and their production of offspring is high. Thus, farmers can synchronize production to match seasonal changes in the availability of feed.

Poultry are the most common livestock in the Third World. Small flocks are found in almost all rural homes, farms, villages, and towns. Picture shows woman in Burkina Faso (West Africa) winnowing pounded cowpeas. Her chickens and guinea fowls scavenge any broken, spilled, or spoiled seeds that would otherwise be wasted. Most Third World poultry flocks live a wary, half-wild existence, scrounging for insects, earthworms, snails, seeds, leaves, and leftovers from the human diet. (IDRC)

## THE INDUSTRIAL CHICKEN

Throughout modern livestock farming the trend is toward more intensive methods, and poultry specialists have set the pace. In many countries, since the 1920s, barnyard fowl have given way to egg and broiler factories. The old-fashioned chicken reared outside on corn stubbles for 5 or 6 months has been replaced by the broiler, mass-produced in controlled-environment houses in 7–10 weeks.

As a result of this revolution in poultry raising, small farmers who once made a comfortable living from a few laying hens have been forced out of business. These economic changes have also forced poultrymen to have larger and larger flocks to survive. The largest broiler-chicken companies even control their own breed development, feed production, house construction, slaughtering, and freezing; many even have wholesale outlets.

The rapid changes in poultry farming methods can be attributed to the application of advanced technology. The development of the incubator to replace the mother hen sitting her seasonal clutch of eggs was the first major step toward intensive poultry farming.

In addition, chickens were the first livestock to receive serious attention from geneticists. Before World War II, it was discovered that crossbreeding selected pure and inbred lines could result in dramatic increases in production. Hybrids tailor-made for egg or meat production quickly ousted the old pure breeds such as the Rhode Island Red, White and Brown Leghorns, Light Sussex, and the various crosses among them. Chicken broilers made by crosses involving parents derived from Cornish and Plymouth Rock have supplanted all others.

This situation now prevails in most industrialized countries. The breeding of commercial stocks is in the hands of a few corporations for each commodity (white eggs, brown eggs, chicken broilers, turkeys) and each has national or even global distribution of its hybrid stocks.

Although poultry contribute substantially to human nutrition in the tropics, it is a small fraction of what it could be. The meat is widely consumed and is in constant demand. An excellent source of protein, it also provides minerals such as calcium, phosphorus, and iron, as well as the B-complex vitamins riboflavin, thiamine, and niacin. Nutritionally as complete as red meat, it is much lower in cholesterol and saturated fats. Poultry eggs are also important sources of nutrients.

## A BREAKTHROUGH IN POULTRY HEALTH

Newcastle disease is endemic in developing countries and is a constant threat to poultry. Farmers dread this virus, first identified half a century ago in northern England, that brings diarrhea, paralysis, and death to most poultry. It is severe, highly contagious, and can cause 100 percent mortality. When it strikes an area, farmers must kill all chickens—even healthy ones—to stop it from spreading.

Only Australia, New Zealand, Northern Ireland, and some Pacific islands are unaffected. But, although the disease is not found in Australia, certain strains of the virus *are* present in Australian chickens. These strains are completely harmless, but Australian researchers have found that they induce antibodies that are effective against Newcastle disease.

In a joint project (funded by the Australian Centre for International Agricultural Research), scientists from Malaysia's University of Agriculture and Australia's University of Queensland* have put this to good use. They have produced a live culture of the harmless virus that farmers can spray onto feed pellets to vaccinate their birds.

Field tests of the new vaccine, carried out in Southeast Asia, have been extremely promising. Simply coating feeds with the virus seems to be enough to immunize some chickens, which then pass the immunity on to the others in the flock as well as to new hatchlings. In Malaysia, which has 49 million chickens and a population willing to pay a premium for tasty village poultry meat, one economist estimates that the vaccine might increase rural incomes by 25 percent.

Conventional vaccines must be stored under refrigerated conditions, which most villages lack. But the Malaysian workers made the Newcastle disease vaccine tolerant of heat. By selective breeding, they now have strains that resist 56°C for at least 2 hours. Thus, even in the tropics, the vaccine remains effective for several weeks without refrigeration. The researchers have also devised methods for coating the vaccine onto pelleted feeds. Because the virus can withstand heat, they use a machine designed for coating pharmaceutical tablets.

At this stage, the project is showing every promise of producing a cheap means of reducing Newcastle disease losses among chickens throughout much of the world. Already inquiries have come from other Asian countries and from Africa, and it is hoped that the vaccine may eventually benefit many countries.

* Peter Spradbrow, Latif Ibrahim, and Rod Cumming.

They are a renewable resource, easy to prepare, and are among the best sources of quality protein and vitamins (except vitamin C).

In spite of their numbers and potential, poultry are rarely accorded primary consideration in economic development activities. All in all, these small birds lack the appeal of large, four-legged livestock. Indeed, most countries have little knowledge of the contribution household birds actually make to the well-being and diets of their peoples. In some countries—even those where birds are widely kept—there is little or no poultry research or extension. And where such programs do exist they usually focus almost exclusively on the production of chickens under "industrial" conditions near cities (see sidebar, page 75).

Most developing countries now have these intensive chicken industries, in which birds are kept in complete confinement. However, these commercial operations provide food for people in the cash economy, not for subsistence farmers. Moreover, grain is sometimes diverted or imported to maintain these operations, perhaps causing food shortages, higher prices, or depleted foreign exchange. Thus, in this section we focus on other, neglected, aspects of poultry production.[1]

The neglect of poultry that scavenge around the rural farmhouses and in village yards is understandable. The birds are scattered across the countryside where extension programs are difficult to implement. Their presence is often so ingrained in traditional village life that they are taken for granted and ignored by the authorities.

Yet village poultry deserve greater attention. As converters of vegetation into animal protein, poultry can be outstanding. In fact, it is estimated that, in terms of feed conversion, eggs rank with cow's milk as the most economically produced animal protein, and that poultry meat ranks above that of other domestic animals.

Most Third World poultry flocks live a wary, half-wild existence, scrounging for insects, earthworms, snails, seeds, leaves, and leftovers from the human diet. From dung and refuse piles they salvage undigested grains, as well as insects and other invertebrates. Often the persons who care for them are women or children. Some keep the birds around the house, penning them at night for protection from predators and thieves.

This almost zero-cost production has, in spite of high losses, a remarkable rate of return. Any improvements that require the purchase of supplies cut severely into the profitability. The first step in improving the production of free-ranging poultry is vaccination against diseases (especially Newcastle disease, fowl pox, and Marek's disease) and a modest, supplemental feeding during times of seasonal scarcity.

[1] Intensive poultry production is also important, however. For example, Mexico had only about 50,000 broilers in 1958—now it produces about 20 million, and chicken meat is consumed by more than half the population. The United States had no "chicken in every pot" in 1932, but now produces billions of the birds each year, and virtually everyone can eat chicken often.

# 5

# Chicken

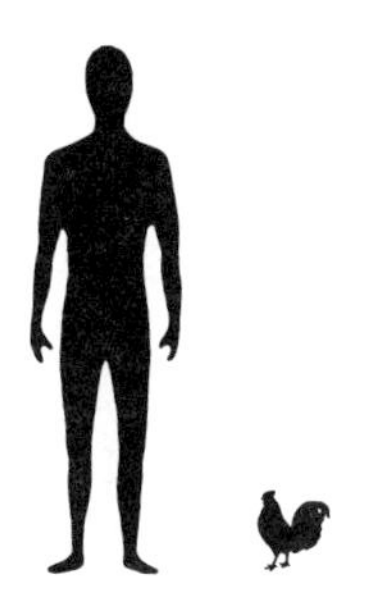

Chickens (*Gallus gallus* or *Gallus domesticus*)[1] are the world's major source of eggs and are a meat source that supports a food industry in virtually every country. There may be as many as 6.5 billion chickens, the equivalent of 1.4 birds for every person on earth.[2]

No other domesticated animal has enjoyed such universal acceptance, and these birds are the prime example of the importance of microlivestock. Kept throughout the Third World, they are one of the least expensive and most efficient producers of animal protein.

To the world's poor, chickens are probably the most nutritionally important livestock species. For instance, in Mauritius and Nigeria more than 70 percent of rural households keep scavenger chickens. In Swaziland, more than 95 percent of rural households own chickens, most of them scavengers. In Thailand, where commercial poultry production is highly developed, 80–90 percent of rural households still keep chickens in backyards and under houses. And in other developing countries from Pakistan to Peru, a similar situation prevails.

Clearly, these chickens should be given far more attention. They represent an animal and a production system with remarkable qualities; they compete little with humans for food; they produce meat at low cost; and they provide a critical nutritional resource.

Scavenger chickens are usually self-reliant, hardy birds capable of withstanding the abuses of harsh climate, minimal management, and

[1] Chickens were domesticated from one or more species of Southeast Asian junglefowl. The actual number is unclear. Taxonomists accepting evidence for descent exclusively from the red junglefowl refer to the domestic form as *Gallus gallus*, and those believing in a descent from this and other species refer to it as *Gallus domesticus*. See sidebar, page 86.

[2] FAO, 1982, *1981 FAO Production Yearbook*, Food and Agriculture Organization of the United Nations, Rome.

For most parts of the Third World, scavenger chickens are a vital part of human existence. (Periscoop)

inadequate nutrition. They live largely on weed seeds, insects, and feeds that would otherwise go to waste.

Unfortunately, however, quantitative information about the backyard chicken is hard to obtain. Few countries have any knowledge of its actual contribution to the well-being and diet of their people. Notably lacking is an understanding of the factors limiting egg production, which is markedly low and perhaps could be raised dramatically with modest effort.

## AREA OF POTENTIAL USE

Worldwide.

## APPEARANCE AND SIZE

Chickens are so well known and ubiquitous that they need no further description. Varying in color from white through many shades of brown to black, they range in size from small bantams of less than

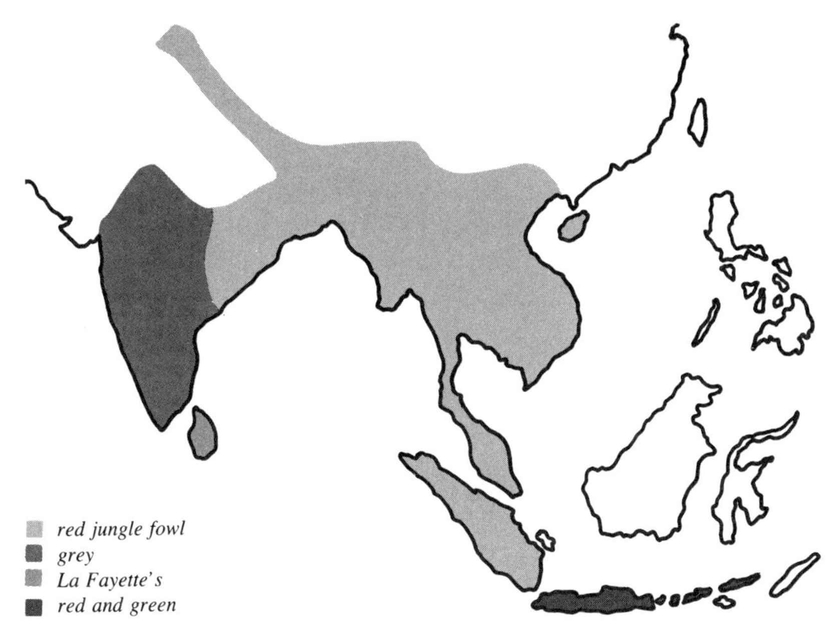


Distribution of the red junglefowl, wild ancestor of the chicken, and its related species. Junglefowls were domesticated in Southeast Asia in prehistoric times. Their domestic descendants had reached the Indus valley by about 2500 B.C. and China by about 1400 B.C. They spread into Central Europe, probably also around 1400 B.C. As it spread, the bird became transformed. For instance, although red junglefowls lay only one egg every two or three months, some modern domestic hens lay eggs daily throughout most of the year.

1 kg to giant breeds weighing 5 kg or more. Scavenger chickens tend to weigh about 1 kg.

The indigenous chickens of Asia are probably descended directly from the wild junglefowl. Those of West Africa are believed descended from European birds brought by the Portuguese in the sixteenth century; those of Latin America probably descend from Spanish birds introduced soon after the time of Columbus.

## DISTRIBUTION

All countries have chickens in large numbers.

## STATUS

They are not endangered, but industrial stocks are replacing traditional breeds to such an extent that much potentially valuable genetic heritage is disappearing.

## HABITAT AND ENVIRONMENT

Although chickens derive from tropical species, they adapt to a wide variety of environments. The modern Leghorn, for example, is found from the hot plains of India to the frozen tundra of Siberia, and from sea level to altitudes above 4,000 m in the Andes. (There are, however, hatching problems at such high altitudes because of oxygen deficiency.) They also occur in desert countries such as Saudi Arabia, which has a vast poultry industry and even exports broilers. (However, the birds need shade and a lot of water where it is hot and dry.)

## BIOLOGY

Chickens are omnivorous, living on seeds, insects, worms, leaves, green grass, and kitchen scraps.

A commercial bird may produce 280 eggs annually, but a scavenger may produce close to none. Commonly, a farmyard hen lays a dozen eggs, takes three weeks to hatch out a brood of chicks, stays with the chicks six weeks or more, and only then starts laying again.

Egg production depends on daylength. For the highest production rate, at least 12 hours of daylight are needed. The incubation period is 21 days. A hen can begin laying at 5 months of age or even earlier, but in scavengers it may be much later. The average weight of the eggs is approximately 55 g from industrial layers and approximately 40 g from scavengers. Hatching success from breeder flocks often exceeds 90 percent. Industrial broilers can be marketed as early as 6 weeks, when they are called "Cornish hens."

## BEHAVIOR

These passive, gregarious birds have a pronounced social (pecking) order. If acclimated, they remain on the premises and are unlikely to go feral. If given a little evening meal of "scratch," they learn to come home to roost at night.

## USES

Chickens have multiple uses. They were probably first used for cock fighting; later they were used in religious rituals, and only much later were raised for eggs and meat. Today, chickens can provide a family with eggs, meat, feathers, and sometimes cash.

## HUSBANDRY

In different parts of the world, people keep scavenging chickens in different ways. The managers are often women and children because they have more time to spend at home to feed the birds and repel predators. Some people leave the birds entirely to their own devices. Many house them at night. Others take the birds each day to the fields, where they may find much more food.

There are many ingenious local practices. In Ghana, for example, farmers "culture" termites for poultry by placing a moist piece of cow dung (under a tin) over a known termite nest. The termites burrow into the dung, and some can then be fed to the chickens each day. Because termites digest cellulose, this system converts waste vegetation into meat.

A ratio of 1 male to 10–15 females is adequate for barnyard flocks. Hens will lay eggs in the absence of a rooster—but of course the rooster is needed if fertile eggs are wanted.

Removing chicks stimulates the hen to lay more eggs. This results in more chicks being hatched, but it requires that the chicks be nurtured and fed until they are old enough to fend for themselves.

## ADVANTAGES

Chickens are everywhere; every culture knows them and how to husband them. They have been utilized for so many centuries that in most societies their use is ingrained. Unlike the case with pork and beef, there are few strictures against eating chicken meat or eggs.

The meat is high in quality protein, low in fat, and easily prepared. In many countries, the village chicken's meat is preferred to that of commercial broilers because it has better texture and stronger flavor. Even in countries with vast poultry industries there is a growing demand for the tasty, "organically grown," free-ranging chicken.

Chickens are more suited to "urban farming" than most types of livestock and can be raised in many city situations.

The birds are conveniently sized, easily transported alive, and, by and large, do not transmit diseases to humans.

## LIMITATIONS

Throughout Asia, Africa, and Latin America, the problems of village chickens are mainly those discussed below.

## High Hatching Mortality

Commonly, a hatch of eight or nine village chicks results in only two or three live birds after a few days. A survey in Nigeria, for instance, showed that 80 percent died before the age of eight weeks. Losses elsewhere are known to be similar. This is mostly because of starvation, cold, dehydration, predators (hawks, kites, snakes, dogs, and cats, for example), diseases, parasites, accidents, and simply getting lost—all of which can be prevented without great effort.

## Chronic and Acute Disease

Poultry diseases can become epidemic in the villages because there are few if any veterinarians. Newcastle disease, fowlpox, pullorum disease, and coccidiosis, for example—all of which are endemic in the Third World—can destroy the entire chicken population over large areas. Lice and other parasites are also prevalent. Scavengers and industrial birds seem to show no differences in their tolerance for such diseases and parasites.

## Low Egg Production

A survey in Nigeria showed that the annual production per hen was merely 20 eggs. Such low production is common throughout the Third World and is caused by a combination of low genetic potential, inadequate nutrition, and poor management. Villagers rarely provide nest boxes or laying areas, so that some eggs are just not found. Some birds have high levels of broodiness, and eggs accumulating in a nest stimulates this. There are indications, however, that some village chickens (for example, some in China) have quite substantial egg-laying potential when provided with adequate feed.[3]

## Low Egg Consumption

In the tropics, many people choose not to eat eggs. Often this is because eggs are the source of the next generation of chickens; sometimes it is because of superstition. Further, eggs do not keep well because most are fertile and, exposed to constant tropical heat, undergo rapid embryo development.

[3] Information from R.W. Phillips.

### Crop Damage

It is often necessary to confine the birds to protect young crops or vegetable gardens.

## RESEARCH AND CONSERVATION NEEDS

Unlike the situation with small cattle, goats, sheep, and pigs, there are few named and recognized breeds of Third World chickens. Yet, nearly every country has at least one kind of village chicken. These have survived there for centuries and are highly adapted to local conditions. In village projects, these unnamed chickens deserve priority attention before other types are sought from elsewhere.

Generally speaking, improving the production of scavenging poultry does not require sophisticated research. Instead, simple precautions are sufficient. These are discussed below.

### Disease Control

At a national or regional level, the initial approach to increasing chicken production in tropical areas should be disease control. There are several outstanding instances of success in this endeavor. For example, the spectacular rise of poultry production in Singapore (from 250,000 birds in 1949 to 20 million in 1957) followed the control of Ranikhet disease. Village flock-health programs, carried out regularly by visiting veterinarians ("barefoot veterinarians"), might be the answer to some of the routine health problems. Today, a prime target should be Newcastle disease, for which there are good chances for success (see page 76).

### Management

The first step in chicken production at the farm level is improved management. With more care and attention, mortality can be greatly reduced. Because incubating and brooding hens must spend the night on the ground, they are extremely vulnerable. Even modest predator controls can be highly beneficial. Building crude and inexpensive nest boxes and constructing a simple holding area around them can substantially raise production by ensuring that more chicks survive.

## THE CHICKEN'S WILD ANCESTOR

Although little known to most people, the red junglefowl has contributed more to every nation than any other wild bird. It is the ancestor of the chicken.

Given its descendant's importance worldwide, the neglect of this bird is baffling. If the cow's wild ancestor, the aurochs, had not become extinct in the 1600s, it would now be worth millions of dollars as the ultimate source of cattle genetic diversity. Yet the world's chicken industry remains virtually unaware of the origin of its source of livelihood.

Like the aurochs, the red junglefowl has a wealth of wild genes, and it deserves more recognition and protection. For one thing, the modern chicken—selectively bred in the temperate zone—is highly susceptible to heat and humidity; the junglefowl, on the other hand, is not. It inhabits the warmest and most humid parts of Asia: Sri Lanka, India, Burma, Thailand, and most of Southeast Asia. It may also be resistant to various chicken diseases and pests.

This is not a rare species. Throughout the wide crescent stretching from Pakistan to Indonesia, junglefowls are still seen in the wild, especially in forest clearings and lowland scrub. Although they are a prized bag for hunters, they survive by fast running and agile flying. They are sometimes sold in village markets, but can easily be mistaken for domesticated chickens, which in this region are often very similar. The wild junglefowl, however, has feathered legs, a down-curving tail, and an overall scragginess.

Junglefowls should be under intensive study. They are easy to rear in captivity and do well in pens, even small ones, as long as they are sheltered from rain and wind. One drawback is their craze for scratching; unless provided plenty of space they promptly tear up all grass and dirt. Another is that junglecocks are violent fighters and must be kept apart. (Cockfighting is probably a major reason why they were initially selected, and thus their aggressiveness is perhaps the reason we have the chicken today.)

These highly adaptable creatures live in a variety of habitats, from sea level to 2,000 m. Most, however, are found in and around damp forests, secondary growth, dry scrub, bamboo groves, and small woods near farms and villages. They are amazingly clever at evading capture and thrive wherever there is some cover.

Other junglefowl species might also provide useful poultry. They, too, can be raised in captivity with comparative ease, as long as the cocks are kept apart. Perhaps they might be tamed

Red junglefowl. (Monte Costa/Waimea Falls Park)

with imprinting and could prove useful as domestic fowl, especially in marginal habitats. They are everywhere considered culinary luxuries and their meat commands premium prices. Moreover, several have colorful feathers, giving them additional commercial value. These other species are:

- La Fayette's Junglefowl (*Gallus lafayettei*). A very attractive bird of Sri Lanka, it is little known in captivity, and only in the United States are there any number in captivity.

- Gray or Sonnerat's Junglefowl (*Gallus sonnerati*). A native of India, this colorful bird produces feathers that are used in tying the most prized trout and salmon flies. Demand is so great that certain populations have declined, and since 1968 India has banned all export of birds or feathers. Nonetheless, there are several hundred in captivity in various countries.

- Green Junglefowl (*Gallus varius*). This is yet another striking bird. The cock has metallic, greenish-black feathering set off by a comb that merges from brilliant green at the base to bright purple and red at the top. Native to Java, Bali, and the neighboring Indonesian islands as far out as Timor, it is found particularly near rice paddies and rocky coasts. This species, too, can be raised without great difficulty, and there are at least 90 in captivity in various parts of the world.

## THE SOUTH AMERICAN CHICKEN

Sheldon Parks

Early European explorers of South America were surprised to discover an abundance of unusual chickens that laid colored eggs and had feathers resembling earrings on the side of the head. While the origin of this bird—commonly called the araucanian chicken and classified as *Gallus inauris*—is debatable, scientists generally agree that it is pre-Columbian. There is archeological evidence that this bird is native to the Americas. It is reported to have occurred in Chile, Ecuador, Bolivia, Costa Rica, Peru, and Easter Island. It still occurs in the wild in southern Chile and on Easter Island.

The araucanian has been called the "Easter-egg chicken" because it lays light green, light blue, and olive colored eggs. It lays well and has a delicious meat. In areas such as southern Chile the eggs are preferred over those of normal chickens because of their flavor and dark yellow yolk. This unusual bird has a high degree of variability; however, specimens of similar genetic background have been grouped to create "breeds" such as the White Araucanian, Black Araucanian, and Barred Araucanian. These are homozygotes and breed true.

The araucanian has been the subject of much public interest; clubs dedicated to its preservation have been formed in the United States, Great Britain, and Chile. Its possible exploitation as a backyard microlivestock deserves serious consideration.

## Nutrition

Improving poultry nutrition is also of prime importance. There are no quantitative data on the quality of a scavenging chicken's diet. Surveys are badly needed so that appropriate, low-cost supplements can be devised.

Chances are that the diet for chicks of scavenging poultry is almost always deficient in available energy. Minimal supplementation in the form of cereals or energy-rich by-products can greatly improve both egg and meat production. However, caution must always be exercised and the supplements given only to chicks. Overfed adults will give up scavenging and stay around the owner's house, without really producing much more meat or eggs.

## Genetic Improvement

Although it seems attractive to replace the scrawny village chicken with bigger, faster-growing imported breeds, it is a process fraught with difficulty. Exotic breeds lack the ability to tolerate the rigors of mismanagement and environmental stress. Many cannot avoid predators, as a result either of being overweight or of having a poor conformation for flight. The local birds, however, probably have a genetic potential that is much higher than can be expressed in the constraining environment. Thus, the environmental constraints should be tackled first.

However, the village birds may have a feed-conversion efficiency that is far less than ideal because they are adapted to a scavenging existence. Modern breeds imported into Ghana, for instance, showed a feed-conversion efficiency of less that 3.5:1 (weight of food eaten: growth and eggs), but the local birds had efficiencies of 11:1.[4]

## Conservation

The need for preserving genetic variability is greater in poultry, especially in chickens, than in any other form of domestic animal. North America, for instance, which years ago had 50 or more common breeds, now relies on only 2 for meat production, and the others have been largely lost. Conservation of germplasm has become a matter of serious concern, and the saving of rare breeds in domestic fowl should not be delayed.

[4] If there is a good carbohydrate source (such as corn), the protein conversion is excellent when this is combined with palm-kernel cake, peanut cake, or other sources of protein that people cannot eat. More high-quality protein is obtained than is in the corn. Information from M.G.C.McD. Dow.

# 6

# Ducks

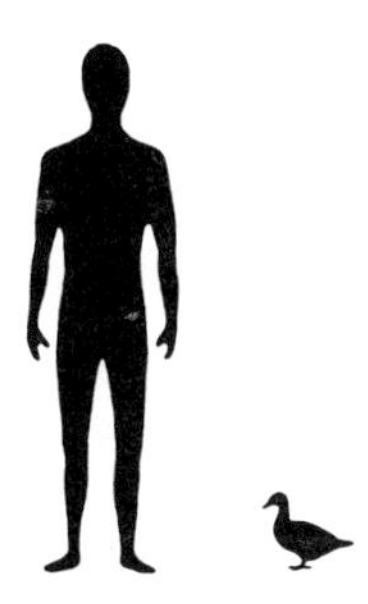

Domestic ducks (*Anas platyrhynchos*)[1] are well known, but still have much unrealized promise for subsistence-level production. Although a major resource of Asia, where there is approximately one duck per 20 inhabitants, they are not so intensively used elsewhere. On a worldwide basis, for instance, they are of minor importance compared with chickens.

This is unfortunate because ducks are easy to keep, adapt readily to a wide range of conditions (including small-farm culture), and require little investment. They are also easily managed under village conditions, particularly if a waterway is nearby, and appear to be more resistant to diseases and more adept at foraging than chickens.

Moreover, the products from ducks are in constant demand. Some breeds yield more eggs than the domestic chicken. And duck meat always sells at premium. A few recently created breeds (notably some in Taiwan) have much lower levels of fat than the traditional farm duck. This development could open up vast new markets for duck meat, especially in wealthy countries, where consumers are both concerned over fat in their diet and eager for alternatives to chicken.

Ducks are also efficient at converting waste resources—insects, weeds, aquatic plants, and fallen seeds, for instance—into meat and eggs. Indeed, they are among the most efficient of all food producers. Raised in confinement, ducks can convert 2.4–2.6 kg of concentrated feed into 1 kg of weight gain. The only domestic animal that has better feed conversion is the broiler chicken.

Raised as village birds and allowed to forage for themselves, ducks become less productive but become even more cost effective because much of the food they scavenge has no monetary value.

[1] This is the commonly used scientific name; others are also used, however. The muscovy duck is covered in a later chapter.

Tsaiya ducks, Taiwan. (Food & Fertilizer Technology Centre)

## AREA OF POTENTIAL USE

Worldwide.

## APPEARANCE AND SIZE

Several distinctive types have been developed in various regions. Most have lost the ability to fly any distance, but they retain a characteristic boatlike posture and a labored, waddling walk. The Indian Runner, however, has an almost erect stance that permits it to walk and run with apparent ease.

Domestic ducks range in body size from the diminutive Call, weighing less than 1 kg, to the largest meat strains (Pekin, Rouen, and Aylesbury, for example) weighing as much as 4.5 kg. For intensive conditions, the Pekin is the most popular meat breed around the world. In confinement it grows rapidly—weighing 2.5–3 kg at a market age of 7–8 weeks. In addition, it is hardy, does not fly, lays well, and produces good quality (but somewhat fatty) meat.

The Khaki Campbell breed is an outstanding egg producer, some individuals laying more than 300 eggs per bird per year.

The Taiwan Tsaiya (layer duck) is also a particularly efficient breed. It weighs 1.2 kg at maturity, starts laying at 120–140 days, and can produce 260–290 eggs a year. Its small body size, large egg weight, and phenomenal egg production make Brown Tsaiya the main breed for egg consumption in Taiwan. More than 2.5 million Brown Tsaiya ducks are raised annually for egg production.[2]

## DISTRIBUTION

The domestic duck is distributed throughout the world; however, its greatest economic importance is in Southeast Asia, particularly in the wetland-rice areas. For example, about 28 percent of Taiwan's poultry are ducks. In parts of Asia, some domestic flocks have as many as 20,000 birds. One farm near Kuala Lumpur, Malaysia, rears 40,000 ducks.

## STATUS

Although ducks are abundant, some Western breeds are becoming rare. Indigenous types are little known outside their home countries and have received little study, so their status is uncertain.

## HABITAT AND ENVIRONMENT

These adaptable creatures thrive in hot, humid climates. However, during torrid weather they must have access to shade, drinking water, and bathing water.

Ducks are well adapted to rivers, canals, lakes, ponds, marshes, and other aquatic locations. Moreover, they can be raised successfully in estuarine areas. Most ocean bays and inlets teem with plant and animal life that ducks relish, but (unlike wild sea ducks) domestic breeds have a low physiological tolerance for salt and must be supplied with fresh drinking water.

[2] Information from Cha Tak Yimp and Chein Tai.

## BIOLOGY

Ducks search for food underwater, sieve organic matter from mud, root out morsels underground, and sometimes catch insects in the air. Their natural diet is normally about 90 percent vegetable matter (seeds, berries, fruits, nuts, bulbs, roots, succulent leaves, and grasses) and 10 percent animal matter (insects, snails, slugs, leeches, worms, eels, crustacea, and an occasional small fish or tadpole). They have little ability to utilize dietary fiber. Although they eat considerable quantities of tender grass, they are not true grazers (like geese), and don't eat coarse grasses and weeds at all. Sand and gravel is swallowed to serve as "grindstones" in the gizzard.

When protected from accidents and predation, ducks live a surprisingly long time. It is not unusual for one to continue reproducing for up to 8 years, and there are reports of exceptional birds living more than 20 years.

Despite large differences in size, color, and appearance, all domestic breeds interbreed freely. Eggs normally take 28 days to incubate; brooding and rearing is performed solely by the female.

Depending on breed, a female may reach sexual maturity at about 20 weeks of age. Most begin laying at 20–26 weeks, but the best egg-laying varieties come into production at 16–18 weeks and lay profitably for 2 years.

## BEHAVIOR

It is generally well known that ducks are shy, nervous, and seldom aggressive towards each other or humans. Skilled and enthusiastic swimmers from the day they hatch, they spend many hours each day bathing and frolicking in any available water. However, most breeds can be raised successfully without swimming water.

Although wild ducks normally pair off, domestic drakes will mate indiscriminately with any females in a flock. In intensively raised flocks, 1 male to 6 females, and in village flocks, 1 male for up to 25 females, results in good fertility.

Most domestic ducks, particularly the egg-laying strains, have little instinct to brood. If not confined, they will lay eggs wherever they happen to be—occasionally even while swimming. To facilitate egg collection, some keepers confine ducks until noon.

Opposite: Duck farm in China. Although ducks are naturally cautious, they quickly become accustomed to familiar humans. (R.N. Matheny, *The Christian Science Monitor*)

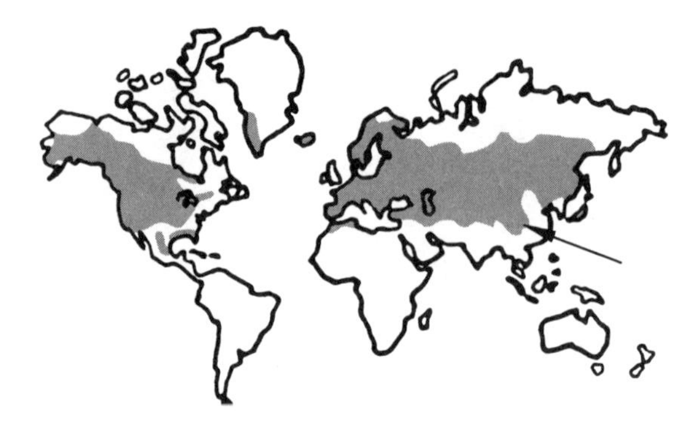

Native distribution of the mallard, ancestor of the domestic duck. The Rouen—a major commercial duck of France—is almost identical in coloration to the wild mallard. From such birds—with their iridescent green heads, brown necks, and gray and green bodies—white mutants were selected in medieval times. Today's big white breeds developed in this way.

## USES

As a meat source, ducks have major advantages. Their growth rate is phenomenal during the first few weeks. (Acceptable market weights can be attained under intensive management with birds as young as 6–7 weeks of age.) Yet, even in older birds, the meat remains tender and palatable.

Eggs from many breeds are typically 20–35 percent larger than chicken eggs, weighing on average about 73 g. They are nutritious, have more fat and protein, and contain less water than hen's eggs. They are often used in cooking and make excellent custards and ice cream. Eggs incubated until just before the embryos form feathers produce a delicacy known as *balut* in the Philippines. Salted eggs are popular in China and Southeast Asia.

Feathers and down (an insulating undercoat of fine, fluffy feathers) are valuable by-products. Down is particularly sought as a filler for pillows, comforters, and winter clothing.

Ducks have a special fondness for mosquito and beetle larvae, grasshoppers, snails, slugs, and crustaceans, and therefore are effective pest control agents. China, in particular, uses ducks to reduce pests in rice fields.[3] Its farmers also keep ducks to clear fields of scattered grain, to clear rice paddy banks of burrowing crabs, and to clear aquatic weeds and algae out of small lakes, ponds, and canals. This

[3] Ducklings, at hatching, are readily "imprinted." Farmers in Asia wave a white flag on the end of a bamboo pole over the ducks as they hatch. The ducks then think that the flag is their mother and follow it to the rice paddy, stay around it all day, and follow it back to the shelter in the evening.

not only improves the conditions for aquaculture and agriculture, it also fattens the ducks.

## HUSBANDRY

In Southeast Asia, droving is a traditional form of duck husbandry, much as it was in medieval Europe. The birds are herded along slowly, foraging in fields or riverbanks as they march to market. The journey might cover hundreds of kilometers and take as long as six months.

This process, however, is generally declining, and most ducks are raised under farm conditions where they scavenge for much of their feed. Throughout Southeast Asia, ducks have been integrated with aquaculture.

Ducks can be raised on almost any kitchen wastes: vegetable trimmings, table scraps, garden leftovers, canning refuse, stale produce, and stale (but not moldy) baked goods. However, for top yields and quickest growth, protein-rich feeds are the key. Commercial duck farms rely on such things as fish scraps, grains, soybean meal, or coconut cake. Agricultural wastes such as sago chips, palm-kernel cake, and palm-oil sludge are being used in Malaysia.[4]

Ducks have a very high requirement for niacin (a B vitamin). If chicken rations are used, a plentiful supply of fresh greens must be provided to avoid "cowboy legs," a symptom of niacin deficiency.

## ADVANTAGES

Of all domestic animals, ducks are among the most versatile and useful and have multiple advantages, including:

- Withstanding poor conditions;
- Producing food efficiently;
- Utilizing foodstuffs that normally go unharvested;
- Helping to control pests; and
- Helping to fertilize the soil.

Also, they are readily herded (for instance, by children).

Excellent foragers, they usually can find all their own food, getting by on only a minimum of supplements, if any. Raising them requires little work, and they provide farmers with food or an income from the sale of eggs, meat, and down.

Ducks can grow faster than broiler chickens if they have adequate nutrients. Like guinea fowl and geese, they are relatively resistant to

[4] Information from Yeong Shue Woh.

## DOMESTICATING NEW DUCKS

Many species of ducks adapt readily to captivity; it is surprising, therefore, that only the mallard and the muscovy have been domesticated so far. Several wild tropical species seem especially worth exploring for possible future use in Third World farms. Because of the year-round tropical warmth, their instinct to migrate is either absent or unpronounced, and the heavy layer of fat (a feature of temperate-climate ducks that consumers in many countries consider a drawback) is lacking. Moreover, because of uniform daylength, they are ready to breed at any time of the year. Candidates for domestication as tropical ducks include:

- Whistling ducks (*Dendrocygna* species). These large, colorful, gooselike birds are noted for their beautiful, cheerful whistle.* They are long-necked perching ducks that are found throughout the tropics. By and large, they are gregarious, sedentary, vegetarian, and less arboreal than the muscovy—all positive traits for a poultry species.

The black-bellied whistling duck (*D. autumnalis*) seems especially promising. It is common throughout tropical America (southwestern United States to northwestern Argentina) and is sometimes kept in semicaptivity. Occasionally, in the highlands of Guatemala, for instance, Indians sell young ones they have reared as pets. When hand reared, the birds can become very tame.** They eat grain and other vegetation, require no swimming water, and will voluntarily use nest boxes. In the wild, they "dump" large numbers of eggs,† so that even if substantial numbers were removed for artificial hatching, the wild populations should not be affected.

- Greater wood ducks (*Cairina* species). The muscovy (*C. moschata*) was domesticated by South American Indians long before Europeans arrived (see page 124). Its counterparts in the forests of Southeast Asia and tropical Africa are, however, untried as domesticates. The white-winged wood duck (*C. scutulata*) is found from eastern India to Java. Hartlaub's duck (*C. hartlaubi*) occurs in forests and wooded savannas from Sierra Leone to Zaire. Both are rare in captivity, but might well prove to be future tropical resources. Both are strikingly similar to muscovies in size and habits, being large, phlegmatic, sedentary, and omnivorous.

* They are known as "pijiji" in Spanish—from the very pretty "pee-hee-hee" sound they make.

** In the 1930s, plant explorer Wilson Popenoe kept some at his home in Antigua, Guatemala. At dawn each morning one would enter his bedroom and tug on the blankets until he got up and threw it into the patio fountain.

† One census of nest boxes in Mexico showed that out of 22,000 eggs laid, 80 percent were not hatched.

disease. They also have a good tolerance to cold and, in most climates, don't need artificial heat.

## LIMITATIONS

Predators are the most important cause of losses in farm flocks. Ducks are almost incapable of defending themselves, and losses from dogs and poachers can be high. Locking them in at night both protects the birds and prevents eggs from being wastefully laid outside.

Ducks do suffer from some diseases, mainly those traceable to mismanagement such as poor diet, stagnant drinking water, moldy feed or bedding, or overcrowded and filthy conditions. Of all poultry, they are the most sensitive to aflatoxin, which usually comes from eating moldy feed. They are also susceptible to cholera (pasteurellosis) and botulism, either of which may wipe out entire flocks. Duck virus enteritis (duck plague) and duck virus hepatitis also can cause severe losses.

If not carefully managed, ducks can become pests to some crops, especially cereals.

As noted, ducks tend to be extremely poor mothers and can be helped by using broody chicken hens or female muscovies as surrogate mothers.

Major limitations to large-scale, intensive production are mud, smell, and noise.

Defeathering ducks is much more difficult than defeathering chickens because of an abundance of small pinfeathers and down feathers.

## RESEARCH AND CONSERVATION NEEDS

These birds already function so well that no *fundamental* research needs to be done. Nonetheless, there are a number of topics that could improve their production.

For example, different types of low-cost systems need to be explored and developed. These must be low-input systems since cash is a limiting factor for most subsistence farmers. One possibility is the integration of duck and fish farming.

A survey of all breeds is needed to determine their status and likelihood of extinction.

One need in countries that already have ducks is to encourage the consumption of duck meat. Indonesia, for instance, has 25 million egg-producing ducks, but little duck meat is consumed.

Research on economically significant diseases is needed.

Brown Chinese Geese

# 7

# Geese

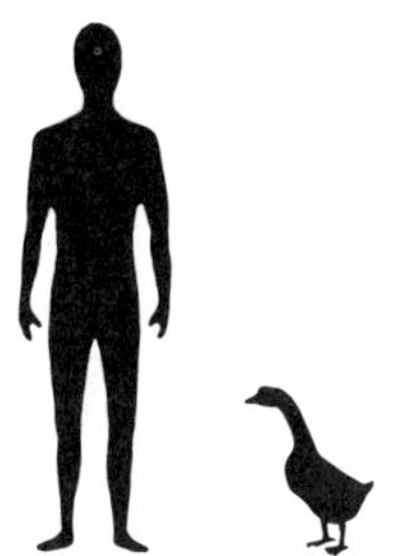

Although geese (*Anser* spp.) were one of the first domesticated animals, they have yet to receive the level of commercial or industrial exploitation of chickens or even ducks. Thus, their global potential is far greater than is generally recognized today.

Domestic geese are easily managed and well suited to small-farm production; they are among the fastest growing avian species commonly raised for meat, and they have immediate application in many developing countries.

These birds are especially appropriate for providing farmers a supplemental income. With little extra work they supply nutritious meat, huge eggs, and rich fat for cooking, as well as soft down and feathers for bedding and clothing. Moreover, their strident voices sound the alarm when strangers or predators approach. They are especially well suited to aquatic areas and marshy lands and are completely at home in warm shallow waterways. Nevertheless, they can thrive away from water. In fact, wherever pasture is available geese readily adapt to captivity.

Geese are grazers, and can be raised almost exclusively on pasture. They are excellent foragers, and on succulent grass can find most or all of their own food. With their powerful bills they pull up grasses and underwater plants and probe soil and water for roots, bulbs, and aquatic animals. Their long necks make them adept at gleaning weeds from hard-to-reach places—such as fence rows, ditches, and swampy areas that baffle larger livestock. They will also feast on vegetable trimmings, garden and table leftovers, canning refuse, and stale baked goods. Like other poultry, they pick up shattered grains of rice, wheat, barley, and other crops, which can reduce the bothersome problem of weeds volunteering in subsequent years.

Geese are available worldwide. In most climates, they require little or no housing. Given reasonable care and protection from predators, mortality can be extremely low.

## AREA OF POTENTIAL USE

Worldwide.

## APPEARANCE AND SIZE

Domestic geese come in an assortment of colors, sizes, and shapes. There are two main types, however. Descendants of the wild greylag goose (*Anser anser*) make up the domestic breeds common in North America and Europe, including the Embden, Toulouse, Pilgrim, American Buff, Pomeranian, Sebastopol, and Tufted Roman breeds. These are generally best suited to temperate climates. On the other hand, descendants of the wild "swan goose" (*Anser cygnoides*) make up the geese of Asia, including the Chinese and "African" types. These breeds seem better suited to hot climates.

In addition to these, many European and Asian countries have their own local breeds and types, and there are even several wild species that show some potential for captive production.

With their long legs and webbed toes, geese are equally at home walking or swimming. Avid walkers, they march long distances to find forage, but return home at dusk. Accomplished and graceful swimmers, geese are able to take to water soon after they hatch. Despite their large size, some domestic breeds—especially the leaner ones—have retained the ability to fly.

## DISTRIBUTION

Geese are found worldwide, but goose farming is nationally important only in Asia and Central Europe.

## STATUS

Domestic geese are not threatened, although much local variation among the breeds is being lost.

## HABITAT AND ENVIRONMENT

Most geese adapt well to hot climates—as long as some shade is

Opposite: Geese being fed on a poultry farm south of Bangkok, Thailand. (Y. Hadar, World Bank)

## DOMESTICATING GEESE (Temperate Zones)

Today's domestic geese are descended from two species: the greylag (*Anser anser*) and the swan goose (*Anser cygnoides*). These were domesticated in Europe and China, respectively. Their domestication occurred in ancient times, long before people knew about genetics, microorganisms, veterinary science, or behavior modifications such as imprinting. Today, armed with such knowledge, more geese may be amenable to domestication. Most of the 15 other wild species adapt to captivity. Compared to most birds, geese spend much time walking and swimming, and are less inconvenienced by pinioning (removing the tip of the wing). Thus, they can be kept outdoors rather than in cages.

Both of the ancestors of today's domestic geese are native to the northern temperate zone. Two more wild species that might make useful domesticates are:

- Canada goose (*Branta canadensis*). North America. People feeding these birds in city parks and wildlife refuges are causing many local flocks to develop. These birds no longer migrate. They are increasing in numbers each year and are well on the way to de facto domestication.
- American swan goose (*Coscoroba coscoroba*). Southern South America. Although most closely allied to swans in shape and physiology, this bird resembles a muscovy (see page 124) in size and behavior. Its calm disposition, as well as its attractive red feet and bill that accent its white plumage, have made it much sought as an ornament for parks.

available. Their waterproof feathers help them adapt well to high-rainfall regions. They also tolerate extreme cold. (For instance, in Canada, geese are wintered outdoors in subfreezing temperatures, with merely a simple shelter from wind.)

For tropical developing countries, the Chinese type, which is widely kept in Southeast Asia, is especially promising. Smaller than most geese (although ganders can weigh over 5 kg), they are the best layers, the most active foragers (making them economical and useful as weeders), the most alert and "talkative," and they produce the leanest meat. Some European breeds, such as Embden and Toulouse, have also been used in the tropics with notable success.

## BIOLOGY

In its diet, the goose utilizes large quantities of tender forage. It can break down plant-cell walls and digest the contents. Although it has no crop for storing food, there is an enlargement at the end of the gullet that serves as a temporary storage organ. Sand and small gravel are swallowed to aid the gizzard in grinding hard seeds and fibrous grasses. Research has shown that geese can digest 15–20 percent of the fiber in their diet, which is 3–4 times the amount that other poultry species can digest.

The natural diet consists of grasses, seeds, roots, bulbs, berries, and fruits, normally supplemented with a little animal matter (mainly insects and snails) picked up incidentally. Most feeding takes place on land. They characteristically feed for prolonged periods, even at night.

Females may lay for 10 years or more. It is generally believed that reproduction is best in the second year and that it remains good until the fifth year. Geese outlive other types of poultry; life spans of 15–20 years are common.

The eggs incubate in 27–31 days. The incubation time is more variable than in most poultry species, perhaps because geese have not been subjected to the selection pressure that is imposed by artificial incubators.

## BEHAVIOR

One of the most intelligent birds, the goose has a good memory and does not quickly forget people, animals, or situations that have frightened it. While personalities and habits vary among individual specimens, there are common behavioral patterns, such as the pecking order, that allow individuals to live peaceably together.

Unless conditions are crowded or there are too many males, geese normally live harmoniously both with themselves and with other creatures. The bond between male and female is strong. Changing mates is difficult, although most geese will eventually accept a new mate after a period of "mourning."

Geese nest on the ground and prefer the water's edge, but they adapt readily to man-made nesting boxes. The gander usually stands guard while the goose incubates the eggs. He then assists in rearing the goslings. Most geese become irritated if intruders approach their nest or goslings, and will even attack people and large dogs.

## USES

As previously noted, these birds provide meat, eggs, fat, and down. The meat is lean, flavorful, and of outstanding quality. The fat accumulates between the skin and the flesh and can be rendered into a long-lasting oil. The eggs are large and taste much like chicken eggs. The "down" (the small, fluffy feathers that lie next to the body of adult birds) is the finest natural insulating material for clothing and bedding, and can fetch a premium price. Worldwide markets exist for both down and other goose feathers. In France, in particular, some geese are raised for their livers (foie gras).

Geese can control many types of aquatic weeds in shallow water as well as grass and some types of palatable broad-leaf weeds on the banks of lakes, ponds, and canals. They can also be used as "lawn mowers" and "weeders" among cotton, fruit trees, and other crops (see sidebar).

Elongated necks not only allow geese to reach many different foods, they also help them keep a watchful eye on the surroundings. With their exceptional eyesight they can see great distances, and the position of the eyes gives them a wide field of vision. Geese are among the most alert of all animals, and strangers cannot calm them into silence. In the high Andes, in Southeast Asia, and in many other locations, they replace guard dogs. In Europe, they are used to guard whiskey warehouses and sensitive military installations (see page 111).

## HUSBANDRY

Methods of caring for adult geese vary according to climate, breed, and people's experiences and needs. Overall, however, the birds cause little trouble and require little expense. They range freely without restriction, feeding themselves and returning home of their own accord. They have strong flocking instincts and can readily be herded from one area to another.

Like all young poultry, goslings are fragile. The highest mortality is caused by predators. Until the goslings are 6–10 weeks old, it is prudent to confine the parents and their young at night in a secure pen or building.

Geese are the only domestic fowl that can live and reproduce on a diet of grass. They cannot remain healthy on coarse dry fodder, but when grass is succulent they need little else other than drinking water. Many legumes also make excellent goose forage.

In the tropics, eggs can be laid year-round. The production seldom exceeds 40 eggs per year, although with feed supplement and simple management, the Chinese breed may yield more than 100 eggs. Geese go broody quickly. To break up broodiness, the goose can be confined for 4–6 days away from, but in sight of, the ganders.

## DOMESTICATING GEESE (Tropics)

Geese of the tropics have seldom if ever been considered for domestication, but they might provide poultry of considerable value. Presumably they are more heat tolerant and lack the layers of subcutaneous fat (which the ancestors of today's geese needed for warmth in the Arctic). They might thus produce lean birds that would fetch premium prices because excessive fat is the major drawback of today's commercial geese. Examples of tropical species that might be domesticable are:

- Egyptian goose (*Alopochen aegyptiacus*). Found throughout the African tropics, this bird is already partly domesticated. However, it is bad tempered and quarrelsome and, so far, this has limited its utility. It has therefore been kept only under semidomestication, without intensive breeding.
- Nene (*Branta sandvicensis*). A native of the Hawaiian Islands, this is one of the most endangered species on earth. So few specimens are in existence that farming enterprises cannot now be envisaged. Yet, should this bird prove amenable and suitable, the possibility of an economic future could boost efforts to build up its now meager populations.
- Bar-headed goose (*Anser indicus*). India and Central Asia. These smallish geese are handsome, dainty, and have a musical horn-like call. They have distinct black bars across the nape, which gives them their popular name. Hand-reared specimens breed well in captivity. Despite heavy hunting, they are still abundant.
- Northern spur-winged goose (*Plectropterus gambensis gambensis*). Tropical Africa (Senegal to Zimbabwe). This large bird is a ground nester, but it has long, bony spurs on the wings that enable it to easily protect its eggs and young from predators.
- Semipalmated (magpie) goose (*Anseranas semipalmata*). Australia and New Guinea. One of the most aberrant and primitive of all waterfowl, this long-legged, sturdy-billed bird has only partially webbed feet. It perches high in trees and has a loud, ungooselike whistling call.

## THE WADDLING WORK FORCE

Because geese relish grasses and shun most broad-leafed plants, some enterprising U.S. farmers in the 1950s began using them to rid cotton fields of grassy weeds, which are difficult to kill with herbicides. The geese were put into the fields as soon as the crop came up. A brace of birds kept an acre of cotton weeded; a gaggle of 12 would gobble as many weeds as a hard-working man could clear with a hoe.

This method of clearing fields was so effective that by 1960 more than 175,000 geese honked their way across the carefully tended farmland, mainly in the Southwest. Seven days a week, rain or shine, the feathered field hands slaved uncomplainingly from daybreak to dusk, even putting in overtime on moonlit nights. Many toiled so diligently that they worked themselves out of a job.

The geese cleared the fields more cheaply than hoe hands. They left the crop untouched and ate only the succulent young weeds. They did not damage crop roots (as hoes or tractors can), and they were safe and selective, unlike many herbicides. On top of all that they spread fertilizer for the farmer, and ultimately provided him meat for the market.

Eventually, farmers found that geese could be used to weed nearly all broad-leafed crops: asparagus, potatoes, berry fruits, tobacco, mint, grapes, beets, beans, hops, onions, and strawberries, for example. Geese were used in vineyards and fruit orchards to eat both weeds and the fallen fruits that could otherwise harbor damaging insects. They were employed in fields producing trees for the forest industry and flowers for florists shops. Some growers turned goslings loose in cornfields to consume the "suckers" (corn, after all, is a grass) as well as the grain left on the ground. This eliminated the problem of corn as a weed when different crops were later planted in those fields.

In the 1970s when cotton acreage dropped and herbicides selective for the troublesome grasses were developed, the use of geese declined. But today, some organic farmers are returning to the practice. From February to June in the Pacific Northwest, fields are resounding once more to the old-fashioned racket of White Chinese geese.

Opposite: Geese keep an Australian citrus orchard free of weeds and of rotting fruits that can attract pests. (Queensland Department of Agriculture)

Goslings grow rapidly and can reach market size as early as 10–12 weeks; most geese, however, are marketed at 20–30 weeks of age, when they may weigh from 5 to 7 kg, depending on type and breed. Some young birds (also called green or junior geese), force-fed for rapid growth, are marketed at 4–6 kg when they are 8–10 weeks old.

If fed a good diet to maximize growth and if slaughtered at, say, 10 weeks, the Embden, Chinese, or African will have a carcass low in fat. However, the carcass normally has much more fat than other poultry.

Geese must have a constant supply of reasonably clean drinking water during daylight hours. Although swimming water is not necessary, it promotes cleaner and healthier birds because they find it easier to care for their plumage.

## ADVANTAGES

Mature geese are independent creatures. When kept in small flocks and allowed to roam the farmyard or field, they require less attention than any other domestic bird with the possible exception of guinea fowl. In areas where grass is green for much of the year, they can be raised on less grain or concentrated feed than any other domestic fowl.

Durability is one of their most attractive features. Along with ducks, geese seem to be the most resistant of all poultry to disease, parasites, and cold or wet weather. They also do well in hot climates as long as drinking water and deep shade are available.

Growth is not only rapid, it is also efficient. If managed properly, goslings can produce 1 kg of body weight for every 2.25–3.5 kg of concentrated feed consumed.

Geese are not usually thought of as prolific layers. However, as noted, some strains of the Chinese breed will yield well over 100 eggs per goose per year. At 140–170 g per egg, that compares favorably with the output of laying chickens.

Opposite: Geese make good "watchdogs." They once saved ancient Rome from the attacking Gauls, and today they help guard modern missiles on military bases in Europe. At the Ballantine bonded warehouses near Glasgow, Scotland, more than 100 geese zealously protect 240 million liters of maturing whiskeys. No matter what the weather or time of day, some of the flock remain awake and alert. Keener of ear and sharper of eye than any dog, they cannot be duped by any blandishments. Since this feathered force was formed in 1959 there has never been a theft at the warehouses. Also, the birds keep the grounds weed free and the grass clipped. And they seem to like being Scotch guards because they are free to fly away at any time. ( Pierre Boulat/Cosmos)

The native origins of geese. The arrows show the probable places where the two main types were domesticated.

## LIMITATIONS

These birds are messy and their loud trumpeting is often irritating. However, unless they have been teased or mistreated or if they are nesting or brooding young, they are not aggressive. But kilo for kilo, they are stronger than most animals, and a harassed or angry adult can express its displeasure effectively with powerful bill and pounding wings.

Excessive concentrations of geese on ponds or along creeks encourages unsanitary conditions, muddies water, hastens bank erosion, and destroys plant life. Where sanitation is poor, salmonellosis can devastate geese and be transmitted, via meat and eggs, to humans. Coccidiosis and gizzard worm are other infections.

Defeathering geese is more difficult than plucking chickens because there are two coats (feathers and down) to remove.

In some situations, geese may need a diet supplement (such as grain) if they are grazing vegetation exclusively. A balance must be struck: too much supplement and they will quit foraging and become too fat; too little and they grow slowly and may suffer malnutrition.

Geese are not fully mature until two years of age. Their overall reproductive rate, therefore, is lower than that of other poultry.

## RESEARCH AND CONSERVATION NEEDS

Poultry researchers worldwide should begin studies to clarify the role that geese could play in helping to feed Third World nations. Studies might include:

- Management practices for tropical areas;
- Breeding and management for increased egg production;
- Incubation techniques;
- Nutrition supplementation (for example, vitamins, minerals, energy, specific amino acids) needed by grazing geese;
- Physiology of digestion and reproduction;
- Clarifying the inheritance of various traits;
- Genetic selection for specific meat, eggs, growth factors, or disease resistance;
- Comparative studies of the relative efficiency (especially of feed utilization) of the various types and breeds for specific climates in underdeveloped countries;
- Weeding tropical crops with geese; and
- Studying diseases and cross-infection with other birds.

# 8

# Guinea Fowl

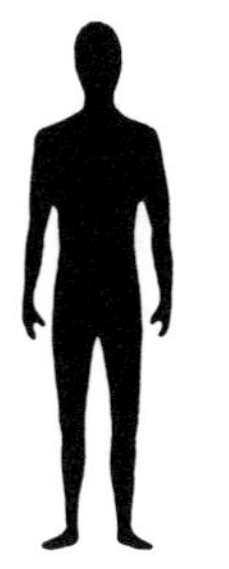

For Third World villages, the guinea fowl (*Numida meleagris*) could become much more valuable than it is today. The bird thrives under semi-intensive conditions, forages well, and requires little attention. It retains many of its wild ancestor's survival characteristics: it grows, reproduces, and yields well in both cool and hot conditions; it is relatively disease free; it requires little water or attention; it is almost as easily raised as chickens and turkeys; and it is a most useful all-round farm bird.

The guinea fowl's potential to increase meat production among hungry countries should be given greater recognition. The birds are widely known in Africa and occur in a few areas of Asia, but they show promise for use throughout all of Asia and Latin America and for increased use in Africa itself. Strains newly created for egg and meat production in Europe—notably in France—show excellent characteristics for industrial-scale production. Also, many semidomestic types in Africa deserve increased scientific assessment as scavenger birds.

Meat from domestic guinea fowl is dark and delicate, the flavor resembling that of game birds. It is a special delicacy, served in some of the world's finest restaurants. Several European countries eat vast amounts. Annual consumption in France, for example, is about 0.8 kg per capita.[1]

Guinea fowl also produce substantial numbers of eggs. In Africa, these are often sold hard-boiled in local markets. In the Soviet Union, they are produced in large commercial operations. In France, guinea fowl strains have been developed that not only grow quickly but lay as many as 190 eggs a year.

Outside Europe, virtually all guinea fowl are raised as free-ranging

[1] Information from P. Mongin.

birds. These find most of their feed by scratching around villages and farmyards. Their cost of production is small, and they yield food for subsistence farmers. In Europe, on the other hand, most are raised in confinement, with artificial insemination, artificial lighting, and special feeding. In the main, this is to produce meat for luxury markets.

Guinea fowl production is beginning to increase all over the world. During the last 20 years, for example, many of Europe's chicken farmers and breeders, wishing to diversify, have switched to this bird. The United States is now studying ways to establish industrial production, and both Japan and Australia are increasing their flocks. Nonetheless, there is still a vast untapped future for this bird.

## AREA OF POTENTIAL USE

Worldwide. This species is robust and resilient and adapts to many climates.

## APPEARANCE AND SIZE

Guinea fowl are somewhat larger than average scavenger-type chickens: adults weigh up to 2.5 kilograms. They have dark-gray feathers with small white spots. Their heads are bare with a bony ridge (helmet) on top, which makes them look something like vultures. The short tail feathers usually slope downwards.

The chicks, known as "keets," resemble young quail. They are brown striped with red beaks and legs. The sexes are indistinguishable until eight weeks of age. After that, the males' larger helmets and wattles and the cries of the different sexes can be identified. Both sexes give a one-syllable shriek, but females also have a two-syllable call.

Like the chicken, the guinea fowl is a gallinaceous species and possesses the characteristic sternum with posterior notches and a raised "thumb."

Among domestic types are pearl, white, royal purple, and lavender. Pearl is the most common, and is probably the type first developed from the wild West African birds. Its handsome feathers are often used for ornamental purposes. The white is entirely white from the time of hatching and has a lighter skin.

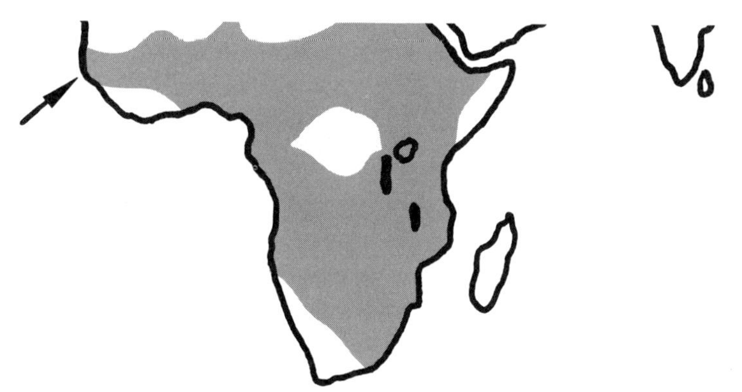

The guinea fowl's native habitat. The arrow shows the probable place where today's main breed originated.

## DISTRIBUTION

Europe dominates industrial production. France, Italy, the Soviet Union, and Hungary all raise millions of guinea fowl under intensive conditions, just as they raise chickens. Elsewhere, guinea fowl have become established as a semidomesticated species on small family farms. Native flocks are found about villages and homes in parts of East and West Africa, and free-ranging flocks can be seen in many parts of India, notably Punjab, Uttar Pradesh, Assam, and Madhya Pradesh. During the slavery era, they were introduced from Africa to the Americas to be used for food. In Jamaica, Central America, and Malaysia, the birds have reverted to the wild state and are treated as game.

## STATUS

Guinea fowl are abundant; in most places even wild populations are not threatened.

## HABITAT AND ENVIRONMENT

Guinea fowl are native to the grasslands and woodlands of most of Africa south of the Sahara where they occupy all habitats except dense forests and treeless deserts. Being native also to temperate South Africa, they appear to have an inherent adaptability to both heat and cold. However, in cool climates, regardless of daylength, they will not begin egg production until temperatures exceed 15°C.

## BIOLOGY

Guinea fowl accept many foods: grains, leaves, ant eggs (for which they will tear anthills open), and even carrion.

Normally, they lay their first egg at about 18 weeks of age. Unlike many wild birds, which produce a single clutch a year, guinea hens lay continuously until adverse weather sets in.[2] Free-range "domestic" guinea hens lay up to 60 eggs a season. And well-managed birds under intensive management lay close to 200. The eggs weigh approximately 40 g. Shells are stronger than those of chickens and are usually brown, but can be white or tinted.

The guinea hen goes broody after laying, which can be overcome by removing most of the eggs. A clutch of 15–20 is common. The incubation period is 27 days.

## BEHAVIOR

These birds never become "tame," but neither do they leave the premises. Although they stray farther than chickens do, they always return. They like to hide their eggs in a bushy corner, often in hollows scratched in the ground. They can fly, although even in the wild they do not fly far. They prefer to roost on high branches and (unless pinioned) can be hard to catch during the day.

Although wild guinea fowl live in groups, they are monogamous by nature and tend to bond in pairs. However, in domestication a single male may serve four or more females.

## USES

As noted, guinea fowl are valuable sources of both meat and eggs. They can also be used to control insect pests on vegetable crops.[3]

Guinea fowl are good "watch animals"; they have fantastic eyesight, a harsh cry, and will shriek at the slightest provocation. Their agitation on sighting dogs, foxes, hawks, or other predators have saved the lives of many a chicken, duck, and turkey. They are brave and will attack even large animals that threaten them.[4]

[2] In West Africa, laying is largely confined to the rainy season, but it can be induced by spraying the birds with water. Information from R. T. Wilson.

[3] In parts of Queensland, Australia, many farmers keep a few "guineas" to assist with controlling grasshoppers in crops and gardens as well as cattle ticks in and around the cattle yards and milking sheds. The birds do no harm to gardens or crops because, unlike chickens, they do not scratch in the ground. Information from A. Hutton.

[4] Information from B. K. Shingari and from A. Hutton, who adds, "I have known them to attack and kill a brown snake (similar to a cobra in venom), and also a taipan."

Guinea fowl on an American farm. (Grant Heilman Photography, Inc.)

## HUSBANDRY

Guinea fowl can be kept in confinement using the methods for raising battery chickens. In this system, breeding stock are housed in cages and artificially inseminated. It gives the best egg production and fertility but requires housing, equipment, and skilled labor.

These birds can also be kept in a semidomestic state in and around the farmyard. In such cases they are penned until they are 12 weeks old. Unaccustomed to foraging for natural food, they constantly return to their artificial food supply. Eventually, however, they learn to subsist by scavenging.

The birds have been called "the worst parents in the world," and are almost incapable of looking after their keets.[5] Because the females are such indifferent mothers, the eggs are best hatched in incubators or under other birds, to avoid the keets' being lost by their natural mothers. In many African countries, eggs are hatched under chickens.

Keets are often kept indoors until they are 3–4 weeks old to protect

[5] According to A. Hutton, "losing half the chicks on a 5-km hike through thick bush the day they hatch is not unusual."

## GUINEA FOWL AND THE ANCIENTS

The earliest reference to guinea fowl can be found in murals in the Pyramid of Wenis at Saqqara in Egypt, painted about 2400 B.C. Aviaries were quite fashionable at the time, and wealthy landowners maintained guinea fowl within their walled gardens. A thousand years later, by the time of Queen Hatshepsut (about 1475 B.C.), the junglefowl (the ancestor of the chicken) had arrived, and from then on it was raised on a substantial scale. Records of this period refer to "walk-in" incubators, constructed of mud bricks and heated by camel-dung fires. The largest could hold up to 90,000 eggs (mainly from junglefowl but some from guinea fowl) and hatching rates of up to 70 percent were claimed.

By 400 B.C., guinea fowl were well established on farms in Greece. Later, they rose to importance in ancient Rome. Pliny the elder (in his *Natural History*, published 77 A.D.) stated that they were the last bird to be added to the Roman menu and that they were in great demand, both eggs and flesh being considered great delicacies. The emperor Caligula offered them as sacrifices to himself when he assumed the title of deity.

The guinea fowl then died out in Europe but was reintroduced by the Portuguese navigators returning from their African explorations in the late 1400s. They gave it the name *pintada* or "painted chicken" and this changed to *pintade* in French, while the name "Guinea fowl" (fowl from Africa) stayed in English, and *gallina de Guinea* in Spanish. Coincidentally, guinea fowl and turkeys were both introduced to England between 1530 and 1550, and the English, smitten with the original French misnomers, were left sorting out "Ginny birds" and "Turkey birds" for the remainder of the century. Both birds were adopted with great enthusiasm, and within 150 years they had utterly displaced the peafowl and swan as the major table birds for festive occasions.

Adapted from R.H.H. Belshaw, 1985
*Guinea Fowl of the World*

them from predators and wet weather. Sexual maturity can be delayed to as late as 32 weeks of age by holding the birds in windowless housing and controlling the lighting. This improves egg size and hatchability and reduces early mortality.

## ADVANTAGES

Compared with the farmyard chicken the guinea fowl's advantages are:

- Low production costs;
- Premium quality meat;
- Greater capacity to utilize green feeds;
- Better ability to scavenge for insects and grains;
- Better ability to protect itself against predators; and
- Better resistance to common poultry parasites and diseases (for example, Newcastle disease and fowlpox).

Surprisingly, this semidomestic bird, which has been farmed for centuries, retains the characteristics (feather morphology, hardiness, social behavior) of its wild ancestor—even when subjected to the most modern intensive-rearing methods employing battery cages and artificial insemination. Thus, it thrives under semicaptive conditions and needs little special care. The birds forage well for themselves and do not require much attention; their meat is tasty and they produce substantial numbers of eggs. Unlike chickens, they don't scratch to get insects out of the soil, so they are less destructive to the garden.

## LIMITATIONS[6]

In backyard production the guinea fowl is supreme, but when produced intensively it costs more to raise than chickens. In Europe, for instance, day-old keets cost about twice as much as day-old broiler chicks. (The major reason is that guinea fowl produce fewer hatching eggs and require a longer feeding period.) Guinea fowl are also more expensive to feed. Their feed conversion (for meat production at the marketing age) is about 3.3–3.6 as compared with a broiler's feed conversion of 1.8–1.9. Moreover, guinea fowl take about twice as long to reach marketable size: they are marketed for meat at age 12–14 weeks, compared with 7–8 weeks for the broilers. Therefore, the selling price of guinea fowl in the Western world is up to twice that of broilers.

Guinea fowl are nervous and stupid. They can be difficult to catch, and when panicking they can easily suffocate their keets.

They are susceptible to some of the common diseases of chickens and turkeys. Salmonella is the most prevalent, but others are pullorum disease, staphylococcus, and Marek's disease.

[6] Most of the information in this section was contributed by A. Ben-David.

## THE GUINEA FOWL'S WILD COUSINS*

The domesticated guinea fowl is descended from just one subspecies of the family's seven known species and numerous subspecies. Some of the others may also have promise as poultry. They, too, generally occur in flocks in bushy grasslands and open forests in Africa. All feed on vegetable matter such as seeds, berries, and tender shoots, and on invertebrates such as slugs. They rarely fly except to roost. They acclimatize well, are easy to maintain in captivity, and can survive long periods away from water.** Their disposition is tame and nonaggressive, and they mix well with other birds.

Wild subspecies closely related to the domestic guinea fowl that might make future poultry in their own right include the following:

- Gray-breasted guinea fowl (*Numida meleagris galeata*). This subspecies is the principal ancestor of domestic guinea fowls. It is found throughout West Africa and probably has many valuable genetic traits. There is much variation in the size and other characteristics among the various individuals. People along the Gambia, Volta, and Niger rivers have long traditions of breeding these birds.
- Tufted guinea fowl (*Numida meleagris meleagris*). This subspecies is quite large and has black plumage thickly spotted with white dots. It is the probable ancestor of the birds reared in ancient Egypt and in the Roman empire (see page 120). Hill farmers in the southern Sudan sometimes breed them in captivity.
- Mitred guinea fowl (*Numida meleagris mitrata*). Probably the most popular game bird in East Africa, this type has a bright blue-green head and red wattles. It was once a common sight in the wild but it has now been decimated by overhunting. It is now most numerous in the Masai lands of Kenya and Tanzania. It has been kept in a semidomesticated form in Zanzibar for several centuries. Zoos and aviaries around the world have imported it, and it has bred well for them.

Wild guinea fowl that are different species from the domestic one but that are still worth considering as potential poultry include the following:

- Black guinea fowl (*Phasidus niger* or *Agelastes niger*). This bird of the tropical rainforests of West and Central Africa is the size of a small chicken. It has sooty black plumage, a naked head, and a pink or yellow neck. It is seldom hunted because the meat tastes dreadful, but this is probably because of a particularly pungent fungus they eat in the forest. Raised on fungus-free forages, these birds are probably very palatable.

• Crested guinea hen (*Guttera* spp.). Three species. These strange-looking birds have a thick mop of inky black feathers above their black, naked faces. Widely distributed in the thickly forested areas of sub-Saharan Africa. Unlike the other species, they prefer the rainforest. They have a musical trumpeting call. At least one species has bred well in Europe. For example, a flourishing colony has been established in the Walsrode Bird Park in Germany.

• Vulturine guinea fowl (*Acryllium vulturinum*). The largest of all guinea fowl, this species is found in parts of Ethiopia, Somalia, and East Africa. One of the most striking looking of all birds, its head is bare and blue, its body black with white spots, and its breast bears long bright cobalt-blue patches on either side. This has been reared as an aviary bird in both Europe and America and might make a useful domesticate.

---

* This section is based largely on information given in *Guinea Fowl of the World*, R.H. Hastings Belshaw, 1985.

** At least some of these birds commence feeding long before dawn and obtain their moisture from dew on the leaves. Moreover, like the camel, they appear to utilize metabolic water very efficiently.

## RESEARCH AND CONSERVATION NEEDS

Agencies involved in international economic development should undertake guinea fowl assessment trials, evaluations, and coordinated introductions to stimulate programs for small farmers and for industries in dozens of countries.

Breeders have been working to improve guinea fowl only since the 1950s. There is a need for more information on growth rate, health, egg production, feed conversion, body weight, carcass yield, laying intensity, fertility, hatchability, and egg weight—especially under free-ranging conditions.

Husbandry research should also be directed towards feeds and feeding systems for growing and breeding stock. Other efforts are needed to increase the hatchability of eggs under natural conditions (under guinea hens or surrogate mothers), and to identify the best lighting regimes (both sexual maturity and rate of lay are influenced by changes in daylength).

The guinea fowl that has become an important domesticated bird throughout the civilized world is descended from just one of seven known species in the family. These birds generally occur in flocks in bushy grasslands and open forest in Africa and Madagascar, and some of the others may also have promise as poultry (see sidebar opposite).

# 9

# Muscovy

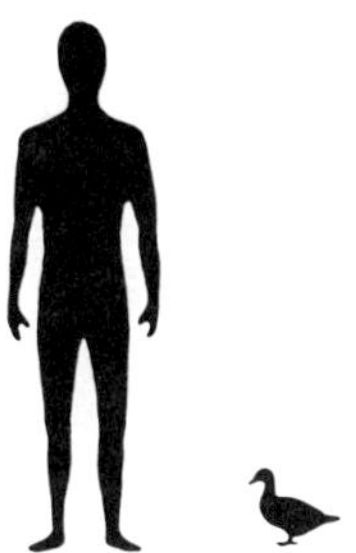

The muscovy[1] (*Cairina moschata*), a unique ducklike species of the South American rainforest, belongs to a small group of waterfowl that perch in trees. In poultry science, however, it is normally grouped with domestic ducks for lack of a better classification.

Except in France, Italy, and Taiwan, muscovies have received little modern research. But their promise can be judged from the fact that they account for 50 percent of the duck meat consumed in France—about 60,000 tons per year—and they are often consumed in Italy and Taiwan as well.

For Third World subsistence farming, muscovies have excellent possibilities. There is probably no better choice for a meat bird that requires minimal care and feed. Tame, quiet, and able to forage for much of their keep, they are inherently hardy, vigorous, and robust. They have heavily fleshed breasts and are highly prized for their meat, which is dark, more flavorful, and less fatty than that of common ducks. An average muscovy gives more meat than a chicken of the same age, and it also survives hot, wet environments better. In addition, muscovies are better parents than the domestic duck. Females are probably the best natural mothers of any poultry species, as measured by their success at incubating their eggs and caring for their young.

All in all, this bird deserves more attention than it has received so far in Third World livestock projects. Dispersed around the warm and hot regions of the world, muscovies already exist in small numbers in backyards and villages, much like the domestic chicken in previous centuries. Despite a lack of research, the present unimproved stocks are already impressive meat yielders. Used more widely and more intensively, they could contribute much to poor people's meat supplies.

[1] Most commonly, this bird is known as the muscovy *duck* or Barbary *duck*. In Latin America today, it is called the criollo duck.

Muscovies and chickens in a village in Benin, West Africa. (WFP/FAO photo by Banoun/Caracciolo)

Crossing the muscovy with the common duck produces a hybrid that combines many of the advantages of both. This cross, known as "mulard," or "mule duck" in English, is raised in France for its liver and meat and is produced in quantity in Taiwan (see sidebar, page 132). It, too, has a major future role.

## AREA OF POTENTIAL USE

Muscovies are suitable for use almost anywhere that chickens can be kept. Moreover, their tropical ancestry and inherent robustness give them an advantage in hot and humid climates.

## APPEARANCE AND SIZE

Although a muscovy somewhat resembles a goose, it is one of the greater wood ducks of tropical South America. It was domesticated in pre-Columbian times, most likely in the rainforests of Colombia. Related wild types, looking very much like the muscovy, still occur in South American wetlands, particularly mangrove swamps.

Males have mature live weights of 5 kg and females about 2.5 kg. Both have broad and rounded wings. The adults have patches of bare skin around the eyes, rather than feathers. Much of this is covered in "caruncles," which superficially resemble warty outgrowths. The feet have sharp claws. Both sexes raise a crest of feathers when alarmed.

There is much color variation among the various muscovy populations including types that are called white, colored, black, blue, chocolate, silver, buff, and pied.[2] The most common types (they are not considered breeds) are the white and the colored. The white produces a cleaner looking carcass, but the colored is the most popular meat type in France. Its plumage is an iridescent greenish black, except for white forewings.

## DISTRIBUTION

The native range of the muscovy's probable wild ancestor covers much of Central America and northern South America. The domestic form also occurs over most of Latin America—from southern Chile to the northern limits of traditional culture in lowland Mexico—including

[2] In general, the wild type is black, the highly bred commercial is white, and others are all grades in-between.

the Caribbean, where it was present shortly after Columbus landed.[3] The birds can be observed among the domestic fowl in the high Andes, for example, and are feral in southern coastal areas of the United States.

Carried across the Atlantic, probably in the early 1500s, the domesticated muscovy spread quickly in Europe, and thence to North America, Asia, Africa, and Oceania. Today, it finds favor with the food-loving French as "canard de Barbarie," and France has the greatest concentration of muscovies in Europe.

Down the centuries the muscovy became popular in tropical Asia (especially the Philippines and Indonesia) and in China and Taiwan. Throughout Indonesia (where it is known as "entok") it is popular with villagers for incubating eggs from ducks, geese, and chickens. It is now spreading into Oceania, and has recently gained particular favor in the Solomon Islands.

Muscovies are also known in Africa and can be found in many villages, especially in West Africa.

## STATUS

Not endangered.

## HABITAT AND ENVIRONMENT

Wild muscovies occur mainly along tropical jungle streams, but domestic muscovies are found in many environments from the heat of Central America to the cold of Central Europe. They also tolerate dry conditions, but they thrive best where climates are both hot and wet.

## BIOLOGY

Muscovies utilize high-fiber feeds better than chickens and common ducks, and eat larger quantities of grass. They also consume other green vegetation and readily snap up any insects they can find. If quality forage is available, only a small daily ration of grain or pellets is required for them to reach peak production.

Muscovy females normally hatch and raise large broods efficiently.

[3] The Spaniards first met the domesticated bird at Cartagena, Colombia, in 1514, where, according to Oviedo, the Indians kept it in domestication and called in "quayaiz." He describes the warts about the head and makes the identity clear, showing also that the color had already been affected by domestication. It was extremely abundant in Peru, whence the Spaniards exported it under the name of "pato perulero" to Central America, Mexico, and Europe. (J. C. Phillips, 1922, *A Natural History of Ducks*, Boston and New York: Houghton Mifflin Co., volume 1, page 66.)

The wild muscovy's natural range covers a vast area of tropical America. As a result, their domesticated descendants are well adapted to many conditions, including both tropical and temperate climates. The arrow indicates the area where this bird was probably first domesticated.

It is not unusual to see them with a dozen or more fragile ducklings in tow—many of them adopted from other species. They bravely protect their young and have been known to beat off cats, dogs, foxes, and other marauders.

Normally, muscovies are healthy and live and breed for many years. They suffer few diseases, especially when free ranging. However, they seem to be more susceptible to duck virus enteritis (duck plague) than common ducks.

## BEHAVIOR

While appearing to be slow and lethargic, muscovies can be quick and agile when one tries to catch them. Females are strong fliers and readily clear a standard fence. Males frequently become so ponderous that they cannot get airborne without an elevated perch or the aid of a strong wind. Although they forage over a larger area than chickens, they generally neither decamp nor wander as far as common ducks.

Domesticated muscovies are either solitary or live together in small family groups, but sometimes in winter they flock together on bodies of water. They swim and dive well.

These birds seldom make loud noises. A drake's voice resembles a muffled "puff"; females are almost mute. However, both can hiss or make a soft sound not unlike that of sleigh bells.

## A BETTER FLY TRAP

B.D. Glofcheskie

The muscovy is a voracious omnivore that is particularly fond of insects. For years, some Canadian farmers have sworn that a few muscovies took care of all fly problems on their farms. In 1989, Ontario biologists Gordon Surgeoner and Barry Glofcheskie (see Research Contacts) decided to put this to the test.

Starting with laboratory trials, the entomologists first put a hungry five-week-old muscovy into a screened cage with 400 living houseflies. Within an hour it had eaten 326. Later, they placed four muscovies in separate cages containing 100 flies each. Within 30 minutes over 90 percent of the insects were gone. It took flypaper, fly traps, and bait cards anywhere from 15 to 86 hours to suppress the populations that much.

Moving to field tests, the researchers placed pairs of two-year-old muscovies on several Ontario farms. Videotapes showed the birds snapping at houseflies and biting flies about every 30 seconds and being successful on 70 percent of their attempts. With that efficiency, they achieved 80–90 percent fly control in enclosures such as calf rooms or piggeries. The birds were given only water and had to scavenge for all their food. Females seemed to eat about 10 percent more flies than males, and individuals of any age between eight days and two years were equally effective.

The birds fit the practical needs for farmyard fly control. They stayed close to piglets and calves, to which flies are particularly attracted. They even snatched flies off the hides of resting animals without waking them up. On one farm, the birds huddled between sleeping piglets and were accepted by the sow lying beside them. This was noteworthy because most fly-catching devices (chemical, electrical, or mechanical) must be kept far from animals.

To the Canadians, the economic advantages are clear. A 35-cow dairy needs $150–$390 worth of chemicals for controlling flies during the fly season; muscovy chicks, on the other hand, cost less than $2 each, eat for free, and can be sold for a profit of 200–400 percent.

The researchers point out that employing muscovies does not eliminate all need for insecticides, but it reduces the amounts required. And muscovies are biodegradable, will not cause a buildup of genetic resistance, and taste better than flypaper. Indeed, their meat is excellent, and the naturally mute birds seldom make any noise.

Reportedly, muscovies are kept in some houses in South America to control not only flies, but also roaches and other insects.

---

The muscovy is polygamous (a young male will try to mate with almost any fowl, including chickens). Mating can occur on land or in water. Males are pugnacious and tolerate no opposition. Because of this, they do not do well in close confinement.

## USES

The muscovy is generally raised only for its meat, which is of excellent quality and taste. In stews it is hard to distinguish from pork; cured and smoked it is similar to lean ham.[4] The fat content is low.

Muscovy eggs are as tasty as other duck eggs, and a muscovy female can supply a large number if she is kept from sitting.

These birds are useful for clearing both terrestrial and aquatic weeds.[5]

Down feathers are used, like those of other ducks, in clothing and comforters.

## HUSBANDRY

Muscovies may be raised like common ducks. An ideal grouping is one male to five or six females.

[4] Indeed, perhaps because of this, Israel is a big muscovy producer.

[5] National Academy of Sciences, 1976, *Making Aquatic Weeds Useful: Some Perspectives for Developing Countries*, Washington, D.C.: National Academy Press, p. 49.

## THE MULE DUCK OF TAIWAN*

In parts of Europe, hybrids between muscovies and common ducks are reared for fattening. However, Taiwan has made the most outstanding use of this "mule duck." Thanks in part to this muscovy hybrid, Taiwan's duck industry has grown rapidly in the last decade. The total value of duck products now exceeds $346 million per year. Much of the boom in duck production is due to improved feeding, disease control, and management systems, but much is also due to the performance of the mule duck.

Taiwan. Mule-duck chicks. (Food & Fertilizer Technology Centre)

This hybrid is now Taiwan's major meat-duck breed, and about 30 million are consumed each year. Indeed, the duck industry has been so successful that Taiwan is increasingly exporting frozen duck breast and drumstick meat to Japan. It now provides 24 percent of the duck meat eaten in Japan—most of it coming from mule ducks. Also, Taiwan is exporting partially incubated mule-duck eggs throughout Southeast Asia. And mule ducks supply most of the raw material for Taiwan's large feather industry.

Taiwan farmers have been producing mule ducks for 250 years, but the recent jump in production is due to the use of artificial insemination to overcome the natural reticence of the different species to mate. Fortunately, artificial insemination is well developed and is a standard part of farming practice in Taiwan.

Mule ducks are successful because they have less fat than a broiler chicken and they grow faster. Indeed, they can reach a market weight of 2.8 kg at 65–75 days of age, depending upon the weather, season, and management. In part, this fast growth is because they are sterile and waste no energy in preparing for a sexual existence or in laying eggs.

The usual cross employs a muscovy male and a domestic-duck female. The domestic breeds most employed for mule-duck production are White Kaiya (Pekin male x White Tsaiya female), Large White Kaiya (Pekin male x White Kaiya female) and colored Kaiya. Both sexes of the hybrid offspring weigh about the same.

Crosses between a muscovy hen and a domestic drake are much rarer (traditionally, this was because of the different mating behavior of the two species, but even with artificial insemination available they are not much used) and the males of these hybrids are much heavier than the females. Females of this cross do lay eggs, but the eggs are small (about 40 g) and their embryos do not develop.

There are almost 300 duck-breeding farms in Taiwan, annually producing more than 600,000 female domestic ducks for use in producing mule ducks. Some farmers combine duck raising with fish farming. The excreta of 4,000 ducks on one hectare of pond can provide 30,000 tilapia with 20 percent of their feed. It helps the farmer get rid of waste as well as giving him fresh fish to sell.

---

* Information in this section from Chein Tai.

Except in Taiwan, France, Hungary, and a few other European countries, they exist predominantly in small flocks in farmyards and village ponds. However, they can be reared under intensive conditions in a shed or pen that is well lighted and equipped with low roosts and bedding. Under such conditions, they may be fed diets recommended for rearing common ducks or given coarse feeds, including whole grain. If chicken rations are used, fresh greens must be provided to avoid "cowboy" legs, a symptom of niacin deficiency.

Although they thrive in areas where there is abundant water, they do not require access to swimming water. They prefer to nest under cover and will use nesting boxes. A normal clutch size is 9–14 eggs; however, clutches of up to 28 can occur. There may be 4 clutches annually, and (when the hen does not have to brood the ducklings) some muscovies have laid 100 eggs in a year.[6]

The egg weight, which increases with the female's age, ranges from 65 to 85 g. The eggs require 33–35 days to hatch, a week longer than the common duck's. Hatching success of 75 percent or more is common.

Compared with domestic ducks early growth is slow, which is perhaps why muscovies have not enjoyed wider industrial use. However, after the slow period they grow rapidly and, because they forage on a broader range of vegetation than common ducks, they can scavenge a large proportion of their diet at little or no cost.

When raised intensively, females average 2 kg and males 4 kg at 11–12 weeks of age. Females may reach sexual maturity by 28 weeks of age; males require a month more.

## ADVANTAGES

As noted, the muscovy is an extremely good forager and thrives under free-ranging conditions. Unlike other ducks, it grazes on grass and leaves and will maintain itself on pasture. Apparently, it can digest bran and other fibrous feeds better than common ducks can.

The males are larger than all but the largest strains of table duck. They have exceptionally broad, well-muscled breasts and provide one of the leanest meats of any waterfowl.

The muscovy is apparently more resistant to diseases that regularly decimate other poultry. This is one reason why villagers favored them: when chickens die, muscovies often survive.

The female's strong parental qualities help assure the survival of ducklings with a minimum of human intervention. Her ability to

[6] However, to get 100 or more eggs per female per year is only possible with efficient management, and then only with ducks in their first or second laying year. If the female is allowed to hatch her own eggs and keep her ducklings, she is likely to produce only two broods per year. Information from W. F. Hollander.

incubate and hatch most other poultry eggs is an added advantage to small farmers who have neither the capital to buy, nor the knowledge to operate, artificial incubators.

Unlike other ducks, muscovies are not easily alarmed, and fright does not affect their egg production and laying. Indeed, they are so phlegmatic that automobiles can be major causes of death.

## LIMITATIONS

Because they are a tropical species, these birds are much less tolerant of cold than common ducks and require more protection from freezing weather.

The muscovy's feed conversion is not as good as the chicken's. Also, compared with some other meat-duck breeds, muscovies have a slower rate of growth and require about 4–6 weeks longer to attain maximum development of breast muscles.

Muscovies can be difficult to handle. If their legs are free, the handler may be badly lacerated by the claws.

Although adults have a fair homing ability, muscovies may wander away when local forage is sparse, and young birds may be carried long distances downstream, never to return.

Because they feed on greenery, they can devastate gardens if the plants are very young.

Muscovies can be unsuspected carriers of poultry diseases, so that healthy-looking muscovies may infect the other species.

## RESEARCH AND CONSERVATION NEEDS

Poultry scientists should unite in efforts to advance technical knowledge and public appreciation of this bird. Governments and researchers should begin evaluations of local varieties and their uses and performances. The experiences of France, Italy, Eastern Europe, and Taiwan should be gathered and made available for a worldwide readership.

It is important that the many muscovy varieties within the countries of Latin America—where the bird has a centuries-long history of domestication—be maintained and studied. Many superior varieties and specimens may be awaiting discovery.[7]

The muscovy's nutritional requirements, range and confined systems of management, and disease vulnerability are poorly understood and need study. Especially needed are ways to increase growth rate.

[7] A muscovy-research project of this type has already begun at La Molina University in Lima. Specimens have been collected from all parts of Peru, with researchers selecting those that grow larger more quickly and have other valuable traits.

# 10

# Pigeon

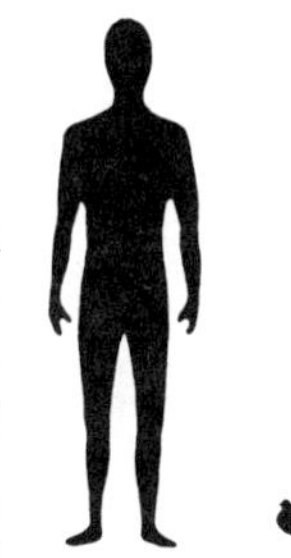

Pigeons (*Columba livia*)[1] are durable birds that can be raised with little effort. Able to survive in inhospitable climates, they fend for themselves—often ranging over many square kilometers to locate seeds and edible scraps. They have been raised for centuries, especially in North Africa and the Middle East. In parts of North America and Europe, they are produced as a delicacy for the gourmet market. But raising pigeons for food is not nearly as widespread as it could be; indeed, in modern times its potential has hardly been touched. Farmed pigeons are particularly promising as urban microlivestock because they require little space and thrive in cities.

Young pigeons (squab) grow at a rapid rate. Their meat is finely textured, has an attractive flavor, and is often used in place of game fowl. Tender and easily digested, it commands premium market prices. In many areas, the continuing demand is unfilled.

Pigeons are traditionally raised in dovecotes—"houses" that protect the birds from the elements and from predators. This system allows free-ranging flight and requires almost no human intervention. Dovecotes are a good source of both squab and garden manure, and they continue to be used, especially in Egypt. On the other hand, pigeons can also be raised in confinement—usually in enclosed yards—with all their needs supplied by the farmer. There are, for example, pigeon farms in the United States with up to 35,000 pairs of breeding birds.

Pigeon production may never rise enough to compete with commercial poultry as a major source of food, but for Third World villages these birds could become a significant addition to the diet as well as a source for substantial supplemental income.

[1] The pigeon family includes both doves and pigeons and has more than 300 species. Some of these other species live in association with man and also have potential as microlivestock.

## AREA OF POTENTIAL USE

Worldwide.

## APPEARANCE AND SIZE

Pigeons have small heads, plump, full-breasted bodies, and soft, dense plumage. They weigh from about 0.5 to nearly 1 kg. A few large breeds (Runts, for instance, which commonly weigh 1.4 kg) are the size of small domestic chickens.

Many breeds have been developed for meat production. They produce squab that grow quicker and have larger breasts than unselected birds.

## DISTRIBUTION

The wild ancestor of the common pigeon—domestic, wild, or feral—is thought to be the rock pigeon or rock dove of Europe and Asia. Today its domestic descendants are bred in virtually every country, and those that have gone feral (reverted to the wild) occur in most of the world's cities and towns.

## STATUS

They are abundant. However, as with most other domestic species there is concern over the decline and loss of certain breeds. Societies have been organized (notably in the United Kingdom) to preserve rare types.

Natural habitat of the rock dove, ancestor of the domestic pigeon.

## HABITAT AND ENVIRONMENT

The domestic pigeon can be raised equally well in temperate and tropical zones. Indeed, this adaptable species can be kept anywhere that wild pigeons exist, including arid and humid regions. It should be noted, however, that cold climates do not favor squab production and hot climates promote vermin and disease.

## BIOLOGY

The pigeon's natural diet consists mostly of seeds, but includes fruits, leaves, and some invertebrates. Feral pigeons consume a wide array of materials, including insects, bread, meat scraps, weed seeds, and many kinds of spilled grains at mills, wharves, railway yards, grain elevators, and farm fields.

For the first four or five days of life, the young are fed "crop milk." This substance, common to pigeons and doves,[2] is composed of cells from the lining of the crop and is very high in fats and nutritional energy. The phenomenal growth rate of young squab has been attributed to crop milk and to its early replacement (within 8–10 days) with concentrated foods, regurgitated by both parents. The parents feed the squab for about four weeks before pushing it out of the nest to prepare for the next clutch.

In domestic birds, sexual maturity (as measured by age at first egg) is reached at 120–150 days. Life span can be 15 years, although growth and egg production decline rapidly after the third year.

## BEHAVIOR

Wild pigeons often nest on cliff sides. Domestic pigeons prefer to nest around buildings, in nooks and shelves and under the eaves—that is, in "pigeonholes."

In domestic varieties, the pair-bond often lasts until severe illness or death. Sometimes, however, a vigorous male will "invade" a nest and mate with the females there. Both sexes take nearly equal part in nest building, incubation, and caring for the young. Typically, there are two eggs to a clutch. Eight clutches a year is not uncommon for a breeding pair. The incubation period is 17–19 days.

Unlike most birds, pigeons drink by inserting their beaks into water and sucking up a continuous draft.

[2] Flamingos, which belong to a different order than pigeons, are the only other birds to produce a similar material.

Courtship is characterized by cooing, prancing, and displays of spread, lowered tail feathers. "Bow and coo" exhibitions are unique to pigeons and doves and differ among species.

## USES

Pigeons are usually raised exclusively for meat. The squab are harvested just before full feather development and before the youngster has started to fly, usually at 21–30 days of age. At this time the ratio of flesh to inedible parts is highest; once flying begins, the meat becomes tougher. Weight depends upon breed, nutrition, and other factors, but usually ranges from 340 to 680 g.

Pigeons are extensively used for scientific research, notably in physiology and psychology. They are also widely kept as pets for plumage and for racing. The pigeon's unique homing ability was recognized in Roman times, and the birds have been trained to return to the dovecote from as far away as 700 km. Even today, homing pigeons are used to carry messages, especially during war.

## HUSBANDRY

Pigeons are easily trained to recognize "home." The wing feathers are clipped and the birds are fed close to the dovecote; by the time they refledge, their homing instinct has been developed. Alternatively, newly captured pigeons may be trained by confining them to the dovecote for at least one week. At first, a little grain is provided in the morning (this is to ensure the birds will return to the coop). The birds can obtain the rest themselves.

Any waterproof house that is easy to clean is suitable for keeping pigeons. Many traditional dovecotes are built of earthenware pots. In Asia and Europe, wooden pigeon towers are generally used.

Unlike chickens, pigeons do not prefer communal roosts. Instead, they prefer nesting shelves, of which there should be two for each breeding pair. The shelves are usually placed in dark corners and are fitted with low walls to keep eggs from rolling out.

Grit is important in the diet, both to provide minerals and to allow the birds to grind feed in their gizzards.

Commercial squab breeds are often kept permanently in pens, a process that requires care and experience. Growers expect an average of 12–14 squab per pair per year, although much depends on environment and management.

Opposite: Afghanistan. Boy and his pigeons. (B. Pendleton, courtesy of the Committee for a Free Afghanistan)

## CARRIER PIGEONS

Most people consider message-carrying pigeons to be a quaint anachronism. But in a few countries (both developed and developing) carrier pigeons are making a comeback, and in the future they may be used routinely once again.

New techniques are making this process far more practical than before. For example, in the past the pigeons would be flown in one direction only. They were transported away from home and at the appropriate time released to find their way back. That was very limiting. But it has since been found that pigeons can be trained to carry messages in two directions: flying from one point to another and then back again. They will do it twice a day, and with almost perfect reliability. The key is to place the feeding station at one end and the nest at the other. This limits the pigeon's range, but they still can handle round-trip distances up to 160 km.

With a little ingenuity, there is no need for a person to monitor the stations to receive the messages as the bird arrives. One simple technique is to arrange the station with one-way doors—one opening inwards, the other outwards. Placing a bar across the outward door means that the bird cannot get out until someone releases it. Thus the message can always be retrieved.

This system has been employed in Puerto Rico and Guatemala, but it could be used almost anywhere. In many parts of the Third World, in particular, there are remote areas with no phones and with hilly, rough terrain where delivering messages can take hours of strenuous travel. Some locations are subject to unexpected isolation by natural calamities or military or terrorist actions.

In Puerto Rico, for instance, we kept pigeons in a village 32 km from the capital. The pigeons could get downtown in 20–30 minutes. It took us 1.5–2 hours each way by road. What was easy for the birds was a major trip for us. Pigeons carried the villagers' requests for certain foods and medicines. Our contact in the city then sent up the supplies by bus. The birds never let us down.

Carrier pigeons are useful for more than just flying far and fast. They have been bred for racing and their large pectoral muscles make them excellent meat producers—much better than the common pigeons normally raised for food. A pair of carriers typically will raise 12–16 young each year, and those not needed for message carrying can be butchered at 28 days of age—yielding meat that is nutritious and considered a delicacy in many countries.

David Holderread

Every day on the northwest coast of France, Petit Gendarme, a black and white carrier pigeon, flies on average 23 km between hospitals on his blood delivery route. Trussed in tiny hand-sewn harnesses, he and a flock of carrier pigeons set out (except during the hunting season) with little red tubes of blood secured to their breasts.

"It's a simple, effective, and cost-saving transport system," said Yves Le Hénaff, head of the Avranches Hospital laboratory, Cotentin, France, a central blood-testing center that serves a number of isolated medical centers along the coast.

The service becomes particularly valuable during the summer tourist rush, when travelers flock to the seashore to visit nearby Mont Saint Michel, crowding the small country roads and increasing the risk of traffic accidents.

The birds' average flight time between the hospitals of Avranches and Granville, for example, a distance of about 27 km, is 20 minutes, including the time for harnessing up. And with a favorable western wind, their best time can reach 11 minutes.

While gasoline costs the equivalent of $0.75 a liter in France, hospital officials say that a few grains of corn is all it takes to run this operation. According to Le Hénaff, who supervises the carrier pigeon operation, the hospital is saving up to $46 a day on gas and auto maintenance.

The 40-year-old Le Hénaff got the idea five years ago from an article in a scientific journal describing a similar experiment in Britain. A year later, he and an associate called on the local seamstress to design a light harness that could hold a tube, which, when filled, weighs approximately 39 g.

The flock now consists of 40 veteran fliers and 20 carrier pigeons in training.

And what if the winged creatures stray en route? Le Hénaff has devised a fall-back option: two pigeons carry two different test tubes containing the same blood sample.

The birds fly every day of the year with the exception of the three-month autumn hunting season. Since the beginning of the experiment four years ago, there have been two casualties. Le Hénaff believes the birds probably met their fate in some Normand's oven.

Sometimes weather is a factor, and heavy fog can keep the delivery team grounded.

So far, the new job seems to benefit the pigeons, too. Unlike sickly city pigeons, whose average life span is about four years, well-cared-for carrier pigeons can live up to 15 years, Le Hénaff said. "And how many people do you know who are willing to stay with the same outfit for that long?"

Sabine Maubouche
*The Washington Post*
December 2, 1986

The birds need fresh water daily and water for bathing at least weekly. Since they feed their young by regurgitation, the adults must have a continuous supply of clean drinking water. Orphan squab can be fed egg yolk until old enough to consume adult feeds.

Like all poultry, confined pigeons must be provided enough supplemental feed to ensure a balanced diet. A mixture of whole grains can be fed for maximum production. It is important that grains be dry and free of mold (pigeons will not thrive on mash). Peas, beans, or similar pulses make good supplements.

## ADVANTAGES

Under extensive conditions—where the birds are released each day to feed themselves—almost no land is needed. Under intensive conditions, where the birds spend their lives in confinement, a mere half hectare can be enough space to raise 2,000 pairs.

Free-ranging pigeons forage over a wider area than most domestic fowl because they fly out to find their feed. Nutrient requirements[3] are similar to those of chickens and other fowl (making allowance for the energy needed for flying), so commercial feed and other supplements—if needed at all—are generally available.

In dovecote culture, pigeons require little or no handling. They brood the young with little intervention. Although not continuous, the production of meat from these fast-growing, rapidly reproducing birds is more sustained than with most livestock.

Almost nowhere are there taboos against consuming pigeon meat. Prices received for squab are normally high, and in most places the demand is constant. The only limitation in some areas is the absence of an effective market, which is usually easy to create.

Squab contains a larger proportion of soluble protein and a smaller proportion of connective tissue than most meats and is therefore good for invalids and people with digestive disorders.

As many hobbyists can testify, raising pigeons can be gratifying.

## LIMITATIONS

Pigeons are subject to few diseases. However, worms, lice, diarrhea (coccidiosis), canker (trichomoniasis), and salmonella (paratyphoid) occur at some time in most domestic breeds. Salmonella exists in low levels in most flocks and will flare up if birds are stressed. Treatments recommended for domestic chickens are usually suitable for pigeons.

[3] Protein, 13.5–15 percent; carbohydrates, 60–70 percent; fat, 2–5 percent; fiber, 5 percent. Clemson University, 1982.

By flying over a wide area and eating grains and other foods, pigeons can cause conflicts with farmers. Indeed, in the 13th century the aristocracy's pigeons became a major grievance of the peasants who saw their seed devoured. On the other hand, "croppers" (breeds with large crops) were developed to steal grain from the lord's fields. The pigeon returned home and his crop was emptied of the grain, which was used by the peasant to make bread.

The birds can become nuisances. They leave droppings in annoying places, some people find them too noisy, and a few people are severely allergic to "pigeon dust."

Every conceivable type of predator can be expected; therefore, precautions must be taken. The dovecote must be well protected against rats, which are the principal enemy of the eggs and the squabs.

Nesting birds need a high-protein diet to raise squab at the high rates of gain that are possible.

## RESEARCH AND CONSERVATION NEEDS

Poultry researchers should study the increased role pigeons might play in Third World economic development. Nothing comparable to the sophisticated selection employed with the domestic chicken has so far been attempted. Given such attention the gains could be great.

Among pressing research needs are:

- Breeding. This needs to be better understood. For example, the effects of hybridization and inbreeding need clarification.
- Environmental limits. Little work has been done outside the temperate regions.
- Diseases. These deserve increased attention.

There is also the potential of "dovecotes" for wild pigeons. Numerous local species are well adapted to local conditions, and these deserve to be tested for "domestication."[4] Many wild species quickly lose their fear of man, and in time they can even become too fat to fly. Wild pigeons are already found throughout the humid tropics and are trapped for meat and rearing in New Guinea and other places. They are already an important food source for many subsistence farmers and shifting cultivators, and with some dovecote management could provide a greater, more dependable source of food and income. The potential for domesticating local pigeon species, especially those suited to the tropics, deserves exploration.

[4] An example is the stock dove (*Columba oenas*) of southern Europe. This migratory and wild bird (in olden days called the tower-pigeon) can be domesticated. Information from G. Peña.

# 11

# Quail

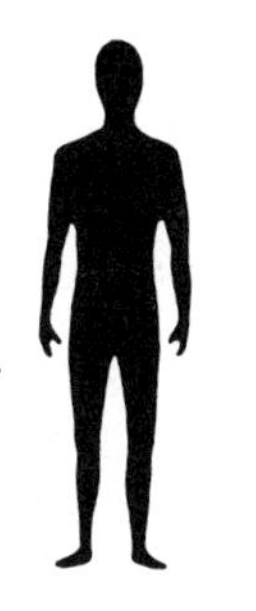

Native to Asia and Europe, quail[1] (*Coturnix coturnix*[2]) have been farmed since ancient times, especially in the Far East. They reproduce rapidly and their rate of egg production is remarkable. They are also robust, disease resistant, and easy to keep, requiring only simple cages and equipment and little space. Yet they are not well known around the world and deserve wider testing.

Quail are so precocious that they can lay eggs when hardly more than 5 weeks old. It is said that about 20 of them are sufficient to keep an average family in eggs year-round. Quail eggs are very popular in Japan, where they are packed in thin plastic cases and sold fresh in many food stores. They are also boiled, shelled, and either canned or boxed like chicken eggs. Quail eggs are excellent as hors d'oeuvres and they also are used to make mayonnaise, cakes, and other prepared foods.

In France, Italy, the United States, and some countries in Latin America (Brazil and Chile, for example) as well as throughout Asia, it is the meat that is consumed. It is particularly delicious when charcoal broiled. One company in Spain annually processes 20 million quail for meat.

Many of the domesticated strains seem to have originated in China, and migrating Chinese carried them throughout Asia. Today, millions of domestic quail are reared in Japan, Indonesia, Thailand, Taiwan, Hong Kong, Indochina, Philippines, and Malaysia, as well as in Brazil and Chile.

[1] This chapter concerns coturnix quail, the quail of commerce. Biologically speaking, quail is a general term that includes many different species and subspecies.

[2] Recently the suggestion has been made that Japanese quail be reclassified from a subspecies of *Coturnix coturnix* to a separate species, *Coturnix japonica*.

A quail in the hand. (Rodale Press)

Commercial production is carried out, as in the chicken industry, in specialized units involving hatcheries, farms, and factories that process eggs and meat. However, quail have outstanding potential for village and "backyard" production as well. It is this aspect that deserves greater attention.

## AREA OF POTENTIAL USE

Worldwide.

## APPEARANCE AND SIZE

Quail come in various sizes. The smaller types are used for egg production, whereas the larger ones are better for meat. Adult females of improved meat strains may weigh up to 500 g.

There arc several color varieties. However, mature females are characterized by a tan-colored throat and breast, with black spots on the breast. Mature males, which are slightly smaller than females, have rusty-brown throats and breasts. All mature males have a bulbous structure, known as the foam gland, located at the upper edge of the vent.

In the United States, the Pharaoh strain is the bird of choice for commercial production. Other available strains tend to be bred more for fancy than for food.

Quail eggs are mottled brown, but some strains have been selected for white shells. These eggs are often preferred by consumers and are easier to candle (the process of holding eggs up to a light to check for interior quality and stage of incubation). An average egg weighs 10 g—about 8 percent of the female's body weight. (By comparison, a chicken egg weighs about 3 percent of the hen's body weight.) Quail chicks weigh merely 5–6 g when hatched and are normally covered in yellowish down with brown stripes.

## DISTRIBUTION

The ancestral wild species is widely distributed over much of Europe and Asia as well as parts of North Africa. Although domestic quail are now available almost everywhere, Japan is probably the world leader in commercial production; quail farms are common throughout its central and southern regions.

## STATUS

Not endangered.

The natural distribution of the quail (*Coturnix coturnix*). The subspecies identified are: (1) European quail (*C. c. coturnix*); (2) Ussuri quail (*C. c. ussuriensis*); (3) Japanese quail (*C. c. japonica*); (4) African quail (*C. c. africana*). Only the Japanese quail is now widely used in domesticated form.

## HABITAT AND ENVIRONMENT

Quail are hardy birds that, within reasonable limits, can adapt to many different environments. However, they prefer temperate climates; the northern limit of their winter habitat is around 38°N.

## BIOLOGY

A quail's diet in the wild consists of insects, grain, and various other seeds. To thrive and reproduce efficiently in captivity, it needs feeds that are relatively high in protein.

The females mature at about 5–6 weeks of age and usually come into full egg production by the age of 50 days. With proper care, they will lay 200–300 eggs per year, but at that rate they age quickly. The life span under domestic conditions can be up to 5 years. However, second-year egg production is normally less than half the first year's, and fertility and hatchability fall sharply after birds reach 6 months of age, even though egg and sperm production continue. Thus, the commercial life is only about a year.

Crosses between the wild and domestic stocks produce fertile hybrids. Repeated backcrossing to either wild quail or domestic quail is successful.

## BEHAVIOR

Only females hatch the eggs and raise the chicks. Males go off and court other females when their partners begin the nesting process.

## USES

Quail eggs taste like chicken eggs. They are often served hard boiled, pickled, fried, or scrambled. Because of their size they make attractive snacks or salad ingredients. They provide an alternative for some people who are allergic to chicken eggs. On frying, the yolk hardens before the albumen.

Quail meat is dark and can be prepared in all of the many ways used for chicken. The two meats are similar in taste, although quail is slightly gamier.

Because of its hardiness, small size, and short life cycle, quail are now commonly used as an experimental animal for biological research and for producing vaccines—especially the vaccine for Newcastle disease, to which quail are resistant.

## QUAIL IN JAPAN

Shigeo Kogure

In some areas of Japan, quail are widely raised for their eggs and meat. However, Japanese originally valued the quail as a songbird. Tradition has it that about 600 years ago people began to enjoy its rhythmic call. In the feudal age, raising song quail became particularly popular among Samurai warriors. Contests were held to identify the most beautiful quail song, and birds with the best voices were interbred in closed colonies. Even photostimulation was practiced to induce singing in winter.

Around 1910, enthusiastic breeders produced the present domestic Japanese quail from the song quail. It was created as a food source and became a part of Japanese cuisine. During World War II it was almost exterminated, but Japanese quail breeders restored it from the few survivors and from birds imported from China. The original song quail, however, were lost. In the 1960s, commercial quail flocks rapidly recovered and Japan's quail population again reached its prewar level of about 2 million birds.

## A QUAIL IN EVERY POT: AN OLD DELICACY FINDS A NEW PUBLIC

For centuries, quail were considered a great delicacy: a dish that only eminent chefs would cook and diners with an appreciative palate could enjoy. These small migratory birds, which are found in one variety or another throughout the world, were available until recently almost exclusively to hunters in the wild.

But now quail are in danger: in danger of becoming commonplace.

In the last few years, quail have gone from being rarefied to a supermarket specialty item. They are on menus in the most elegant restaurants and the most casual cafes and bistros.

Why so much interest?

Quail are now available semi-boneless, which makes them faster and easier to cook, and easier to eat as well. The breastbones are removed by hand before the birds are packaged and shipped to stores. The bones in the wings and legs remain.

A stainless-steel V-shaped pin—invented and patented by a restaurant chef who wanted a way to keep quail flat for grilling—is inserted into the breast. The pin can be left there throughout cooking and removed just before serving.

While whole quail might require 45 minutes to cook, the semi-boneless variety can be grilled in less than 10 minutes, or pan-roasted, braised or sauteéd in less than 20 minutes.

The flavor of farm-raised quail has also helped bring them into the mainstream. Most farm-raised quail have tender meat like the dark meat of chicken, whose flavor is enjoyed by many people.

And at a time when people are searching for foods, specifically animal protein, with low fat and cholesterol, quail fills the bill. The Agriculture Department says that quail skin has about 7 percent fat, about the same as dark meat of roasted chicken without the skin.

Judith Barrett
Adapted from *The New York Times*
June 21, 1989

Many fanciers and hobbyists have also become interested in raising this adaptable species as a pet. Science teachers find it an excellent subject for classroom projects.

## HUSBANDRY

It is necessary to keep quail in battery cages on wire floors because males secrete a sticky foam (from the foam gland) with their feces; on a solid floor, this adheres to the feet and collects dung, leading to crippling and breakage of eggs.

Adult quail can live and produce successfully if they are allowed 80 $cm^2$ of floor space per bird. However, for reproduction about twice that is needed to allow for mating rituals.[3] If properly mated, high fertility rates and good egg hatchability can be expected. To obtain fertile eggs, one male is needed for roughly six females.

Eggs hatch in about 17 days. Chicks require careful attention. Brooding temperatures of between 31°C and 35°C are needed for the first week and above 21°C for the second week. From the second week on, chicks can survive at room temperature. (These temperatures are similar to those required for common chickens.) In cold climates, supplemental heat may be needed as well as protection from cool drafts.

Clean water must be provided at all times, with care taken to prevent the chicks from drowning in their water troughs. Shallow trays, jar lids, or pans filled with marbles or stones may be used.

## ADVANTAGES

Quail production can be started with little money. These easy-care birds can be housed in small, simple, inexpensive cages.

As noted, they are resistant to Newcastle disease.

## LIMITATIONS

Although generally disease resistant, quail are affected by several common poultry diseases, including salmonella, cholera, blackhead, and lice. They also suffer epidemic mortality from "quail disease" (ulcerative enteritis), which can, however, be controlled with antibiotics.

Quail seem to require more protein than chickens, and produce best when given feed that is fairly high in protein.[4] However, they also

[3] Information from H. R. Wilson.

[4] A level of 27 percent crude protein and 2,800 kcal ME per kg ration is recommended up to 3 weeks of age and 24 percent crude protein with the same energy level during a 4–5 week growing period. Layer quail should be fed 22 percent crude protein and 2,900 kcal ME per kg ration. Information from B. K Shingari.

## SIX OF ONE, HALF A DOZEN OF THE OTHER

Z.A.b.M. Noor

The fact that chickens can be crossed with quail has been known for some time, but there has been little attempt to develop the fertile hybrids. Now Malaysia has begun a project aimed at producing a new poultry bird—a cross between a cockerel and a hen quail. Zainal Abidin bin Mohd Noor, of the Department of Veterinary Services in Kuala Lumpur, is creating a strain that produces eggs of good quality and meat with the flavor of both parents. The new bird is intermediate in size between chicken and quail, which is convenient because it is about right for an individual helping.

The crossbreeding is done through artificial insemination. The progeny exhibit a range of appearances, sizes, and plumage colors, depending on the strains of cockerel and quail hens used. In the Malaysian research, cockerels have been local Ayam Kampung, Bantam, Hybro, and Golden Comet hybrids. The quails have been local inbred Japanese quail (IJQ) and imported meat strain quail (IMSQ).

The trials show that the hybrids derived from the IMSQ flocks grew faster and bigger than those from the IJQ cross. The best have been the Hybro x IMSQ crosses, which weigh 475 g at 10 weeks of age. The best of the IJQ group weighed 290 g during the same period.

This type of "tropical game hen" might be a way to introduce hybrid vigor into poultry production.

The researchers who developed the hybrid have named it the "yamyuh."

perform satisfactorily when fed rations designed for turkeys. They have high requirements for vitamin A, which they do not store.

Quail are not suitable as free-ranging "scavengers." They must be kept confined, which is a major constraint. Unlike chickens or pigeons, they have no homing instinct and will not remain on a given site; if released, they will be lost. In addition, since they nest on the ground, they are highly susceptible to predation; they must be protected, especially where certain animals, the mongoose for example, are common.

Artificial incubation is essential. Natural incubation using the female is futile; the females do not go broody and rarely incubate their eggs. The shells are extremely thin, but the eggs can be incubated under a small chicken hen, such as a bantam.[5] The eggs are also subject to minute fractures. However, the shell membrane is extremely tough and unfertilized eggs are generally unaffected, but the cracks cause fertilized embryos to dehydrate and die. This is a serious limitation. Whenever quail husbandry is introduced, artificial incubation should be included.

## RESEARCH AND CONSERVATION NEEDS

Quail deserve to be included in all poultry research aimed at helping the Third World. Through its international scientific program, Japan, in particular, could apply to developing nations its vast experience with quail farming.

Experiences with quail in the tropics (for example, Japanese farmers in the Amazon Basin) and in tropical highlands (for instance, in India, Nepal, or Central Africa) should be collected and assessed to improve understanding of the environmental limits to Third World quail farming.

Cooperation between commercial and laboratory quail breeders should be encouraged. Mutants found at the commercial level would be useful for laboratory work. Conversely, introducing new stocks could help the farmer. In both cases, more genetic diversity might also lead to the production of hybrid vigor, and genetic variability would be conserved.

Sex-linked genes, if they can be found, would be useful to the commercial quail breeder for the rapid sexing of newly hatched chicks. This could lead to more efficient production techniques, like those in the chicken industry.

Although virtually all work to date has been on the Japanese quail, other species and subspecies warrant research and testing.

[5] Information from S. Lukefahr.

# 12

# Turkey

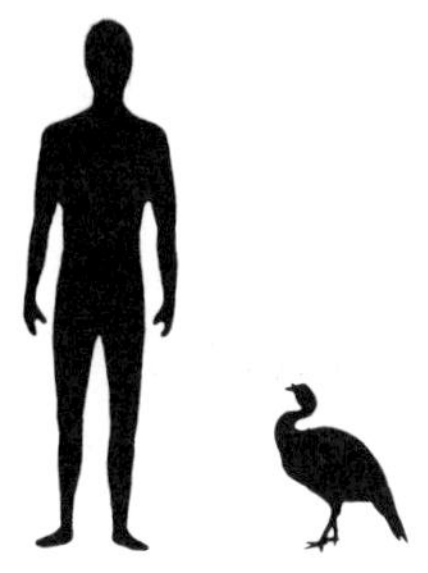

The turkey (*Meleagris gallopavo*) is well-known in North America and Europe, but in the rest of the world, especially in developing countries, its potential has been largely overlooked. Partly, this is because chickens are so familiar and grow so well that there seems no reason to consider any other poultry. Partly, it is because modern turkeys have been so highly bred for intensive production that the resulting birds are inappropriate for home production.

Nevertheless, there is a much wider potential role for turkeys in the future. There are types that thrive as village birds or as scavengers, but these are little known even to turkey specialists. These primitive types are probably the least studied of all domestic fowl; little effort has been directed at increasing their productivity under free-ranging conditions. However, they retain their ancestral self-reliance and are widely used by farmers in Mexico. That they are unrecognized elsewhere is a serious oversight.

Native to North America, the turkey was domesticated by Indians about 400 BC, and today's Mexican birds seem to be direct descendants.[1] Unlike the large-breasted, modern commercial varieties, they mate naturally and they retain colored feathers and a narrow breast configuration. Their persistence in Mexico after 500 years of competition with other poultry highlights their adaptability, ruggedness, and usefulness to people.

These birds complement chicken production. They are able to thrive under more arid conditions, they tolerate heat better, they range farther, and they have higher quality meat. Also, the percentage of edible meat is much greater than that from a chicken. Turkey meat is so low in fat that in the United States, at least, it is making strong inroads into markets that previously used chicken exclusively.

[1] In Spanish, these New World natives are called "criollo" turkeys to distinguish them from improved, reintroduced types. They are also known as "guajolote" or "pipil" in some areas.

Turkeys are natural foragers and can be kept as scavengers. Indeed, they thrive best where they can rove about, feeding on seeds, fresh grass, other herbage, and insects. As long as drinking water is available, they will return to their roost in the evening.

Appreciation for the turkey could rise rapidly. Interest already has been shown by several African nations. A French company has created a strain of self-reliant farm turkeys and is exporting them to developing countries.[2] Researchers in Mexico are displaying increased interest in their national resource. And as knowledge and breeding stock continue to be developed, it is likely that village turkeys will become increasingly popular around the world.

## AREA OF POTENTIAL USE

Worldwide.

## APPEARANCE AND SIZE

Modern turkey breeding has been so dominated by selection for increased size and muscling that commercial turkeys have leg problems and cannot mate naturally (they are inseminated artificially). These highly bred birds are adapted for large-volume intensive production, and must be raised with care. As noted, this chapter emphasizes the more self-reliant, less highly selected turkeys found in Mexico and a few other Latin American countries. They do not require artificial insemination, and with little attention can care for themselves and their young.

Fully grown "criollo" turkeys of Mexico are less than half the size of some improved strains. Males weigh between 5 and 8 kg; females, between 3 and 4 kg.[3] They vary in color from white, through splashed or mottled, to black. The skin of the neck and head is bare, rough, warty, and blue and red in color. A soft fleshy protuberance at the forehead (the snood) resembles a finger. In males it swells during courtship. The front of the neck is a pendant wattle. A bundle of long, coarse bristles (the beard) stands out prominently from the center of the breast.

## DISTRIBUTION

The unimproved domestic turkey is essentially limited to central Mexico and scattered locations throughout nearby Latin American

[2] Information from ADETEF, 1980.
[3] Information from M. Cuca.

countries. Some village birds are also kept in India, Egypt, and other areas, but these are descended from semi-improved strains exported from North America and Europe in earlier times. Generally speaking, few turkeys are found in tropical countries outside Latin America.

## STATUS

Domesticated turkeys are not endangered; there are estimated to be about 124 million in the world. However, the wild Mexican varieties, ancestors to the first domesticated turkeys sent to Europe, may now be endangered since their distribution in southwestern Mexico has been greatly reduced. Certainly, some primitive domestic strains in the uplands of central Mexico are also being depleted. A separate type, independently domesticated by the Pueblo Indians of the southwestern United States, seems to have disappeared entirely.

## HABITAT AND ENVIRONMENT

Turkeys can be reared virtually anywhere. Their natural habitat is open forest and wooded areas of the North American continent, but

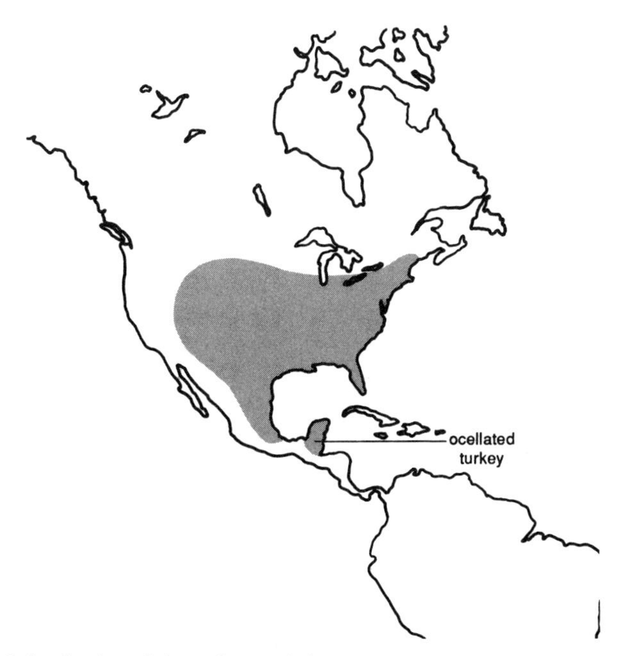


The original distribution of the turkey and the ocellated turkey.

in Mexico they are raised from sea level to over 2,000 m altitude, from rainforest to desert, and from near-temperate climates to the tropics.

## BIOLOGY

The range of diet is broad. Turkeys eat greens, fruits, seeds, nuts, grasses, berries, roots, insects (locusts, cicadas, crickets, and grasshoppers, for example), worms, slugs, and snails.

Reproduction is generally seasonal and is stimulated by increasing daylength. (A minimum daylength of 12 hours is required.) The birds can reach sexual maturity at six months of age and may start breeding at this time. Ten days after first mating, the hen searches out a nest and commences laying. Industrial birds in temperate climates lay, on average, 90 eggs a year. The nondescript type of turkey in the tropics seldom lays more than 20 small eggs (weighing about 60 gm) before going broody.

## BEHAVIOR

Domestic turkeys walk rather than fly, and find almost all their food on the ground. They can, however, fly short distances to avoid predators.

The commercial birds have lost many abilities for survival in the wild; they can no longer exist without human care. However, village types can do well with little management.

Turkeys prefer to make their own nests but can be induced to lay in a convenient spot if provided with nest boxes.

## USES

These birds are raised almost exclusively for meat. In many countries, they are a treat for holidays, birthdays, and weddings. In their native range of Mexico and Central America, the "unimproved" birds are usually produced as a cash crop for market. They receive little care or feed, and thus they are almost all profit—providing a significant income supplement to many rural homes.

Opposite: Criollo turkeys being raised in a backyard on the outskirts of Mexico City. (G. Hettel)

## HUSBANDRY

The principles of turkey management (nutrition, housing, rearing, and prevention of disease, for example) are basically the same as those for other poultry.

In Mexico, turkeys are usually kept under free-ranging conditions around houses and villages. Some shelter and kitchen scraps are occasionally provided. A number of them, however, are confined in backyards as protection from marauders and for shelter against rain and wind.

One malc can service up to 12 females. Roomy nests are needed. (As a rule, turkeys require three times the space occupied by chickens.) Most range turkeys are corralled when they begin to lay, so as to protect them from predators. Eggs may be gathered to prevent broodiness and thereby increase production. The eggs may be kept for several days (cool, but not refrigerated) if turned daily, and then may be placed under a chicken hen. (A setting chicken can be used this way to hatch up to nine eggs at a time.) Hatching takes 28 days.

As in other birds, newly hatched turkeys (poults) must be kept warm during the first weeks of life. Until they begin foraging and have full access to pasture they are usually fed broken grain or fine mash, as well as finely chopped, tender green feed.

---

### "AN INCOMPARABLY FINER BIRD"

*The turkey was domesticated in Mexico some time before the Conquest. It is the one and only important domestic animal of North American origin. When the Spanish arrived, they found barnyard turkeys in the possession of Indians in all parts of Mexico and even in Central America. However, the Aztecs and the Tarascans, originating in west-central Mexico, seemed to have achieved the highest development of turkey culture, and it is probable that turkeys were domesticated in the western highlands, perhaps in Michoacan. Wild turkeys of that region are morphologically very similar to the primitive domestic bronze type. Both the Aztecs and Tarascans kept great numbers of the birds, including even white ones. They paid royal tribute to their respective kings in turkeys, according to the Relación de Michoacán. The Tarascan king fed turkeys to the hawks and eagles in his zoo. The economy of some highland tribes was based on the cultivation of corn and the raising of turkeys.*

A. Starker Leopold

---

## THE INDUSTRIAL TURKEY

The modern domesticated turkey is thought to be descended from two differing wild subspecies, one found in Mexico and Central America and the other in the United States. The southern type is small, whereas the U.S. native is larger and has a characteristic bronze plumage.

Mexican turkeys were exported to Europe soon after the Conquest, and spread rapidly. In the 17th century, some were returned to North America, where they interbred with the eastern subspecies of wild turkey, producing a heavier bird, which was then re-exported to Europe.

These types underwent little change until this century, when the Englishman Jesse Throssel bred them for meat quality. In the 1920s, he brought his improved birds to Canada, where their large size and broad breasts quickly made them foundation breeding stock. Crossed with the narrow-breasted North American types, these heavily muscled meat birds quickly supplanted other varieties.

About the same time, the U.S. Department of Agriculture began the scientific development of a smaller meat turkey derived from a more diverse genetic base. By the 1950s, the Beltsville Small Whites predominated in the home consumption market in the United States.

Although free-ranging turkeys are simple to raise, confined turkeys require more complex management. The birds need uncrowded, well-ventilated conditions and should be on a wire or slatted floor to reduce parasitic infections. Any feeds recommended for chicks are suitable, but the protein content should be somewhat higher; that is, about 27 percent. They can be fed mixed grains, corn, and chopped legume hay. It may be necessary to provide vitamin supplements and antibiotics and take steps to prevent coccidiosis.

## ADVANTAGES

The birds are efficient and generally take care of themselves. They tolerate dry, hot, or cold climates and forage farther than chickens. They are large, fast growing, highly marketable, low in fat, and tasty.

## THE TURKEY'S TROPICAL COUSIN

Okapia

The ocellated turkey (*Agriocharis ocellata*) occurs in Yucatan, Guatemala, and Belize. It is much like the common turkey in size, form, and behavior; however, unlike the common turkey, which in Mexico lives in the high mountain pine and oak forests, the ocellated turkey inhabits bushy, semiforested lowlands. This splendid bird lacks the kind of beard sported by the common turkey gobbler, is generally more metallic in appearance, and has brighter coppery colors. The chief character is a neck and head that are bare, blue, and profusely covered with coral-colored pimples. It also has a yellow-tipped protuberance growing on the crown between the eyes.

This species is worthy of investigation by poultry researchers because it might prove to be domesticable. It was possibly domesticated by the Mayas, whose ruins often include appropriately sized stone enclosures whose soil has elevated levels of phosphorus and potassium. Even today, in the rural Petén area of Guatemala, ocellated turkeys are sometimes kept around houses as scavengers.

## LIMITATIONS

Young birds are readily affected by temperature changes and must be protected from the sun as well as from sudden chills, such as may occur at night. They are particularly susceptible to dampness, especially if associated with cold. One peculiarity is the turkey's aversion to any change in feeding routine or the nature of the food.

Young turkeys are susceptible to parasitic infestation as well as to the same type of bacterial and virus diseases as chickens (for example, fowlpox and coccidiosis). Blackhead, a devastating disease of young turkeys, is carried by a common parasitic nematode, and can be contracted from chickens. Medicines are available to prevent or treat most disease and pest problems.

## RESEARCH AND CONSERVATION NEEDS

Turkey development is almost nonexistent in the Third World (and much of the rest of the world, too). Although commercial turkeys are highly developed in some countries, little or no research has been conducted on the criollo turkey. Research on physiology, disease, and husbandry of the criollo turkey should be given high priority.

The need for conservation of genetic variability is perhaps more critical in this species than in almost any other domesticated animal. The unimproved types in Mexico should be collected and assessed, and a program to conserve the stocks should be initiated. An analysis should also be made of the traditional management and performance of these birds. In addition, the four or five recognized turkey subspecies should be evaluated for their potential as seed stock for Third World countries.

# 13

# Potential New Poultry

Several preceding chapters have discussed the possibilities of domesticating certain wild birds.[1] Here, briefly, are highlighted other wild species with qualities that might make them suitable for sustained production. It should be understood that their practical use in the long run is pure speculation; they are included here merely to guide those interested in exploring the farthest frontiers of livestock science.

Collectively, poultry have become the most useful of all livestock—and the most widespread. Yet only a handful of species are employed. Of the 9,000 bird species, only a few (for instance, chickens, ducks, geese, muscovies, pigeons, and turkeys) have been domesticated for farm use. Strictly speaking, all birds are edible—at least none have poisonous flesh—so it seems illogical to conclude that these are the only likely candidates. Perhaps they are not even the best.

At first sight there may seem to be little need for new species, but poultry meat is in ever increasing demand and there are many niches where the main species are stricken by disease, or are afflicted by heat, humidity, altitude problems, or other hazards. For these areas, a new species might become a vital future resource. Perhaps some could even become globally important. The modern guinea fowl, for example, is a relative newcomer as a worldwide resource (see page 120).

The birds now used as poultry were domesticated centuries ago by people unaware of behavior modification, nutrition, genetics, microbiology, disease control, and the other basics of domestication. Today we can tame species that they couldn't. In particular, the new understanding of "imprinting" may make the domestication of birds easier today than ever before.

[1] These include junglefowls (see page 86), ducks (98), geese (104 and 107), guinea fowl (122), and ocellated turkey (164).

In this highly speculative concept, the birds described on the following pages are worth considering. They all eat vegetation and tend to live in flocks, which makes them likely to be easy to feed and to keep in crowded conditions. Most are sedentary, nonmigratory, and poor fliers. All but three (tinamous, sand grouse, and trumpeters) are gallinaceous.

Gallinaceous birds are already the most important to people. The best known are chickens, turkeys, quail, and guinea fowl. But there are about 240 other species. Most are chickenlike: heavy bodied with short, rounded wings, and adapted for life on the ground. Although some are solitary, many are sociable. Basically vegetarian, they also eat insects, worms, and other invertebrates. The young birds are extremely precocious, walking and feeding within hours of hatching. All of these are advantageous traits for domestication.

Game birds are also emphasized here. Many today are considered gourmet delights, and this should give them a head start in the marketplace. Indeed, some are already being raised in a small way on game farms and are at least partly on the way to domestication.

## CHACHALACAS

These brownish birds (*Ortalis vetula* and nine other species) are found throughout Central and South America, and, given research, could possibly be raised on a large scale. A sort of "tropical chicken," they tame easily, live together in dense populations, and protect their chicks extremely well. They commonly scavenge around houses and people often put out scraps to feed them.[2] The chicks are easily hatched, grow fast, and can be fed standard chicken rations.[3]

There is already considerable demand for these birds. Everywhere they are found, they are prized as food. In some areas they constitute the single most important game species, and are heavily hunted to supply local communities. Although they have less meat than a chicken, it is tastier and darker.

Chachalacas are very adaptable. They occur mainly around forest edges and thrive in the thickets that appear after tropical forests have been felled. They do well close to humans, and their populations are not threatened, despite much hunting. Indeed, they seem well adapted to existence around villages and towns. Although not strong fliers,

[2] During winter months, when natural foods become scarce, small flocks frequently solicit handouts of everything from potato chips to popcorn from delighted humans. Information from W. R. Marion.

[3] In 4–5 months of growth they approach the adult size of 500–600 g. Information from W. R. Marion.

they are one of the few tree-roosting gallinaceous species. Primarily fruit eaters, they also consume tender leaves, twigs, and buds, and they scratch up the ground, presumably for insects.

Although excitable and noisy, chachalacas become remarkably tame when fed by people. In a few cases, full domestication has almost been reached. Farmers like to have chachalacas around and have even used them to guard domestic chickens. These very raucous and fearless birds will take on all potential threats, even weasels.[4]

## GUANS

Close relatives of the chachalacas, guans[5] are glossy black birds about the size of small geese. They are highly gregarious and perhaps could be raised in larger numbers. They commonly live around houses, farms, and settlements in their native region of tropical America.

Unlike most game birds, guans are chiefly tree dwellers, but they also feed on the ground. Some 12 species are known. All are relentlessly hunted for food and sport—their tameness and inability to fly far or fast making them easy targets. The rapid destruction of tropical forests threatens their populations in some parts of their range. Conservation projects and specific plans of action are being proposed for the most threatened species. Perhaps for the other species, game-ranching projects or even outright domestication might provide just the right incentive for their protection and multiplication.

## CURASSOWS

Curassows are also relatives of guans and chachalacas, but they are even larger—up to 1 m tall and 5 kg in weight. At least seven species are found over the vast area from northern Mexico to southern South America.[6] Among them are Latin America's finest game birds.

It might be possible to produce curassows in organized farming or ranching. They are commonly called "tropical turkeys" because they look like and run like turkeys. Indeed, Latin Americans normally refer

[4] L. Griscom has related a story of a chachalaca living in Ocos on the Pacific coast of Guatemala. It was allowed to move freely around the village. Its chosen task was to keep peace and order among the domestic fowl. Whenever two cockerels began fighting, it raced up and separated them. The cockerels would run away as soon as the chachalaca "cop" appeared on the scene.

[5] These birds fall into several genera, but the guans proper are *Penelope* species.

[6] Examples are the great curassow (*Crax rubra*), which is found from Mexico to Ecuador; the helmeted curassow (*Pauxi pauxi*) of the mountains of Venezuela and Colombia; and the razor-billed curassow (*Mitu mitu*) of the Amazon.

## FOREST BIRD RANCHING

This report has intentionally focused on intensive farming—the type where people bring feed to animals in captivity. However, where this normal type of farming is of marginal value, "ranching" free-ranging birds may often be a more effective option. In this, the farmer simply monitors and improves the condition of the range and devises methods to harvest the birds on a sustainable basis.

"Bird ranching" may today have outstanding merit, particularly in tropical rainforests. Hence, in this chapter we emphasize birds of the jungle. These might help make standing rainforests profitable producers of income, and thereby provide economic incentives to stop felling trees for cow pastures. Indeed, forest birds might become part of a whole new "salvation farming" that makes forests more valuable than fields. It is a technique that may contribute to preserving both bird life and its vitally valuable habitat.

to them as "pavos" or "pavones," as if they were the real thing. Their plumage ranges from deep blue to black, invariably with a purple gloss, and all have rather curly crests on their heads. They are not good fliers and spend most of their time on the ground.

Curassows are increasingly hunted; their tropical forest habitat is shrinking, and the subsequent loss of populations is a calamity. They are special targets, not only because they are large but also because their light-colored flesh makes exceptional eating.

There is hope that these large wild fowls can be raised and managed in organized programs. Even now, people commonly keep them around their farms and villages. For example, on a number of Venezuelan ranches, yellow-naped curassows can be seen wandering around the cattle yards as if they were chickens.[7]

## MEGAPODES

Megapodes (family Megapodiidae) include some of the world's most interesting birds. They have temperature-sensitive beaks and employ nature's own heat sources as incubators. The best-known species build piles of leaves and use the heat of decomposition to incubate their

[7] Information from F. Wayne King.

eggs. The species of Papua New Guinea and Indonesia, however, take advantage of sun-warmed sand or even geothermal activity.

People have long revered these birds. Aborigines in Australia, Melanesians in New Guinea, and many Micronesians all protect the bizarre nesting sites, and "farm" them for eggs. Local people consider the large eggs special delicacies, and sometimes the egg-laying sites are owned and exploited for generations without a single bird being killed for food.[8]

Programs that provide sustainable supplies of eggs have been established in Papua New Guinea. One is near Mt. Tovarvar, a simmering volcano on the island of New Britain. Here, megapodes gather in large numbers to lay eggs in the hot sands. They dig until they locate sand that is exactly 32.7°C, before laying their huge (more than 10 cm long and 6 cm wide) pink eggs. Each year the villagers dig up some 20,000 eggs, which are an important source of protein and cash income. The government now regulates the harvest in a way that protects the bird population while supplying a nourishing food.

Megapodes are found in only a few parts of the world, but projects such as those in Papua New Guinea provide hope and guidance not only for the sustainable "ranching" of megapodes, but also for other species elsewhere. Many wild birds yield locally important products—down, colored feathers, eggs, meat, and skins, and they make excellent songbirds and pets, for example. Their management on a sustainable basis may in certain cases be the key to turning local people into the most dedicated conservationists of all.

## PARTRIDGES AND FRANCOLINS

Partridges include many small game birds native to the Old World. They are robust, precocious, and larger than quails. Some lay many eggs—the European partridge, for example, lays up to 26 in a clutch. Newly hatched chicks are soon able to feed themselves and can fly within a few weeks, sometimes even within the first few days.

Species that may make useful poultry include:

- The European (or gray) partridge (*Perdix perdix*);
- The rock partridge (*Alectoris*), bantamlike birds of Africa; and
- The chukar (*A. graeca*).

A native of the vast area from southeastern Europe to India and Manchuria, the chukar is stocked as a game bird in many countries.

[8] On some Pacific islands, starving Japanese garrisons nearly or completely wiped out the colonies during World War II.

## MALAYSIA'S MOBILE MOUSETRAP

*Although the report emphasizes microlivestock as food suppliers, it should be realized that small animals—even wild ones—can have other important uses as well. The following interesting example, with possible worldwide implications, comes from recent experiences in Malaysia.* *

Certain rodents are major pests on farms and plantations. Now, however, Malaysian zoologists are finding that owls, particularly barn owls (*Tyto alba*), can help control them. An owl pair and its chicks annually consume 1,500 or more rodents. This is not new knowledge; indeed, on farms throughout the world, the barn owl has always been a welcome guest. What is new is that Malaysians are showing how outstandingly effective this process is, and they have initiated major projects to attract and maintain these feathered friends.

Barn owl. The shape of the face and the position of the eyes allow this bird to hear, see, and pinpoint even tiny rodents on the darkest nights. (R. White, *New Zealand Herald*)

Barn owls are found in many parts of the world, but were formerly almost unknown in Peninsular Malaysia. In 1969, however, a pair began nesting in an oil-palm plantation in Johore State. Since then, these birds have steadily increased in numbers and have spread throughout most of the peninsula. Today, the population is increasing remarkably quickly as more and more managers erect nest boxes for the owls to live in.

The owls are proving to be a good way to remove rats and are notably effective in plantations of oil palm. They perch on fronds and fly under and between the rows of trees. A cost of $1–$2 per hectare per year is all that is required to install nest boxes, a negligible outlay for the control of such a serious and expensive problem.

It is believed that the barn owls hunt mainly in plantations and other agricultural areas and not in the rainforest. Barn owls are, after all, primarily adapted to open spaces and not dense forest.

Perhaps this experience can be replicated and adopted in other locations and with other crops. Grain crops—notably rice—are particularly prone to the ravages of rodents, and one trial has commenced in Selangor State in a rice area. The concept of using owls for rodent control is also catching on in the United States. Indeed, owl nest boxes are being erected in Central Park in the heart of New York City.

* Based on information provided by Christopher M. Smal.

It is now produced routinely under poultrylike husbandry in many parts of the United States, not only for hunting clubs, but also for expensive food markets. The birds are generally raised on turkey rations and dress out at about 500 g after 18 weeks. They sell for more than broiler chickens and are a profitable sideline for increasing numbers of poultry farmers.[9]

A group of closely related birds are the francolins (genus *Francolinus*), of which there are 34 species in Africa and 5 in West and South Asia. These adaptable birds are sturdy, live in a variety of habitats, and tend to be rather noisy. Basically, they are partridges with leg spurs. They are highly regarded as a food source and are hunted and trapped wherever they are found.

Francolins are much like quail, but are several times larger. Arabs introduced one of the most beautiful species (*Francolinus francolinus*) into southern Spain, Sicily, and Greece during the Middle Ages.

[9] Information from A. Woodard.

However, it was hunted so heavily that it soon became extinct in Europe. More recently, francolins have been introduced to the Soviet Union.[10]

Francolins inhabit steppes, savannas, primeval forests, and mountains. They thrive in cultivated land with much cover. The clutch consists of 6–8 hard, thick-shelled eggs. In recent times at least one program to domesticate them for food has been started in Africa.[11]

## PHEASANTS[12]

One pheasant, the red junglefowl, gave the world the chicken (see page 86). The other 48 species may have some potential, too. These are rarely seen forest birds; all but one are confined to Asia.[13] Because they are prolific they can sustain heavy predation, and many species, notably the ring-necked pheasant (*Phasianus colchicus*), are constantly hunted.

People in several countries have learned to exploit pheasants on farms and estates. As a result, there is a vast amount of information on how to rear and manage these birds. So far, however, it has been applied only to sport hunting in wealthy societies; the potential of raising pheasants for the mass market should now be seriously addressed.

The most dramatic-looking pheasant, the peacock (*Pavo cristatus*), is raised as a poultry species in Vietnam. The meat of the young birds is considered outstanding. In fine restaurants in New York, a peacock dinner is reputed to cost $150. Common peafowls are considered sacred in many parts of India, where they have become so tame that they are essentially domesticated. They also control snakes.

## QUAIL

Domestic quail have been previously described (page 146), but dozens of wild quail species and subspecies occupy many different habitats and ecological niches in almost all parts of the world. Out of

[10] In 1932, three cocks and two hens were released in Agri-tschai Valley in the Nucha area of Kachetis (Caucasus). By 1947 francolins were all over this valley, and had also settled the Alasan River valley more than 100 km away.

[11] D. F. Adene and D. Akande, 1978, A diagnosis of coccidiosis in captive bush fowl (*Francolinus bicalcaratus*) and identification of the causative coccidia, *East African Wildlife Journal*, 16:227–230.

[12] The term pheasant is usually reserved for the large, colorful, long-tailed members of the Phasianidae family, subfamily: Phasianinae.

[13] The exception is the extraordinary and beautiful Congo peacock, whose discovery in Central Africa as recently as 1936 created an ornithological sensation.

all this genetic wealth only one species—the Japanese quail—is widely used. Yet many other species seem easy to raise, becoming exceedingly tame after about the sixth generation.

The management and even perhaps intensive production of these various local quails might provide long-term benefits for many developing nations. Quail meat ranks among the finest.[14] Some of these lesser-studied birds are more meaty than the Japanese quail or have other possibly useful traits. Much is known about rearing a few of them because they are used in sport hunting or laboratory research. The possibility of domestication, therefore, is not farfetched.

Particular quail that might be considered for domestication are the lesser-known subspecies of *Coturnix coturnix*. These subspecies are found in various places, including the following:

- Europe (*C. c. coturnix* breeds in the area ranging from northern Russia to North Africa and from the British Isles to Siberia. In winter it migrates to tropical Africa, Asia, and southern India.)
- The Azores (*C. c. conturbans*)
- The Azores, Madeira, and the Canary Islands (*C. c. confisa*)
- Cape Verde Islands (*C. c. inopinata*)
- East Africa (*C. c. erlangeri*)
- Tropical Africa, southern Africa, Madagascar, and Mauritius (*C. c. africana*)
- Japan (*C. c. japonica*, the most probable ancestor of the domesticated quail)
- China (*C. c. ussuriensis*, a possible ancestor of the domesticated Japanese quail)

## TINAMOUS

Tinamous are quail-like birds of Central and South America's forests and grasslands. They are, however, much larger than quail and resemble small chickens, with plump bodies and no visible tail. There are more than 40 species, and all are much sought for food because their meat is tender and flavorful. The breast is surprisingly large, and its flesh is pale and translucent. One species, the great tinamou (*Tinamus major*), has been called "the most perfect of birds for culinary purposes." Frozen tinamous from Argentina were formerly sold in the United States under the name "South American quail."

Tinamous are found mainly in tropical areas, but are also widely distributed in Argentina and Chile. They dwell in varied habitats:

[14] The domestic bird has all dark meat, but at least some of the wild ones (the northern bobwhite, for instance) have both white and dark.

rainforests, thickets, bushlands, savannas, and grasslands up to 5,000 m altitude in the Andes. Some species sleep in trees, others on the ground. They spend their days creeping about in heavy cover, flying only when forced.

At least some species tame readily. Indeed, during the nesting period males become so tractable that they can be picked up off the nest. At the turn of the century, many tinamous were raised as game birds in France, England, Germany, and Hungary. However, for reasons unknown, subsequent attempts to settle them in Europe have failed. Tinamous have been raised in Canada without undue difficulty; they showed little or no stress under captivity and there were few losses.[15]

Tinamous may also prove suitable for egg production. They lay clutches of 16–20 spectacular-looking shiny eggs that seem to be made of sky-blue and bright-green ceramic.

## SAND GROUSE

Sand grouse (mainly *Pterocles* species) are highly adapted to life in arid regions—desert, dry grasslands, arid savanna, and bushveld. Their entire body (including most of the bill and feet) is covered with dense down, which in the desert insulates them from the burning heat of midday and the freezing cold of night. It also protects the nostrils against blowing sand and dust.

These pigeonlike birds are found throughout the drier regions of Africa and Asia—for instance, the Sahara, Kalahari, Namib, Arabian, and Thar deserts. They live mainly on small seeds, and sometimes flocks of thousands may be seen at waterholes, flying in for a drink from up to 80 km away. For peoples of the driest spots on earth, these birds may make a useful food species: for one thing, they are not endangered. Indeed, they are proliferating as drought and overgrazing is increasing the amount of dry, desolate rangeland that they prefer. The bore holes provided for livestock have both boosted their populations and afforded a place where these wide-ranging birds can be easily captured. When nesting, sand grouse are highly vulnerable to foxes, jackals, mongooses, and other predators. Protection of the nesting sites may be the key to maintaining their populations if harvesting schemes are introduced.

## TRUMPETERS

Trumpeters (*Psophia* species) might prove to be a useful species for sustainable production within tropical forests. As "tree poultry," these

[15] Information from P. Thiessen.

relatives of cranes could help provide meat without destroying the trees, as is now done to raise cattle.

These chicken-sized birds inhabit South America's jungles.[16] They are nonmigrating, ground-dwelling, and are often kept as pets, notably by Amerindians. Under human protection, trumpeters become very tame. They recognize strangers and challenge them with a loud cackle.[17]

Fully adapted to the forest environment, they can run fast, but fly poorly. In the wild, this makes them easy targets for hunters. Because of this and the fact that they make excellent eating, they are approaching extinction in some areas.

No attempts have been made to rear these birds in numbers, but this should be tried. They feed mainly on plant materials, particularly berries of all kinds. They also relish grasshoppers, spiders, and centipedes, and are particularly fond of termites.

Trumpeters require trees; they completely avoid cultivated land. Thus, as the destruction of forests in South America continues, their habitat is shrinking. Although their existence is not as yet threatened, the long-term prognosis is bleak. If managed in "forest-ranching" programs, however, they might be saved from extinction and thriving populations built up.

[16] They are part of the type-fauna of the Amazon region where they are called jacamins.

[17] It has been said that trumpeters, like cats, are predisposed to domestication. They hang around human habitations, even when they have a choice.

*Interest in rabbits continues to increase. It is now widely recognized that the raising of small animals in developing countries has great potential as a means of improving human nutrition and economic security. The famines in Africa, Latin America, and Southeast Asia starkly illuminate the need for maximum efficiency in food production to maintain the quality of human life. Rabbit raising contributes to meeting these needs.*

P.R. Cheeke, N.M. Patton, S.D. Lukefahr, and J.I. McNitt
*Rabbit Production*

*Rabbits are especially well adapted to backyard rearing systems in which capital and fodder resources are usually limiting factors in animal production. When rabbits are reared according to the techniques appropriate to the environment they can do much to improve the family diet of many of the most needy rural families, while at the same time supplying them with a source of income. With more advanced technology rabbit production can also help to supply big city meat markets.*

Food and Agriculture Organization
*The Rabbit: Husbandry, Health, and Production*

# Part III

# Rabbits

Contrary to popular opinion, the domestic rabbit is a substantial part of the world's meat supply. Annual production of rabbit meat is estimated to be one million metric tons, and the total number of rabbits is approximately 708 million.[1] However, rabbits are now intensively raised for food only in temperate, mostly industrialized, nations. France, Italy, and Spain, for example, have long consumed rabbit meat; West German production was 20,000 tons each year; Hungary raises rabbits in large numbers (two of its commercial rabbitries have more than 10,000 does each); and the United States raises almost 8.5 million rabbits each year for consumption in homes and restaurants.[2]

In most developing countries, on the other hand, rabbits are not well known—at least compared with other livestock. But they have great promise there, and in recent years there has been a dramatic increase in interest. For those developing countries where information is available, rabbit meat production almost doubled between 1966 and 1980. For instance, several African countries—among them Ghana, Kenya, Malawi, Mauritius, Mozambique, Nigeria, Sudan, Tanzania, Togo, and Zambia—now have national rabbit-raising programs. A number of Asian countries—such as the Philippines, Indonesia, India, and Vietnam—are also encouraging rabbit farming. And some Latin American countries—Mexico, Costa Rica, and El Salvador, for instance—are actively promoting rabbits for subsistence farmers.

Ghana is also extensively promoting rabbit farming. Although able to produce all the cereals its population needs, it cannot produce enough meat to satisfy demand. In response, the government organized "Operation Feed Yourself." The National Rabbit Project was created

[1] Lukefahr, 1985.
[2] Bennett, 1975.

## FAST FOOD

The order Lagomorpha includes more than 60 small, quick-maturing, and rapidly reproducing species. It seems illogical to think that only one is useful as microlivestock. In principle, any rabbit, hare, or pika could be raised in captivity. All are clean, fast growing, and rapid breeding. They are opportunistic feeders and can digest fibrous vegetation. Their meat tastes better than chicken and does not carry the stigma of rodent. The animals are small, inoffensive, efficient at foraging, and generally tolerant of difficult environments. In theory, at least, they could be raised on vegetation not used by people or by many domesticated livestock.

Species worthy of exploratory research include the following.

### Hares

The common hare (*Lepus europus*) has not been domesticated, but it is nevertheless a major cash crop of several countries. In Argentina, for example, there is a booming, million-dollar enterprise that exports hundreds of thousands of carcasses, mainly to Germany where they are sold as game meat. For Argentine campesinos, many of whom have few sources of livelihood, trapping hares provides a vital income. In New Zealand, too, hare has become an export item.

A closely related species (*Lepus capensis*) is native to Africa, and perhaps could be "ranched" in the same fashion.

### Rainforest Rabbit

The forest rabbit, or tapeti (*Sylvilagus brasiliensis*), is commonly eaten in its native habitat, which extends from southern Mexico to southern Brazil. It occurs in various hot and humid areas of Central and South America and probably within the Amazon Basin itself. Thus, this creature seems a possible candidate for a "tropical rabbit" that can be raised under sweltering conditions, perhaps even in rainforest regions. Although it seems to be heat resistant, it has an especially fine fur.

Little is now known about the tapeti. It is rather secretive and its natural history and even its range are still uncertain. However, its populations appear stable and it is not threatened with extinction. It produces litters of 1–3 young after a 44-day gestation, and may bear 4 litters a year. This may seem a lot, but compared with other wild rabbits, the litter size is small and the gestation period long.

under this program to provide farmers with breeding stock and practical information on rearing rabbits. (To qualify for the purchase of new breeding stock, would-be rabbit raisers are required to take an intensive three-day course in rabbit husbandry, which is provided at no charge.) With both official and popular support, the rabbit's potential for Ghana has been enhanced through media campaigns complete with radio jingles (examples: "Get the bunny money!" "Grow rabbits—grow children." "Get into the rabbit habit!"), television spots, and large posters. Already, rabbit breeding is included in school curricula and rabbit meat is available in school lunches.

Other countries have mounted similar campaigns. In Mexico, for instance, teachers raise rabbits in rural schools as a way of training students; scores of government officials have taken to breeding rabbits in their homes; and several army units are raising rabbits as mess-hall substitutes for costly beef, pork, and chicken. In Nigeria, farmers can now acquire rabbits from 18 government rabbit-breeding centers, which distribute thousands of animals each year. In Costa Rica, the government has similarly established a series of breeding, distribution, and rabbit-farming training centers. And in El Salvador, the technology of rabbit production is being transferred to farmers via the army.

Although rabbits are ideal microlivestock in a general sense, rabbit rearing has many problems and limitations. Poor management is a common difficulty. Unlike the traditional method of keeping scavenger animals, rabbits have to be contained and cannot be left to find their own food. Raising rabbits requires more skills, more time, and much more effort than raising barnyard chickens or other familiar scavengers.

For all that, rabbits produce more food than scavenging animals; they are less likely to damage crops because they are kept confined; they live exclusively on forage, which tends to grow vigorously in tropical zones; and they generally produce a more valuable product. The rabbit's potential is far from exploited, and rabbit farming will have to increase enormously before its promise for the small farms of the world is realized.

There is, however, an increasing concern over a recent outbreak of an exceptionally virulent viral rabbit disease—hemorrhagic tracheopneumonis, which attacks the lungs and lung tissue, killing 48 hours after the onset of symptoms. The virus, which has ravaged the animals in parts of Asia and Europe, was identified in China five years ago in Angora rabbits imported from Germany. It spread to Korea in 1986, and in early 1988 moved through southern and eastern Europe and spread as far as Egypt. It has also been identified in Mexico. Vaccination may become a future prerequisite of rabbit rearing in many countries.

# 14

# Domestic Rabbit

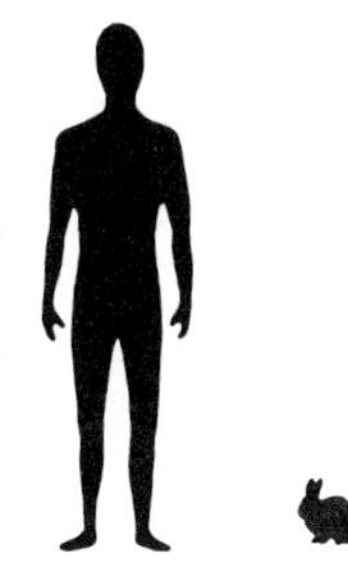

The domestic rabbit (*Oryctolagus cuniculus*)[1] is suited to small-scale production and backyard farming. It is easily maintained, requires scant space, makes minimal demands on the family budget, and thrives on plant materials that are usually disdained by humans. It utilizes forage efficiently, even coarse vegetation that is high in fiber, and under ideal conditions it can grow so rapidly that its rate is only slightly lower than that of broiler chickens.[2]

The rabbit's capacity for reproduction is legendary. In theory, a single male and four females can produce as many as 3,000 offspring a year, representing some 1,450 kg of meat—as much as an average-sized cow.[3] The meat is pink, delicately flavored, and is usually considered a premium product that provides variety in the diet. It has more protein and less fat and calories per gram than beef, pork, lamb, or chicken.

Some breeds are raised for their wool. The long-haired Angora, for instance, yields a luxury fiber that makes a soft, lustrous fabric. It sells at high prices and makes these animals very valuable.

Rabbit pelts also bring cash. They are used in fur coats and other luxury garments. In addition, rabbit feet and tails are used in good-luck charms and many curios.

## AREA OF POTENTIAL USE

Worldwide.

[1] The European rabbit is the ancestral form of all domestic breeds. Initially, rabbits were classified as rodents—members of the order Rodentia. However, because of an extra pair of incisor teeth, they are now classified in a separate order, Lagomorpha, and are not considered close to the rodents at all.

[2] In one trial, under exceptional conditions, rabbits gave 2 kg of meat at age 10 weeks with a feed conversion of 3.5:1; broilers gave 1.8 kg at age 6–8 weeks with a 2.0 to 1 feed conversion. Information from T.E. Reed.

[3] Information from T.E. Reed.

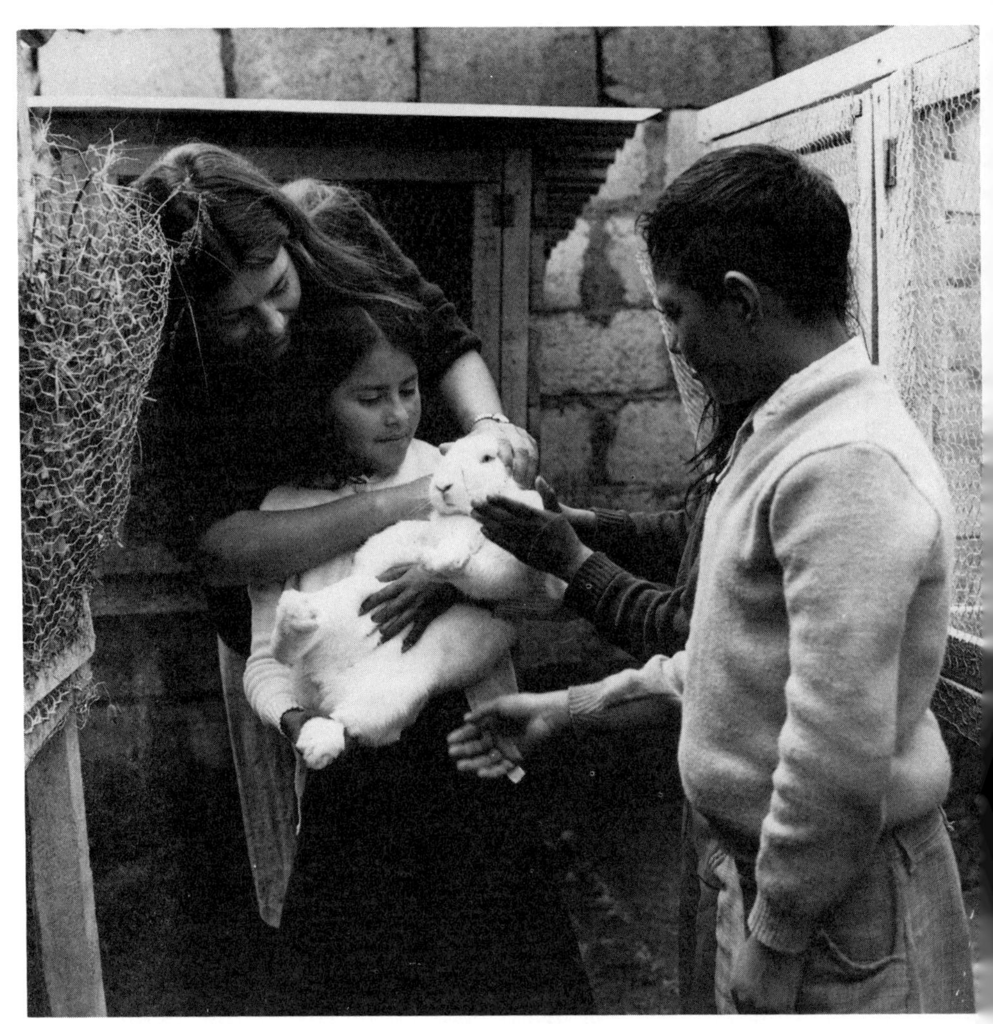

Tigualo, Ecuador. Rabbit raising can be done in inexpensive facilities at home. The operation shown here was organized by Peace Corps volunteers as part of a $600 project to supply protein to about 150 Quechua Indians in the Ecuadorian highlands. This tiny grant (from the U.S. Agency for International Development) introduced rabbits and guinea pigs to enhance the nutrition of, and provide some income for, Indian families. (U.S. Peace Corps)

## APPEARANCE AND SIZE

Rabbits are well known for their long ears, fluffy tails, and long hind legs. Many commercial breeds are white, although colored types are sometimes chosen because of special qualities in their meat or pelts.

There are many breeds and much genetic diversity within and between breeds. (Almost 160 varieties are recognized by the American Rabbit Breeders Association.) However, in both North America and Europe, the New Zealand White has traditionally displaced most other breeds for commercial meat production. This medium-weight breed bears large litters, is a good milk producer, and has good mothering ability. It reproduces best under intensive farming and, among purebreds, yields the most meat. A full-grown New Zealand White weighs 4–5 kg, giving about 2 kg of meat at 8–10 weeks of age. Large breeds include the Flemish Giant or the Checkered Giant, which weigh more than 6 kg at maturity.

Hybrids are rapidly replacing purebreds in Europe for commercial production. Specific crosses of breeds have been shown experimentally to be more productive overall compared with purebred New Zealand Whites.

Different meat breeds are preferred in various countries. For example, in Ghana the most popular are Flemish Giants, New Zealand Whites, Yellow Silvers, and Checkered Giants; in Tanzania and Nigeria, New Zealand Whites and Dutch are preferred; in China, Chinchillas and Japanese Large Whites are the most widely consumed. Some smaller breeds—for instance, the Polish—are also valuable for husbandry.

Some Third World strains have already evolved. They show high tolerance to local conditions (for example, the Baladi—the main strain of the Sudan and the Near East—and the Criollo of Mexico). The Baladi has a small body and relatively low production, but it is hardy and tolerates harsh conditions.

Specialized breeds have been developed for wool, fur, and laboratory research. The Angora wool breed has already been mentioned. The Rex breed produces a high-quality pelt used in furs.

## DISTRIBUTION

The wild ancestor of the domestic rabbit was originally restricted to Spain and Portugal. Today, its descendants are found worldwide.

## STATUS

Plentiful.

## HABITAT AND ENVIRONMENT

Domestic rabbits are best suited to temperate climates, but they do well in tropical and subtropical conditions if hutches are constructed and sited to take advantage of shade and cooling breezes. Ventilation is important (but care must be taken to avoid direct exposure to cold drafts). Prolonged exposure to temperatures higher than 30°C reduces both fertility and growth. Apparently, all breeds tolerate heat equally well. However, heat is shed through the ears, and the longer the ear, the more heat a rabbit will tolerate. Lop-eared varieties withstand heat poorly.

## BIOLOGY

Rabbits eat fibrous vegetation. In addition to normal feces, they produce special droppings called cecotropes. Softer and smaller than the regular fecal pellets, they are excreted in clusters and are swallowed as soon as they are eliminated. Cecotropes are rich in bacterial protein, and this double digestion (coprophagy) enables the animals to utilize the fermentation products formed in the cecum. This process is rather like that of ruminants, and rabbits are sometimes called pseudoruminants.[4]

Breeding begins at 4–6 months of age and may continue up to age 4, occasionally to age 6. Gestation takes 28–32 days. Females can conceive within 24 hours after giving birth and can produce a second litter merely 4 weeks later. With good feed and early rebreeding, 9 or more litters a year are possible. (Such a rate is only achievable under exceptional management, however.) Litter size depends on breed and body weight. Small breeds average 4 young per litter; large breeds 8–10. Births occur at any time of the year, but production slackens when the weather is exceptionally cold or hot, when feed is scarce, or when days are short. Extremes of heat or cold can also affect the survival of the young.

Rabbits raised under subsistence conditions are likely to produce 4 or 5 litters a year, with an average of 5–8 young per litter, depending on management and feed quality.[5] Annual production of about 20 weaned offspring per female per year under tropical and subtropical conditions is common. The young remain in the nest until they are 2–3 weeks old. Their eyes open at approximately 10 days of age. About 4 months are required to produce a 2-kg market rabbit under subsistence conditions.

[4] Although they utilize fiber, they do it less efficiently than cattle.
[5] Information from K. Mitchell.

Opposite: Raising rabbits can foster reforestation. Here in eastern China, a peasant woman feeds rabbits in her home with foliage from edible trees planted to reforest nearby hill slopes. (Shi Panqi, Xinhua News Agency).

## BEHAVIOR

Rabbits that receive human handling are very gentle and can be trained to live inside people's houses and even use a "litter box."

## USES

Rabbits are multipurpose animals yielding the following products:

- Meat. Delicious hot or cold, fancy or plain, it can be breaded and fried, broiled, baked, or barbecued.
- Wool. The fineness of rabbit hair is an asset in the production of wool, which is the plucked or shaved hair of the long-haired Angora breed. It is usually mixed with fine Merino sheep wool to give more substance and to improve its wearing quality. An average Angora rabbit produces about 850 g of wool each year. (Some specimens produce as much as 1,000 g.)
- Fur. The fur is dense.
- Leather or vellum. Rabbit hide has the tension and strength required for tiny drive-belts in tape recorders and other delicate machines.
- Fertilizer. Rabbit manure often contains high proportions of nitrogen, phosphorus, and potash, and it comes in convenient dry-pellet form.
- Tourist charms. In many societies, rabbits are connected with good luck. Feet and tails are used for car decorations, key chains, charms, and mementos that appeal to tourists.

Rabbits are also used in biochemical and physiological research.

## HUSBANDRY

Rabbits can be housed in hutches ranging from sophisticated commercial cages to simple packing crates with a few ventilation holes and rough troughs for food and water. In all cases, watertight roofing is essential. A floor space of only 0.25 $m^2$ is sufficient for one rabbit, but about 1 $m^2$ is recommended for a female and her young.

Starting small-scale rabbit production is generally inexpensive. An almost infinite variety of backyard feeding and drinking equipment can be made from various scrap items, such as old bottles. The main criteria are that cleaning should be easy and spillage minimized.

In practice, diets can be based largely on herbage: grass, leaves, legumes, crop residues, and kitchen scraps. However, the diet must be wholesome, and caged rabbits fed on forage usually need some grain or agricultural by-products (rice bran, for instance) as a dietary supplement. Supplementation is particularly important for newborns and lactating females, whose diet must contain about 16 percent protein and at least 18 percent fiber. When "noncommercial" feeds are used, salt must be added to prevent salt deficiency.

Because of higher protein content, legumes (for instance, alfalfa, cowpea, vetch, or pea) are better than grass. Alfalfa is particularly valuable, and in the Sudan and Mozambique it is already grown extensively for feeding rabbits. On diets consisting of alfalfa and rye grass, weaned New Zealand Whites have demonstrated growth rates of 38–39 g per day in animals weighing up to 2 kg.

## ADVANTAGES

Rabbits, as mentioned, can utilize almost any type of edible vegetation. Also, despite their diminutive size, they can collectively produce as much meat per unit of forage as large livestock, or even more (see page 183).[6]

There is much genetic diversity. Differences in growth rate, fertility, maternal ability, milk production, disease resistance, heat resistance, and other features have been noted. This is useful, since a wide genetic base enhances the likelihood of success of selection programs.

Rabbits are easy to handle and can be raised under primitive conditions. They require little financial investment and their husbandry is easily accomplished in the home by women and children.

The animal's rapid reproduction is a big advantage.

## LIMITATIONS

Tropical conditions produce special problems. There, rabbits must be protected from heat and rain. Stress brought on by high temperatures, high humidity, and wet conditions can lead to respiratory disorders and even sudden death.

Most diseases are caused by poor management. Dirty or wet cages lead to diarrhea, sores, mites, and ringworm, all of which can cause serious losses. Enteritis (diarrhea) often kills 20 percent or more of all rabbits before they attain market age and weight. A major disease problem in most countries is coccidiosis, which is particularly harmful to young rabbits. Again, damp and unsanitary conditions increase the susceptibility; better management can control it.

[6] Cheeke and Patton, 1979.

## HOW RABBITS WERE DOMESTICATED

For 30 or 40 million years the wild species *Oryctolagus cuniculus* lived only in the area that today is Spain. Caves there contain Stone Age drawings of it. Phoenician traders landing on the Iberian Peninsula in about 1100 B.C. found huge numbers of these wild rabbits. The little animals were unknown to them and they mistook them for the hyraxes they had seen in Africa. (Although small and rabbitlike, the hyrax is actually related to elephants.) Since the Semitic name for hyrax was shaphan ("one who hides"), the Phoenicians named the peninsula I-shepan-im, from which the Latin name Hispania developed. Thus, "Spain" actually means "island of hyraxes," even though these African animals have never occurred there.

Given the rabbit's reproductive powers and adaptability it is surprising that it hadn't spread beyond Spain, but dense forests covered most of Europe after the last Ice Age. The rabbit, which is suited to open country, only spread rapidly after man had cleared most of the trees. Even then, the natural spread was slowed by the Pyrenees mountains blocking the way into the rest of Europe.

Ancient Romans became acquainted with rabbits after they invaded Spain, and they eagerly added wild rabbit meat to their banquets. The meat was so popular that around 1 A.D. Roman voyagers released a pair of rabbits onto the Balearic Islands. In time, these produced so many offspring that the islanders had to appeal to the Roman emperor for help. They even asked to be moved to another country if the emperor could not get rid of the plague of rabbits!

Eventually, Romans in Italy, France, and other parts of the European mainland began raising rabbits for meat. They kept them in special cages called leporaria. Their rabbits were probably not truly domesticated; instead they were netted in the wild and caged for fattening before being prepared for the table. The Romans had little incentive to domesticate an animal that could be so easily captured.

The rabbit was the last farm animal to be domesticated. It seems likely that this did not begin until the Christian era when monks in French monasteries began taming rabbits. In those days, rabbit embryos and newborn young were considered delicacies, called "laurices." In 600 A.D. the Pope declared that laurices were "not meat," and permitted them to be eaten during fasts and in monasteries of strict discipline where meat was forbidden. Within a few years, the animal was domesticated.

In some countries—notably Australia and New Zealand—escaping rabbits have become a serious menace and have destroyed crops and grazing lands. Because of this threat, it is illegal to import rabbits into some countries.

At present, many people are unaccustomed to eating rabbit. Indeed, where commercial ventures have been established in areas with an otherwise plentiful meat supply, there have been financial failures. However, where rabbit meat is familiar, there is usually great demand for it. Also, in poorer areas where animal protein is in short supply, the tasty pink meat is widely appreciated.

## RESEARCH AND CONSERVATION NEEDS

Government-sponsored rabbit-research stations and programs are found in France, Germany, Italy, Spain, the United States, and the United Kingdom.

Rabbit husbandry is well known, but much basic research is needed; for example, specific nutrient requirements, breed comparisons, disease control, reproductive management, and efficient housing and equipment. There is a particular need to reduce the labor required for feeding, breeding, caging, and cleaning.

With the increasing number of rabbit programs in Africa, Asia, and Latin America, there is a need to share information and ideas among the various countries. The exchange of experiences with rabbit breeding, health and nutrition, and the practical experiences and field studies could be of great value.

Further research in rabbit nutrition is necessary to identify nutrient requirements more precisely. Moreover, links between nutrition and disease should be clarified.

Further research into the cause and prevention of enteritis is needed. (At present, this condition is prevented by maintaining a fiber level in the diet of at least 18 percent and keeping the energy level relatively low.)

Legume shrubs could be an answer to the feed problems in the dry season. Deep-rooted shrubby legumes, such as gliricidia or leucaena, remain green well into the dry season and have high protein contents. Rabbits find the leaves of leucaena palatable (see companion report *Leucaena: Promising Forage and Tree Crop for the Tropics*. National Academy Press. Washington, D.C. 1984), and they are fairly resistant to mimosine (a sometimes toxic amino acid found in leucaena foliage). More research on this promising approach is needed.

As noted earlier (page 181), a killer virus has recently appeared. Studies into its epidemiology and control are most important.

*Outsiders who hope to improve conditions in underdeveloped areas, sometimes . . . introduce new food avoidances to the communities they came to help [if outsiders] show repugnance toward consuming goats, . . . rats, . . . crows, insects, intestines and blood, then the people they are educating may likewise give up those . . . foods and lose valuable proteins.*

Calvin W. Schwabe
*Unmentionable Cuisine*

*You can count on the fingers of one hand the domestic animals that produce virtually all of mankind's meat and milk—a selection made more than 10,000 years ago by our Neolithic ancestors. Yet the earth teems with thousands of species of animals; why limit ourselves to cattle, pigs, goats, and sheep? Given the world's shortages of energy and water and arable land, why not try to domesticate wild animals? The effort would save many species from extinction, provide the world with more food, and introduce gentle farming to fragile environments.*

N.D. Vietmeyer

# Part IV

# Rodents

Rodents are the world's most widespread, adaptable, and prolific group of mammals. They reproduce well, grow fast, learn quickly, and adapt to a wide variety of local conditions. Many convert vegetation into meat efficiently, digesting some fiber, even though their stomach, like man's, is a simple one.

It seems probable, therefore, that some species would make suitable microlivestock—a notion supported by the previous domestication of the guinea pig, laboratory rat and mouse, gerbil, and hamster. Indeed, "ranching" rodents might be an effective way to increase food supplies in remote areas. It could also be a mechanism to ensure the survival of rare rodents whose natural habitats are being rapidly destroyed.

## RODENTS AS FOOD

Rodents are already common foods in many countries and are valued items of commerce. It has been estimated that 42 of 383 cultures eat rodents.[1] But the fact that they are a major meat source is almost unrecognized. This is due in part to cultural misunderstanding. Rodents suitable for human food or other products do not live in filth, like common rats. They are clean and vegetarian. Like rabbits, they eat grass and grains.

In some regions of the world, cooked rodent meat is regarded as the epitome of dining. In many countries, local rodent species are the

[1] H. Leon Abrams. 1983. "Cross Cultural Survey of Preferences for Animal Protein and Animal Fat." Paper presented at Wenner-Gren Foundation Symposium No. 94, 23–30 October, Cedar Key, Florida.

most eagerly sought meats. City markets in different parts of Latin America carry guinea pig, paca, capybara, and vizcacha. Markets in Asia may carry rice rats, cloud rats, and bandicoot rats. Those of rural Africa are filled with "bushmeats"—usually including grasscutters, giant rats, and several other rodent species. These are often preferred to the meat of domestic stock and fetch higher prices than beef. And the amounts of rodent bushmeat available are not minor. In one year, for example, hunters in Botswana have brought to market 3.3 million kg of meat of the rodent called springhare (see page 278).

Fondness for rodent meat is not restricted to the tropics. In the United States, squirrel was once a much sought treat. Fat, nut-fed gray squirrels went into Brunswick stew, which has been called the most famous dish to emerge from the campfires and cabins of Colonial America. Thomas Jefferson liked it. Today, squirrel is the country's number two game animal (after deer), and many are still eaten.

Ancient Romans kept fat dormice in captivity, serving them as a delicacy. "The fat dormice are fattened up in barrel-like pots like those in country houses," wrote Varro (116–27 BC). "One feeds these animals large amounts of acorns, chestnuts, or other nuts."[2] This small rodent remains a prized food in Europe and still appears on tables in certain areas. The meat is regarded as a delicacy because it tastes of almonds and other nuts. Often it is roasted, broiled, and cooked with its cracklings.

Rodents have seldom been included in livestock programs or economic development plans. Yet human appetite has actually caused the extinction of a number of species. Caribbean Indians ate several endemic rodents (one of which was as big as a bear), and may have caused several species to become extinct just before the time of Columbus. Others may soon follow the same dismal route, including the beautiful cloud rat of the Philippines, the hare-like mara of Argentina, the vizcacha of southern South America, and the gentle hutias of the Caribbean.

The guinea pig is described in a later chapter, but as it is the epitome of a rodent microlivestock species, some historical background is given here. It was domesticated for food use at least 7,000 years ago, probably in what is now the central highlands of Peru and Bolivia. With only llama and deer available, the prehistoric Andean peoples had few readily available sources of meat. They adopted wild cavies, and found that these rangeland rodents (which are more closely related to porcupines than to rats or mice) were gentle, manageable, and easy

[2] At some meals, scales were set up so that dormice could be weighed, and notaries certified the weights of the dormice eaten. To have raised the fattest dormice added to the prestige of the host.

to rear. By the time the Spaniards arrived in the 1500s, the "cuy" (pronounced "coo-ee," like the faint cry it makes) was a major food from Argentina to the Caribbean.

This impressed the conquistadores, who introduced cuys into Europe, where they also became a delicacy.[3] Within a century, these easily transported animals began to appear on tables in many parts of the Spanish empire. Guinea pigs are now reared in campesino huts in the mountains of central Mexico, in the Philippines, and in several African nations, along with other areas of the world.

Elsewhere, guinea pigs came to be used only as house pets and laboratory animals. Although during World War II Mussolini's government urged Italians to keep them to supplement their meager meat rations, their use as food was largely ignored in most parts of the world.

## DOMESTICATION

The idea of domesticating rodents may seem radical, but domestication projects are already under way with capybara in Venezuela (see page 206), paca in Panama (see next page and page 262), giant rat in Nigeria (see page 224), and the grasscutter in Ghana (see page 232). Rodent husbandry is not complicated and the animals' environmental requirements seem relatively simple and easy to satisfy. Moreover, rodents are not usually fastidious feeders, and being essentially vegetarian will readily accept a wide variety of commonly available foodstuffs.

## LIMITATIONS

As with most animals with which man is in close contact, rodents can transmit human diseases.[4] With care, however, managed rodents need not be any more dangerous to care for or to eat than pigs or horses—both of which are worldwide food resources.

[3] Some of the ships from South America stopped in West Africa for water and supplies, which is a possible reason why the animal came to be called "guinea pig" in English. Another explanation is that they cost an English gold "guinea," and yet a third is that the name "guinea" merely refers to something foreign.

[4] In parts of tropical Africa, for instance, lassa fever has become a serious problem in recent decades. It is transmitted to people when they handle or prepare for cooking mouselike rodents that are infected with the lassa virus. This virus is not known to be carried by any of the species in this report.

## DOMESTICATING RODENTS

*To domesticate the paca (see page 262) would seem to be impossible. These large rodents of Central and South America are nocturnal and fiercely territorial; they have low fecundity and take 10 months to reach weaning, and they have tender skin that is easily damaged. Most researchers have written them off as candidates for domestication. But at least two have undertaken to beat the odds. We present the findings of one of them here to show that, using modern techniques, even species that normally fight each other to the death on sight are potential farm animals.*

Through years of studying pacas in Panama, Smithsonian biologist Nicholas Smythe has found that with care and planning the aggressive behavior can be so radically altered that the animals become calm. Indeed, some become almost loving.

Nicholas Smythe and some of the newest of the world's domesticated creatures. (Smithsonian Tropical Research Institute)

Newborns, Smythe found, undergo "imprinting," and when he places them with docile adults or with humans, the fierce territoriality never develops. He nurses newborns on "surrogate" mothers that have been imprinted on people. The youngsters then welcome human company and, if turned out of the cage, return there voluntarily. "It's difficult to imagine a more manageable animal," Smythe said. "Technically speaking, they are behaviorally indistinguishable from traditional domestic animals."

As of this writing, Smythe has three generations totalling about 50 individuals, and has several "families" of gentle pacas living together in harmony. He has observed that they lose their nocturnal habit and, although they live mainly on fruits in the wild, they readily eat leafy vegetables and other foods in captivity. His captive specimens have recently begun to breed. The offspring remain docile, but they have so far averaged only a little more than two young per female per year.

"If we can just double the reproduction rate, then raising pacas can compete economically with raising cattle," Smythe explained. "The potential for a bigger brood is all in the animal's anatomy, and if successful, pacas in the wet tropics could produce as much protein as cattle."*

Pacas need the shade and protection of the forest. Thus, paca raising might provide an alternative to cutting down rainforests for cattle raising. Instead of toppling trees and planting pastures, people could farm pacas in the forest, and perhaps make as much or more money at the same time. In tropical America, the ready acceptance of paca meat is a near guarantee that all they produce will be snapped up at premium prices. In the past, many territorial and aggressive species have been dismissed as being impossible to domesticate or manage. But Smythe has demonstrated that with imprinting and other methods of behavior modification, these need be dismissed no longer.

Indeed, the paca may already be becoming a new domesticated species. In the first stage of his experiments, Smythe had to train his captive-born pacas to be social and nonaggressive. Subsequent generations, however, need no training, adopt the new behavior patterns of the parents, and do not revert to aggressive asocial behavior. By the third generation, they have become as accepting of, and indifferent to, people as cattle or sheep.

---

*Smythe points out that they can possibly do so now. On former Amazon forestland, a cow produces approximately 180 kg of meat in 4.5 years. A group of 5 female and 1 male pacas will, in the same time, produce 45 young, each yielding 4.2 kg of meat (if slaughtered at age 5 months), for a total of 189 kg.

# 15

# Agouti

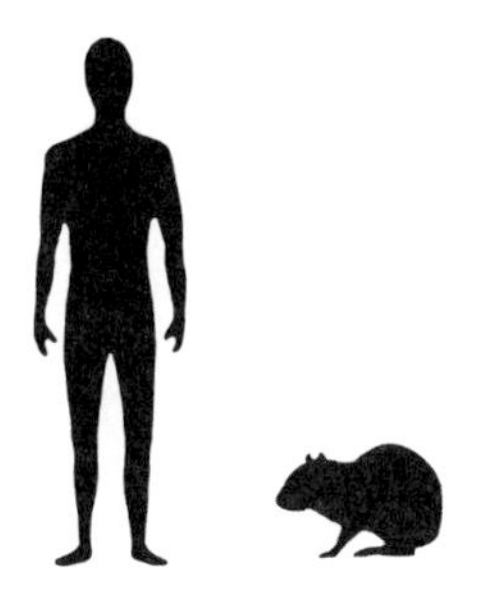

Among the best known of all animals of the American tropics, agoutis (*Dasyprocta* species)[1] are prolific rabbit- or hare-sized rodents that are probably easily farmed. They are valued for food and are hunted throughout most of their range. Indeed, agouti meat, once common in Latin markets, is now difficult to find because of indiscriminate killing. Agouti hunting is already prohibited in Brazil; restaurants in Belem, for example, once offered a variety of "cotia" (agouti) dishes at prices equivalent to those of choice filet mignon, but since the early 1970s they have been banned from serving it. Other countries will probably have to institute similar bans.

Agoutis are active, long-legged, and high-strung. They flee in panic at the slightest alarm. They do not climb but they do burrow occasionally, being essentially specialized ground-dwellers that live in tropical forest regions.

There have been no organized scientific attempts to raise these swift, shy animals in captivity, but Latin Americans sometimes keep them as "domestics," especially in parks and large gardens. (Agoutis are well known, for instance at the Goeldi Museum in Belem, Brazil.) These animals seem to tame easily, and could perhaps be mass-produced on a large scale like rabbits or guinea pigs. They make affectionate pets, sometimes refusing to return to the wild. A research project on captive breeding of two local agouti species (*Dasyprocta mexicana* and *D. punctata*) for food is already under way in Tuxtla Gutierrez, Mexico.

## AREA OF POTENTIAL USE

Most of lowland, tropical Latin America and the Caribbean.

[1] There are about 11 *Dasyprocta* species. This is not the rodent called agouti in West Africa. That is the grasscutter (see page 232).

## APPEARANCE AND SIZE

Agoutis are delicately built, graceful, nimble, and beautifully proportioned. They have slender bodies, short ears, and look somewhat like a rabbit that has been "jacked up" in back. Generally, adults are 40–60 cm long and weigh 2–5 kg. Some are even bigger.

They run well and are good jumpers. From a standing start an agouti reportedly can leap as high as 2 m or as far as 6 m; however, as long as they are well fed, there is little problem keeping them behind a wall only 1 m high.[2] Reportedly, they sometimes climb easy-sloping trees to collect green fruits, but researchers studying Central American agoutis report that they are strictly terrestrial.[3] They swim well.

The body hair is thick, coarse, and glossy: pale orange to black on the back, and white to yellow on the belly. Some species have faint stripes, and some have a rump that contrasts with the rest of the back. The short tail is partially concealed under the long body hair.

## DISTRIBUTION

Agoutis occur over a vast area from southern Mexico to Paraguay, including many islands in the Caribbean.[4]

## STATUS

Because they occasionally damage sugarcane plantings and because the meat is particularly tasty, people hunt agoutis relentlessly, especially near cities and towns. Now, in the 1990s, they are becoming rare because of excessive hunting and habitat destruction. Many Latin Americans have never heard of them. In Mexico, for instance, there are few places where agoutis survive, and *Dasyprocta mexicana* may become extinct if habitat destruction and overhunting continue in its restricted range. In Costa Rica and Panama, agoutis occur only where there is little or no hunting or human interference.

[2] Information from D. Butcher.

[3] Information from D. Janzen and W. Hallwachs.

[4] Agoutis were once imported into the West Indies and released to provide game meat for slaves.

Opposite: Agouti. (W. Hallwachs)

## HABITAT AND ENVIRONMENT

From sea level to elevations of at least 2,500 m, the adaptable agouti lives in many habitats: moist lowland forests, dry upland forests, thick brush, and savannas. However, although they thrive in secondary growth areas, they are mainly forest dwellers. Nonetheless, they often enter fields to forage, and young animals occasionally are seen in open areas such as grassy stream banks and cultivated fields.

## BIOLOGY

Agoutis shelter in hollows among boulders, in riverbanks, or under tree roots. They also hide in heavy brush and sometimes in holes dug out by other species.

These herbivores eat seeds, fruit, stalks, leaves, roots, and other succulent plant parts, as well as occasional insects and fungi.

They seem to mate twice each year. The estrous cycle is variable, but is only about 34 days long. The young are born after a gestation period of 3.5–4 months. Usually, there are twins; however, single births and triplets have been recorded. Newborns are fully developed and are able to run around within hours. They start feeding on solids within a few days. Puberty occurs at about 9 months of age. Life expectancy is 10 years or more.

## BEHAVIOR

In the wild, agoutis are shy and retiring. Every sense seems constantly triggered for instantaneous action and sometimes they become hysterical. If danger threatens, they usually "freeze," but when discovered they stamp their feet as an alarm signal and dash away, nimbly dodging obstacles.

Despite excessive timidity, they can be violent among themselves.

In undisturbed forests, agoutis are diurnal and are often seen. But around villages they become nocturnal, as a means of self-preservation.

For the most part, these rodents live in loosely formed pairs, with previous litters living around their territory as "satellites." There is some "bigamy," some "philandering," and some "divorce."[5].

Despite their long claws, they display much finger dexterity. To eat, they usually sit erect, crouching on their haunches and holding the food in their forepaws. If it has a skin, they carefully peel it before starting their meal. They save some nonperishable foods (nuts, for

[5] Information from D. Janzen and W. Hallwachs.

The native range of the agoutis. Within this area, 11 species are found, but the best known and most widespread is *Dasyprocta punctata*.

instance) by digging holes in scattered locations, dropping each one in a separate location, stamping it down, and covering it over. This behavior helps disperse the seeds of many species of trees so that agoutis benefit tropical forests and reforestation.

## USES

As noted, agoutis are popular game animals. They are often hunted with dogs that even follow them into the burrows. Agouti meat is tasty, although it is usually said to fall short of the meat of the paca (see page 262) because it is leaner and gamier.

## HUSBANDRY

Agoutis adapt well to captivity. With appropriate care they can be bred without difficulty.[6] The nervousness that is pronounced in nature is quickly lost in captivity. The young become tame pets. They can be fed on foods such as leafy vegetables, fruit, potatoes, and bread scraps. Although many wild specimens have become nocturnal, captives readily readapt to daylight.

Being entirely terrestrial, agoutis require no trees, but they do need space. Given enough area, they get on well (with each other and with

[6] At the Lincoln Park Zoo in Chicago, 2 males and 3 females yielded 38 offspring between 1978 and 1982. See Merritt, 1983.

### ACOUCHIES

Close relatives, the green and red acouchies (*Myoprocta acouchy* and *Myoprocta exilis*) also deserve study. These are smaller animals with longer tails, bearing a little plume of white hairs. Although even more delicate and hypersensitive than agoutis, they can be kept in captivity and breed well. They then become less nervous and are easily handled. Acouchies show remarkable intelligence and even some affection for those they trust. They frequent rainforests, but are rare or even absent in disturbed areas. Adults weigh up to 1.5 kg.

The general biology (diet, reproduction, activity rhythm), is almost the same as that of the agouti, but they live in smaller home ranges (0.6–1.2 hectares versus 2.5 hectares for the agouti) and travel singly, although belonging to a well-established family unit. Adult males tolerate the juvenile males. They occur only in Colombia, the Guianas, Ecuador, Peru, and Brazil.

Farming methods would probably be the same as for the agoutis, but acouchies always need plant cover.

different species), and they breed freely. To avoid fighting, it seems necessary to separate females from males at puberty. Probably removing progeny from breeding pens at weaning could also help avoid most of the interpersonal aggression. In large areas with plenty of cover (banana plants, for instance), groups can be kept, but breeding may be disappointing. Husbandry may be most appropriate in large enclosures (50–100 agoutis) with some animals then removed to small cages 0.5–1 $m^2$ for selective feeding.[7]

## ADVANTAGES

Agoutis are appropriately sized: a dressed carcass can weigh 1–3 kg. The meat is good, and large commercial undertakings in urban centers could profit from the ready market that already exists.

The animals are prolific: females can produce up to two litters a year, each litter averaging two offspring. In protected areas, populations may grow fast.

These forest dwellers might provide a source of meat and income without destroying the forests in favor of cattle pastures. Also, they thrive in disturbed areas as long as there is some cover.

[7] Merritt, 1982.

## LIMITATIONS

Experiments in Brazil show that agoutis are highly susceptible to foot-and-mouth disease.

The animals might become pests: they eat the roots, leaves, and fruit of agricultural crops and occasionally damage sugarcane and banana plants. However, current experience suggests that if they escape captivity they are quickly caught by hunters and do not reach pest levels.

Live agoutis have strong-smelling anal glands that may be offensive to breeders or could contaminate the meat if the animals are carelessly handled.

Where the rainforest is destroyed, the agouti population is destroyed. The animals were once well known throughout Latin America, but not anymore. In some areas, therefore, wild breeding stock may not be locally available. Moreover, people may have become sufficiently unfamiliar with them that their value may no longer be appreciated.

In captivity, they can be the prey of large birds such as eagles.

## RESEARCH AND CONSERVATION NEEDS

The taxonomy of agouti species needs clarification.

Husbandry experiments are required, including studies on topics such as:

- Nutrition;
- Growth rate;
- Shelters and enclosures;
- Reproduction; and
- Techniques for catching, moving, marketing, and managing the animals.

One area where agoutis might profitably be raised is in enclosures in palm plantations. Palms such as the babassu provide food, shade, and shelter, while fallen and rotten logs offer secure retreats from predators. This deserves investigation.[8]

Instead of clearing vast areas of rainforest for cattle pasture, as is being done in much of Latin America, people might well "farm" agouties in the forests. Few of the settlers flooding into such regions can afford, let alone raise, beef. Small-scale agouti farming offers a promising and inexpensive alternative that would be gentle on the fragile land.

[8] Smith, 1974.

# 16

# Capybara

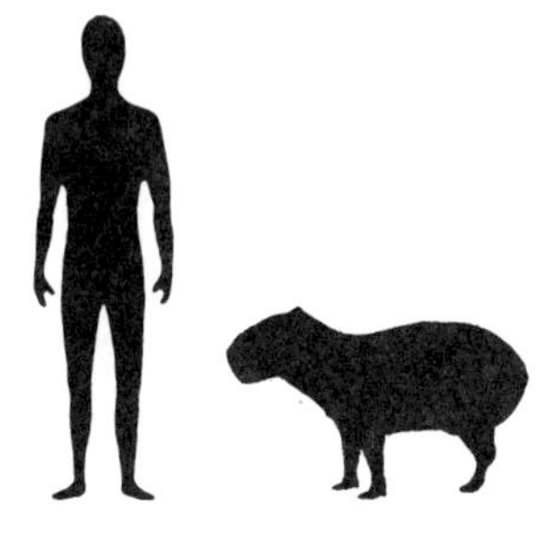

The capybara (*Hydrochoerus hydrochaeris*), the world's largest rodent, can be as big as a sheep and weigh as much as a small person. Its natural habitat is the environs of South America's rivers, marshlands, and swamps, where it feeds on the grasses and reeds that grow near water.[1]

Because of its size, tasty meat, valuable leather, and rapid reproduction, the capybara is a candidate for both ranching and intensive husbandry throughout the hot and humid lowland tropical regions of Latin America. It seems easy to handle. It is commonly raised in zoos or occasionally as a pet, and has, in at least one instance, been proven successful in large commercial herds.

In floodplain ecosystems, capybaras complement cattle because they prefer to graze swamp grasses rather than the dryland grasses on which cattle feed. They have simple stomachs, but are one of the more efficient herbivores. Although they are "selective feeders" that eat lush waterside grasses "preserved in quality" by the water, they also graze pasture, usually selecting new growth that is often too short and scattered for cattle, with their large muzzles, to eat.

## AREA OF POTENTIAL USE

The floodplains of the South American subtropics and tropics where the animal is indigenous.

## APPEARANCE AND SIZE

Although they have blunt, horselike heads, capybaras look like gigantic guinea pigs. They are ponderous, barrel shaped, and have a

[1] Strictly speaking, capybara is probably too big to be a "microlivestock," but we include it here because of its close relationship with other rodents in this section.

tail too small to be seen from a distance. Their skin is tough and covered by sparse, bristlelike hairs: the color above is reddish brown to gray; beneath, it is yellowish brown.

The front legs are shorter than the back. Slightly webbed toes—four on the front feet and three on the back—make them good swimmers. They dive with ease and can stay underwater for up to five minutes. They also move nimbly on land.

The capybara is extremely large for a rodent. In size and color, it looks much like a pig. Often more than 100 cm long and 50 cm high at the shoulder, it can exceed 50 kg liveweight. Indeed, specimens weighing up to 90 kg have been reported.

## DISTRIBUTION

Before livestock were introduced, the capybara grazed widely over riverine regions throughout South and Central America. Today, it is found in the flooded grasslands from Panama to Paraguay. Mainly, it occurs in the watersheds of the Orinoco, Amazon, Paraguay, and Parana rivers. High population densities exist in the Pantanal of western Brazil and on the Llanos floodplains of Venezuela and Colombia.

## STATUS

There are few precise population counts, but capybaras can occur in large numbers.[2] However, in many areas they appear to be on the verge of extinction, being deliberately eradicated by farmers who think they compete with cattle and transmit diseases. Also, in some areas illegal hunting goes on year-round and great numbers are killed. The animals are particularly vulnerable during the dry season, when they concentrate around the diminished river channels and water holes.

## HABITAT AND ENVIRONMENT

As noted, most capybaras live in swampy or grassy areas bordering rivers. However, some are found in other habitats, ranging from open

[2] A census done on one ranch in the Llanos of Venezuela indicated about 47,000 capybaras on 50,000 hectares. Information from R. Lord.

Opposite: Capybara are semiaquatic creatures that feed largely on aquatic vegetation. (D.W. Macdonald)

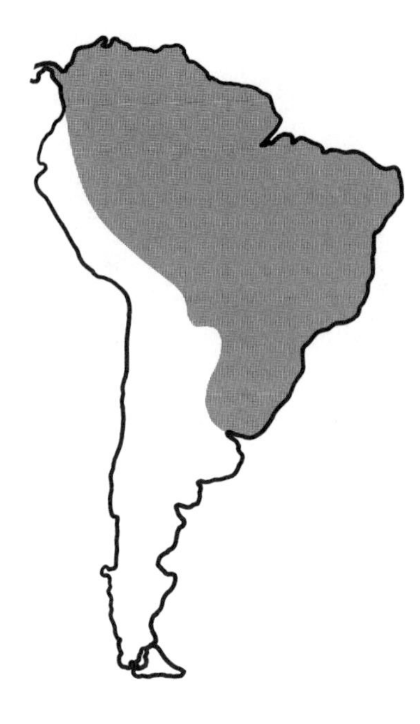

The capybara's native range.

plains to tropical rainforests. But even here they stay near ponds, lakes, streams, and swamps, and never venture much more than 500 m from water.

## BIOLOGY

The capybara, like all rodents, is a simple-stomached animal, but it is a true herbivore. Its digestive system is especially adapted for fibrous materials. The large cecum—the site of enzymatic digestion—serves a function like that of the rumen of sheep, cattle, and goats. It has a digestive capacity similar to that of a sheep's rumen.

Like rabbits and all the rodents, capybaras are coprophagous. That is, during the morning hours when they are resting, soft feces from the cecum are passed a second time through the digestive system.[3]

Contributing to the animal's digestive ability is its efficient mastication. It chews its forage seemingly incessantly, reducing it to extremely small particles before it is swallowed.

[3] Information from R. Lord.

Under natural conditions, the females annually bear 1 or 2 litters, each averaging from 4 to 6 offspring. Birth weight is between 1 and 2 kg, depending on litter size and sex. Both males and females reach sexual maturity when they reach a liveweight of 30 kg or more—usually between the first and second year of life.

## BEHAVIOR

Capybaras are intelligent, shy, inoffensive, and harmless. In undisturbed ecosystems, they are gregarious and live in family groups of up to 30. The young follow the mother about for many months after birth.

Unlike most rodents, they do not construct dens, but the groups have specific resting areas.

The animals are both diurnal and nocturnal and, like many herbivores, they graze at daybreak and dusk, and perhaps also at midnight. They spend the morning resting in weeds on riverbanks, and at noon they cool off by bathing for an hour or so before grazing. They may feed belly deep in water.

Capybaras wallow in mud, allowing it to dry on their skin before bathing again. Mange can develop in captivity when they cannot take a mud bath.

When startled, a capybara barks loudly and dashes away, but after running 200 m or so it tires, slows down, and may lapse into hyperthermia. At that point a hunter can easily catch it. However, if the animal reaches water, it usually eludes the pursuer because it swims so well—especially underwater.

## USES

Capybara meat is white and has qualities and properties (such as high emulsification) that might allow it to compete with pork and other meats in the food industry. Spanish-style sausages, Italian-style mortadellas, frankfurters, and German-style smoked chops have been produced experimentally.[4] However, at present, the meat is mainly consumed only in the dried and salted form. It is particularly popular in Venezuela, where more than 400 tons are sold every year, especially during Easter festivities.[5]

[4] Gonzalez-Jimenez, 1977a.

[5] Centuries ago, Venezuelans and Colombians petitioned the Pope for special dispensation to eat this semiaquatic animal on traditional "meatless" days; approval was granted, and since that time the capybara has been an important food during Holy Week.

The capybara's hide is one of the best for glove making. This luxury product, known in international commerce as carpincho leather, fetches high prices on European markets because it is more heat resistant than most leathers and because it stretches in only one direction. This one-way grain allows gloves to stretch sideways without lengthening and looking sloppy.

## HUSBANDRY

The capybara appears suitable for raising as a livestock animal. Amerindians traditionally collected capybara orphans during the hunting season and raised them until needed for food. Capybara breeding was reported in Brazil as early as 1565.

Modern attempts have been made towards domestication. Researchers at the Institute of Animal Production in Venezuela, for instance, started a breeding program in the 1970s using 20 females and 5 males. Since that time they have continuously kept capybaras in confinement. Through selection and management, they have improved the reproduction of captive animals. The current aim is to get 16 offspring per mother per year. Newborns are weaned after 5 weeks and the mother is returned to the breeding pen.[6] In Colombia, similar work is in progress, and guidelines for raising capybaras on breeding farms have been published.[7] In Brazil, research has been carried out to study capybara nutrition, genetics, management, reproduction, and social behavior in total confinement.[8]

## ADVANTAGES

Throughout South America, the price of beef has increased greatly within the last few years, thereby providing a new incentive for capybara husbandry. It has also forced many campesinos to eat more wildlife, which adds another incentive for producing capybara meat on farms and thereby perhaps helping to relieve pressure on the wild stocks.

When tame, the animals are amenable to handling without physical restraint. They are so tractable that in Surinam a blind man once used one as a guide animal.

[6] Gonzalez-Jimenez and Parra, 1975.

[7] Information from E. Gonzalez-Jimenez.

[8] Conducted by the University of São Paulo at its wildlife research center (CIZBAS). Information from Abel Lavorenti.

Capybaras can be raised on a variety of readily available vegetation: leaves, roots, fruits, and vegetables. They thrive in coarse grasses, if given opportunity to select nutritious parts. Their large incisors allow them to bite off short grasses that many herbivores cannot use. For instance, they eat "capybara grass" (*Paspalum fasciculatum*) that is abundant on river edges in Venezuela and is normally too short for cattle to graze. This makes for low-cost feeding and utilization of a resource that is otherwise unused.

Capybaras are at home in hot, humid environments and are fully adapted to life on the tropical floodplains and seasonally flooded savannas. They thrive in extreme climates where cattle struggle, such as in the parts of the lower Paraguayan Chaco where summer temperatures reach 45°C.[9] An ecological benefit to raising capybaras is that there is no need to alter habitats by introducing exotic forage plants.

They reproduce quickly. Age at first conception for females is about 1.5 years, and the time between parturitions is generally shorter than that of goats or sheep in the tropics. Young capybaras grow so fast that in 18 months they can reach a liveweight of more than 40 kg. In their natural conditions, they are more disease resistant than cattle. The annual productivity is said to exceed that of cattle in many parts of its range.

This species is already so widely eaten in South America that the meat from farmed animals should be readily acceptable.

## LIMITATIONS

Capybaras occasionally raid fields and can harm sugarcane, rice, bananas, sweet potatoes, cassava, corn, and other crops. In many parts of Brazil, they are considered agricultural pests and are shot.

Confining these animals in high density may create serious problems. Infectious diseases and parasite outbreaks seem to be worse than those that occur with conventional livestock. Aggression might prove a limitation to capybara husbandry: it is almost impossible to cage two adult males together or to introduce new animals to an existing group.

The animals may transmit disease to people and livestock. They can harbor foot-and-mouth disease and are known to be susceptible to brucellosis. They also carry a form of trypanosome, *Trypanosoma evansi*.

Compared with cattle, capybara use only a small proportion of the total plant biomass. They are largely selective feeders, and for satisfactory performance must have sufficient area to select the plants they

[9] Information from D.J. Drennen.

need. If placed in a paddock of only coarse grass, most will eventually die. Like goats and gazelles, capybara probably select a diet that is at least 15 percent richer in crude protein than a typical cattle diet.[10]

High mortality has never been observed in Venezuela, but keeping the animals alive on a farm in some areas may not be easy. In one trial, more than half (55 percent) of the capybara died of disease, and a few of septicemia (the result of wounds incurred during fights), but most apparently of trypanosomiasis. Other losses were caused by speeding vehicles (29 percent), poaching (6 percent), and predation, mainly by jaguars (12 percent).[11]

## RESEARCH AND CONSERVATION NEEDS

It is important for researchers to undertake the following:

- Gather specimens from different regions for comparative evaluation.
- Assess experiences of zoos and farms.
- Undertake nutritional trials.
- Initiate captive breeding trials—measurements of growth rates, space requirements, feed needs.
- Characterize the animal's productivity.
- Study the capybara's basic physiology and production potentials.
- Investigate biological factors, such as reproductive physiology, and social behavior (both in the wild and under controlled conditions).
- Determine the factors influencing capybara reproduction, growth, and development.
- Determine the animal's adaptability and economic merit in various farming systems.
- Study the influence of environment on reproduction rate.
- Determine their complementarity with water buffalo or other ruminants that normally use swampy habitats.
- Determine relative causes of mortality (such as diseases specific to capybaras) and predation (especially of the young) by spectacled caiman, crested caracayes, black vultures, and others.

[10] Information from R. McDowell.

[11] G.B. Schaller and P.G. Crawshaw, Jr. 1981. Social organization in a capybara population. *Sonderdruck aus Saugetierkundliche Mitteilungen*, BLV Verlagsgesellschaft mbH Munchen 40, 29. Jbg., Heft 1, Seite 3–16, February 1981.

# 17

# Coypu

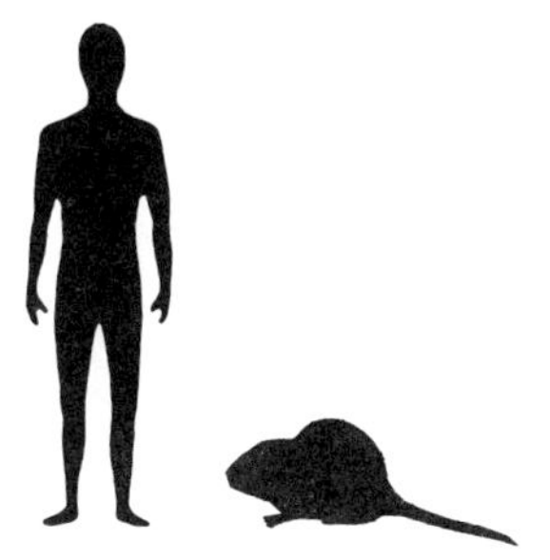

The coypu[1] (*Myocastor coypus*) is an aquatic rodent native to South America. It has been called the "South American beaver," but its size is actually closer to that of a small dog or an agouti. It seems suited to be a microlivestock species because, compared with most rodents, it has a large body size and a relatively high reproduction rate. Moreover, it is easy to manage, and there is much literature on how to raise it in captivity.[2]

Fur is the main item of commercial value. In the late 19th century, it was in such high demand that the animal was nearly exterminated. However, in 1922 Argentineans began raising coypu in captivity and this practice spread through South America and to other regions. In many European countries and in various locations in the United States some specimens escaped or were released, and coypu have become established in the waterways.

Coypu meat is tasty and is consumed in many regions of South America as well as in parts of Europe. Because of the absence of musk glands, the meat is free of the "gamy" flavor found in squirrels and rabbits. It is moist, fine grained, medium light in color, and firm. It is one of the mildest and tenderest of wild meats.

## AREA OF POTENTIAL USE

This animal has been widely distributed, but its area of safest use is within its natural range in South America.

[1] These animals are generally called coypu in English, but their fur is known around the world as nutria. Coypu is the better name to use because throughout South America the word "nutria" refers to an otter. Other names are swamp beaver and quoulyas.

[2] A study in Chile revealed, for example, that 80 commercialized farms were maintaining 48,000 reproducing females and producing 500,000 skins per year. Information from D.L. Huss.

## APPEARANCE AND SIZE

The coypu is adapted to a semiaquatic existence and has webbed feet, valvular nostrils that can be closed to keep water out, and underfur that remains dry even under water. Its long, powerful claws on the forefeet are used for grooming, excavating burrows, and digging up and holding food. The tail is slender; the extremely large incisor teeth are orange-red.

An adult is 40–65 cm long and weighs 7–10 kg. Some occasionally weigh up to 17 kg. Males are larger than females. The pelage is thick, with coarse guard hairs overlying the underfur. The soft dense underfur (the commercially valuable pelt called "nutria") is about 2 cm long on the belly, and 2–5 cm long and less dense on the back. The color is yellowish to reddish brown on the back and pale yellow on the belly.

## DISTRIBUTION

The coypu is distributed through southern Brazil, Paraguay, Uruguay, Bolivia, Argentina, and Chile. It has been widely introduced in North America, Europe, northern Asia, and eastern Africa. As a result of escapes and releases from fur farms, the animals are now feral in Europe, North America, northern Asia, Japan, and East Africa. In the United States, they are abundant in Louisiana, Oregon, Florida, and the Chesapeake Bay region.

## STATUS

In various countries, the animal's status ranges from that of a rarity to that of a pest. Wild coypu are protected by law in Argentina because of overhunting, but there are about 100 producers of farmed coypu.[3] Elsewhere the animals are destroyed en masse to reduce the threat of damage to irrigation ditches, dams, and agricultural crops. In England a decade-long program has eradicated them.

## HABITAT AND ENVIRONMENT

Coypu mainly inhabit the banks of fresh or brackish waterways.[4] They live in temperate zones and are highly sensitive to freezing

[3] Information from D.L. Huss.

[4] They look for water for protection and to cool off on hot days. They stay in tall canes near water during the day and feed on land during the night.

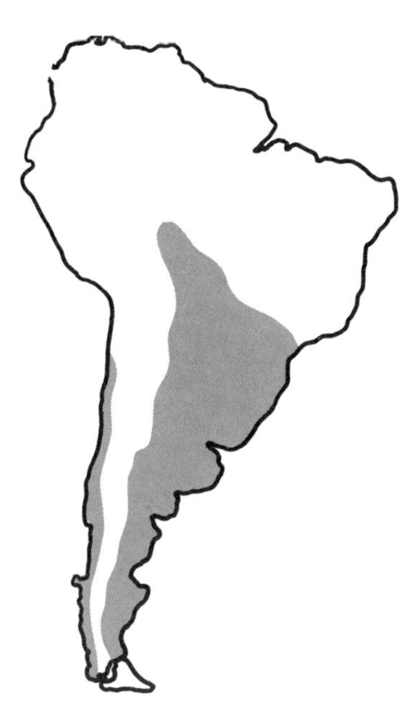

The coypu's native distribution. This is the only region safe for raising this extremely adaptable and potentially pestiferous animal.

conditions. Also, their heat tolerance is poor; lowland tropical regions may be too hot for them.

## BIOLOGY

The coypu feeds mainly at night. The diet consists chiefly of plants, particularly water plants and reeds. Large amounts of fibrous vegetation is decomposed in the cecum, where bacteria break down cellulose particles. Mussels, snails, and other small organisms are also often eaten.

The animals burrow into soft soil or construct nests out of vegetation above ground.

Coypu are relatively fast breeding. Females first give birth at ages ranging from 6 to 15 months. From then on, they produce 2 or 3 litters a year. They are able to mate and give birth at any time of the year, although more young are born during certain seasons. The gestation period is between 128 and 140 days. There are 5 or 6 (sometimes up to 12) young in each litter. Newborns are well developed, able to see, and fully covered with hair at birth.

The female's four or five pairs of mammae are located on the side of the body, an adaptation that permits the young to nurse while floating with their mothers in the water. In captivity, young are nursed for two months, but can survive if weaned at five days. The mean body weight at birth is about 225 g, but growth is rapid during the first five months.

Coypu have a potential life span of more than six years. However, they seldom survive more than three years, and in the wild probably no animals are older than five years.

## BEHAVIOR

These are passive creatures, usually entirely lacking in aggression. They are shy and fearful; the slightest disturbance will send them scurrying to the shelter of water, a burrow, or other hiding place. With their large incisors they can bite viciously, but in captivity they tame down, even to where they can be carried around by hand. Compared with domestic animals, they are very sanitary in their feeding and living habits.

In hot climates they are nocturnal, in cooler climates crepuscular, and in cold weather they become diurnal. Captive animals become conditioned to diurnal activity if fed during daylight hours. Most of the active period is spent feeding, grooming, and swimming. Grooming is done by scratching and "nibbling" the fur, and an oily secretion from glands located near the mouth and anus lubricates the pelage. Secretions from the anal glands are also employed for marking out territory.

Excellent swimmers, coypu spend most of their time in the water. They can remain submerged for five minutes or more. On land, they lumber about with awkward, clumsy movements; however, when the need arises they can run fast and jump short distances.

Although they usually live together in pairs, coypu will form large colonies. They tend to remain in one area throughout their lives: their daily "cruising range" has been measured at less than 45 m.[5] However, drought or freezing weather can induce mass migrations.

The burrows, which are dug in sloping banks, are usually short with no branching tunnels and generally end in a simple chamber.

[5] W.H. Adams, Jr. 1956. The nutria in coastal Louisiana. *Louisiana Academy of Science* 19:28–40.

The coypu. Because the guard hairs are coarse, the coypu is a grizzled, unattractive animal. However, after being dressed, the carcass looks very much like that of a large rabbit. (John McCusker, *The Times-Picayune*)

## USES

In South America, the coypu has a long history of use. Nutria fur was in such high demand and the animal was hunted so avidly at the beginning of the last century that it became rare and had to be protected by government decree. As a result, populations increased dramatically. Nowadays, coypu are protected in many areas, but widespread poaching has reduced their numbers and range.

Many thousands are killed each year just for the guard hairs, which are used in making felt.

Coypu are used in marsh management to reducc infestations of aquatic weeds and to keep waterways open.

## HUSBANDRY

In most areas where it is found, the coypu is trapped by commercial hunters. However, as noted, several countries have coypu farms. In Germany, the animals have been raised on diets consisting chiefly of potatoes supplemented with oats, clover, corn, hay, green forage, legumes, turnips, or cabbage. Elsewhere, feeds generally include such materials as hay, corn, crushed oats, greens, root vegetables, apples, bread, and rabbit feed.

To confine coypu, a wall of stone or concrete or a fence of stout wire netting is necessary. It must be set 1 m deep into the ground and rise 1–1.5 m above ground. Water must be available.

Where selected strains of animals are kept, it is usual to house each female in separate small compartments, complete with pools and shelter boxes. Each is then paired with a male, which is removed after mating to leave the female to rear her brood in seclusion.

## ADVANTAGES

The fur is particularly valuable because the female's nipples are so high that the soft belly fur is unbroken.

This herbivore is much cheaper to feed than the furbearing carnivores such as mink. Furs of coypu raised in captivity fetch a price about three times greater than furs from wild animals.[6]

[6] Information from CORFO, Chile.

## LIMITATIONS

Wherever the coypu has escaped it has damaged embankments and stream banks. The burrows sometimes weaken dikes that protect low-lying areas from flooding. In northern Europe and eastern England, for example, it is considered a serious pest. In rice paddies, coypu could become particularly devastating. They can also damage crops and natural plant communities.

Coypu can carry viruses that result in toxoplasmosis, papillomatosis, rabies, and equine encephalomyelitis; bacteria that cause salmonellosis, paratyphoid, and leptospirosis; protozoans that produce sarcosporidiosis and coccidiosis; and rickettsia. Common diseases of captive specimens are bacterial pneumonia, hepatitis-nephritis, *Strongyloides* infection, and neoplasms.

European winters often cause the coypu's tail (which is hairless) to freeze, but the animal hardly seems to notice. A more dangerous situation arises when lakes, streams, or rivers freeze over; beneath the ice, coypu cannot find their way as easily as beavers, and often drown.

## RESEARCH AND CONSERVATION NEEDS

Little research needs to be done. There is massive literature on farming coypu. Nonetheless, the animal's behavior is little studied, and there are few reliable published observations on its social organization.

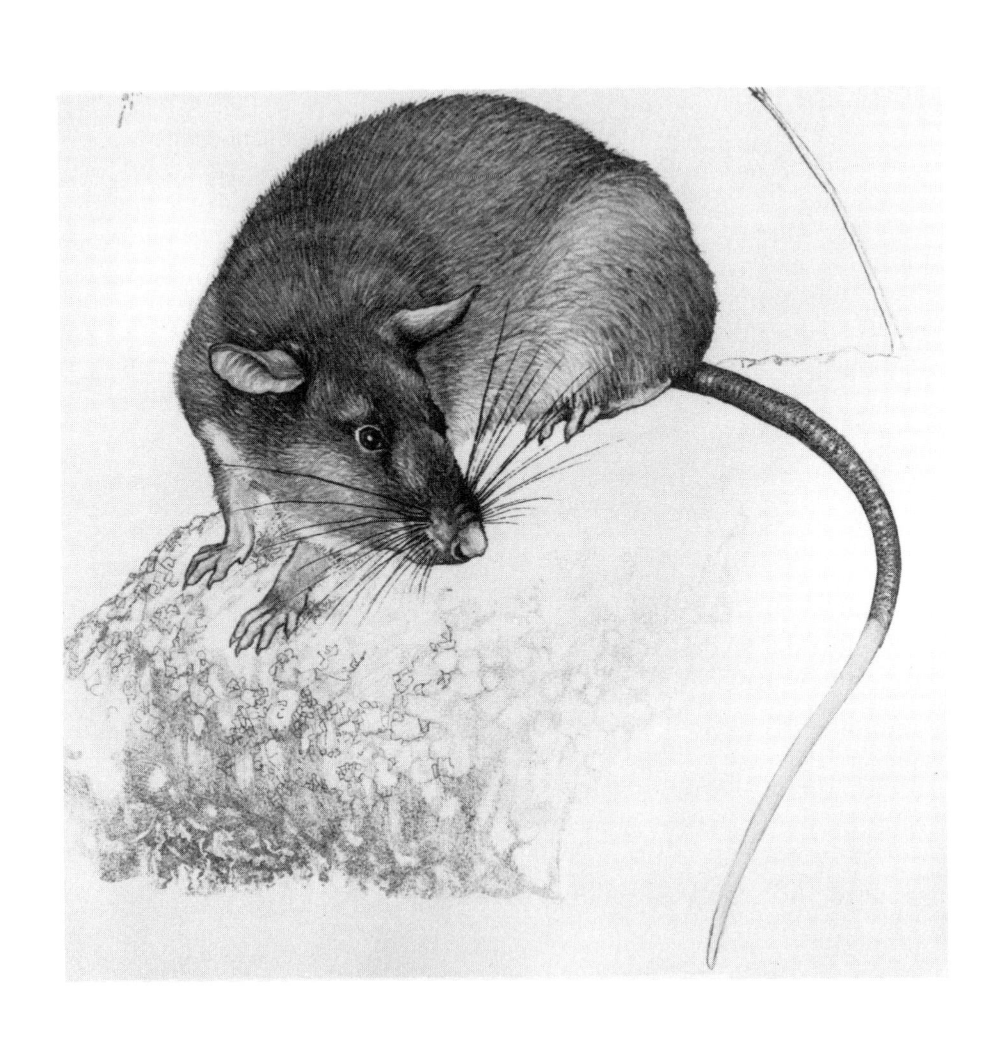

# 18

# Giant Rat

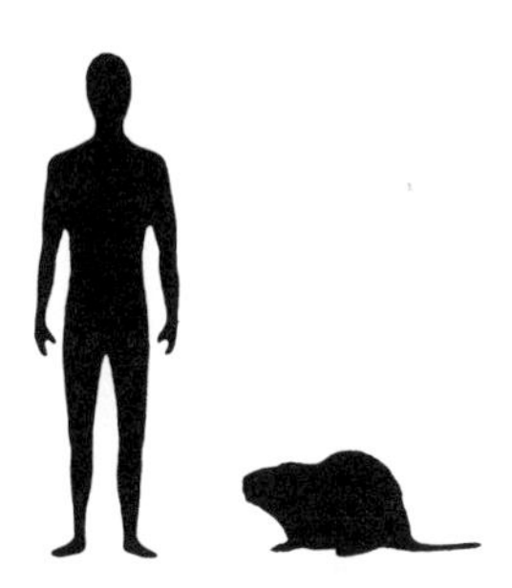

The giant rat, also known as the pouched rat, is one of Africa's largest rodents.[1] Two species have been distinguished: *Cricetomys gambianus*, which lives chiefly in savannas and around the edges of forests and human settlements; and *Cricetomys emini*, which occurs mainly in rainforests. Both are highly prized as food.

These animals are solitary, but they are easy to handle, have a gentle nature, and make good pets. Researchers at the University of Ibadan in Nigeria have been developing techniques for managing them in captivity. Breeding stocks were established in 1973, and since then so many generations have been bred that this small population is considered domesticated. Commercial-scale giant rat farming is now being established in southern Nigeria.

This is a promising development because giant rats are a common "bushmeat" throughout much of Africa. Since these herbivores are well known there, and are acceptable as food, they may have as much or more potential as meat animals than the introduced rabbits that are getting considerable attention (see page 178).

## AREA OF POTENTIAL USE

The intertropical zone of Africa from the southern Sahara to the northern Transvaal.

## APPEARANCE AND SIZE

This species is among the most striking of all African rodents.

[1] Only the brush-tailed porcupine, the springhare (see page 278), and the grasscutter (see next chapter) are larger.

Because of its large size, it often causes amazement—even alarm—when seen for the first time. The body measures as much as 40 cm, and, on average, weighs about 1–1.5 kg. The record for a hand-reared specimen is 1.6 kg.[2]

Apart from its size, the best known species (*Cricetomys gambianus*) is noted for the dark hair around its eyes, a nose that is sharply divided into dark upper and pale lower regions, and a tail that has a dark (proximal) section and pale (distal) section. The overall body color is a dusky gray.

The lesser known species (*Cricetomys emini*) has short, thin, and relatively sleek fur. Its upper parts are pale brown; the belly is white.

## DISTRIBUTION

Giant rats are commonly found from Senegal to Sudan, and as far south as the northern region of South Africa. The main species is mostly found in moist savannas, patches of forests, and rainforests. However, it can also be found in all West African vegetation zones from the semiarid Sahel to the coast. It also exists at high altitudes—up to about 2,000 m in West Africa and 3,000 m in eastern Africa.

The rainforest species occurs in the great equatorial forest belts of Zaire and neighboring Central African countries.

## STATUS

These animals are probably not threatened with extinction. However, they have been exterminated in some areas (such as in parts of eastern Zaire) where the human population is dense, the land fully cultivated, and the wildlife overhunted. Although common, they are not as well known as one might suppose from their bulk and from the fact that they are sometimes found around, and even inside, houses.

## HABITAT AND ENVIRONMENT

Giant rats occur largely in lightly wooded dryland regions or in forested humid regions. They cannot tolerate high temperatures or truly arid conditions. They often live in farm areas and in gardens. Their burrows are commonly found inside deserted termite mounds and at the base of trees. Some have also been found in the middle of cassava fields.

[2] Information from M. Malekani.

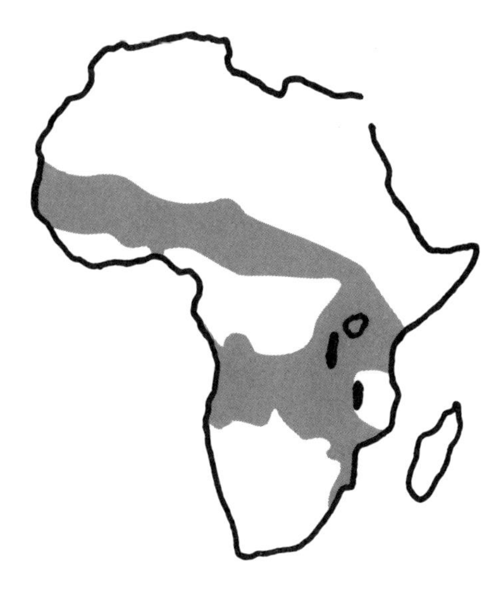

The giant rat's native distribution.

## BIOLOGY

These are herbivores with a tendency to omnivory. They prefer fruits, but also subsist on tubers, grains, vegetables, leaves, legume pods, and wastes (such as banana peels). However, they are not grass-eaters. Giant rats also kill and eat mice, insects (caterpillars, cockroaches, and locusts, for example), and probably many other small animals.[3] They are particularly fond of mollusks (such as snails).

Reproduction is prolific and year-round. The female attains puberty at 20–23 weeks and the gestation period is about 28–42 days. The young are weaned at 21–26 days of age but stay with their mother until 2–3 months of age. So far, the record for the most litters has been 5 in 9 months. It thus seems possible that a female can reproduce 6 times a year. Litter size ranges between 1 and 5, but 4 is most common. Thus, in 1 year a single female could produce 24 or more young.

## BEHAVIOR

These strictly nocturnal animals usually lead solitary lives and forage alone. Mostly, they occupy a burrow by themselves, except when the

[3] Information from M. Malekani.

young are being raised. The burrows can be complex. Below the entrances are vertical shafts leading to a system of galleries and chambers for storing food, depositing droppings, sleeping, or breeding. The home range is individual and limited (1–6 hectares). In the wild, one male "supervises" the home ranges of several females.

In captivity, the animals are often seen sitting up and ramming large amounts of food into their spacious cheek pouches. With full cheeks, they return to their burrows and disgorge the food into a "larder." Food (chiefly hard nuts) is stored there.

They swim and climb well.

## USES

A study carried out in Nigeria showed that the giant rat produces about the same amount of meat as the domestic rabbit.[4] The meat's nutritional value compares favorably with that of domestic livestock, and African villagers know how to preserve it by smoking or by salting.

The giant rat has recently attracted attention as a potential laboratory animal.

## HUSBANDRY

Farmers in Nigeria have traditionally trapped the juveniles and fattened them for slaughter. They usually keep the animals in wire cages and feed them daily with food gathered in the wild as well as with scraps from the household.

As noted, the program at the University of Ibadan indicates that the giant rat can be domesticated. Already, specimens are being bred and reared in an intensive program. They adapt to captivity after about a month. They are subsequently transferred into breeding cages, which are wooden boxes with a rectangular wire-mesh "playroom." Each cage holds a breeding pair or a nursing female with its young. Experimental feeding cages have also been designed.[5]

Food-preference trials show that palm fruits and root crops (especially sweet potato) are preferred to grains and vegetables. Nutritional studies show that the animals can tolerate up to 7 percent crude fiber in their rations. Although largely vegetarian, they eagerly consume dry and canned dog food.

[4] Ajayi, 1975.
[5] Information from S.S. Ajayi.

## ADVANTAGES

These animals have several advantages:

- They are well known and much sought after for food.
- They have adapted to life in lowland tropics.
- They are able to live on locally available plant materials, including vegetable waste.
- They reproduce rapidly.
- They are more tolerant of captivity than the grasscutter (see next chapter). This is largely because omnivorous feeding makes them easier to feed than the grasscutter and other strict herbivores.

## LIMITATIONS

This species could easily become a pest. It is recommended for rearing *only* in areas where it already exists. The crops it damages include cacao, root crops, peanuts, maize, sorghum, vegetables, and stored grains and foods. There is also the possibility that this rodent may transmit diseases to humans.

A project at the University of Kinshasa in Zaire reports problems in getting giant rats to reproduce in captivity. When two specimens were paired they sometimes fought so viciously that copulation was impossible.[6] Special management may be required, such as housing animals in adjacent cages before actually introducing them to each other. Moreover, selection for docility may also be necessary.

The ratlike appearance is not attractive, and a few African tribes have taboos against consuming the meat of these animals.

## RESEARCH AND CONSERVATION NEEDS

Throughout Africa south of the Sahara, giant rat domestication deserves experimentation and trials. Success would open up the potential for supplemental meat supplies in rural and urban areas where meat is now scarce. Tests are needed to determine the factors that favor breeding: temperature, aeration, light, privacy, and size and form of cages. Moreover, diets that are cheap and easy to make from local feedstuffs must be identified.

[6] Information from M. Malekani, who adds that "the rainforest species seems more docile and sociable than the *C. gambianus* in our domestication."

Further research on the domestication of the giant rat might include:

- Identifying husbandry techniques that are applicable at low cost in rural areas;
- Studying food digestibility and setting up various diets;
- Illuminating social behavior: pairing of animals, the best moment for pairing, duration of pairing, age of partners;
- Outlining the basics of husbandry (for instance, capital costs, food conversion ratios, growth rates) and making simple and cheap cages;
- Studying biology (anatomy, physiology, birth records, growth rate); and
- Testing the practical likelihood that this rodent may transmit diseases to people and other animals.

The giant rat has an interesting commensal relationship with *Hemimerus*, an insect that feeds on secretions in the skin. It seems to cause no irritation or damage, and may even benefit the host by helping to keep the skin clean. Caging these animals results in the general loss of the insect, but attempts should be made to maintain them and to determine their role and life cycle.[7]

The potential of this species as a laboratory animal in nutritional, clinical, and pharmacological research also deserves exploration.

[7] Information from M. Malekani.

# 19

# Grasscutter

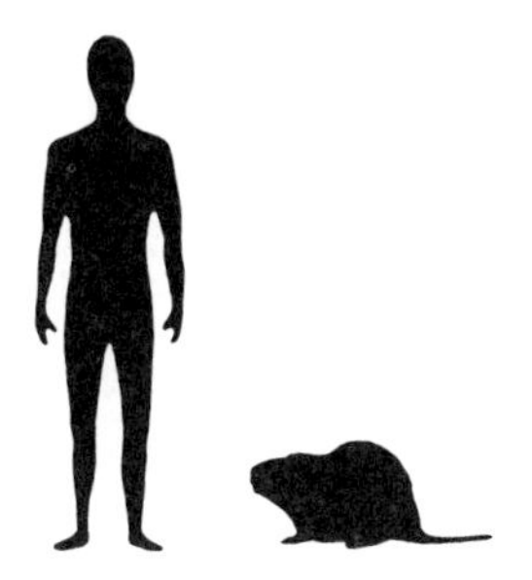

The grasscutter[1] (*Thryonomys swinderianus* and *Thryonomys gregorianus*) is found in many forests and savannas of Africa. Its meat, said to resemble suckling pig, often sells for more per kilogram than chicken, beef, pork, or lamb. It is the preferred, and perhaps most expensive, meat in West Africa. Indeed, in Ivory Coast it sells for about $9 per kg. With prices like that, grasscutter is a culinary luxury that only the wealthy can afford.

If domestication of this wild species were successful in providing meat at a price similar to that of poultry (the second most popular meat), markets would be unlimited.[2] However, as production costs are high, long-term research will be required before grasscutter production can be profitable to the small farmer. This research should now be undertaken.

In an effort to capitalize on the markets for this delicacy, agricultural extension services of Cameroon, Ghana, Ivory Coast, Nigeria, Togo, and particularly Benin are already encouraging farmers to rear grasscutters as backyard livestock. They furnish breeding stock and information, and maintain central offices for records. In addition, a bilateral cooperation project in Benin has started experimental work on improved breeding methods combined with the study of animal responses under domestication.[3]

In future, this vegetarian animal might become the African equivalent of South America's guinea pig, playing an important role in reducing Africa's chronic protein shortage.

[1] Also referred to as the cutting-grass or cane rat; in French-speaking African countries, the grasscutter is referred to as agouti, which simply means an animal from the bush. It is not the true agouti (see page 198).
[2] Baptist and Mensah, 1986.
[3] Baptist and Mensah, 1986.

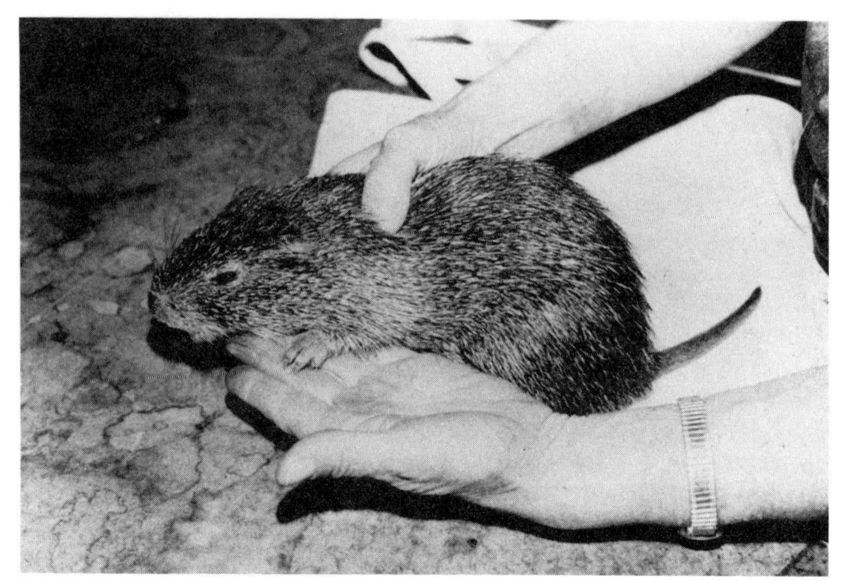

Natal, South Africa. This young grasscutter, only a few months old, already weighs 1 kg. The rounded nose and ears, the spiny fur on the back, and the very short tail all distinguish it from true rats. The handlike forepaws are adept at holding and manipulating grass stalks. (A.J. Alexander)

## AREA OF POTENTIAL USE

Humid and subhumid Africa south of the Sahara.

## APPEARANCE AND SIZE

Grasscutters are robust animals with short tails, small ears, and stocky bodies. Taxonomically, they are more closely related to porcupines than to common rats or mice.

Although many varieties have been described, there are probably only two species. The larger (*Thryonomys swinderianus*) weighs 9 kg or more and has a head-and-body length of up to 60 cm. The smaller species (*Thryonomys gregorianus*) may occasionally reach 8 kg and a body length of 50 cm.

Both species have yellow-brown to gray-brown bodies, with whitish bellies. The fur is extremely coarse, firm, and bristly—reflecting the animal's kinship to the porcupine. The tail is scaly and has short, sparse hairs.

Both species have thick, heavy claws and enormous orange incisors that can chew through even the toughest vegetation. (Grasscutters have been known to tear holes in corrugated iron fences.) Nevertheless, they do not bite when handled, although their claws sometimes cause injuries.[4]

## DISTRIBUTION

Grasscutters occur in grassland or in wooded savanna throughout the humid and subhumid areas of Africa south of the Sahara. They often live in forest-savanna habitats where grass is present. They do not inhabit rainforest, dry scrub, or desert, but they have colonized the road borders in forest regions. Distribution is determined by availability of adequate or preferred grass species for food. Specifically, *Thryonomys swinderianus* occurs in virtually all countries of west, east, and southern Africa. *Thryonomys gregorianus* occurs in savannas in Cameroon, Central African Republic, Zaire, Sudan, Ethiopia, Kenya, Uganda, Tanzania, Malawi, Zambia, Zimbabwe, and Mozambique.

## STATUS

Despite heavy hunting, these animals are not threatened with extinction. Nonetheless, many individual populations are well below carrying capacity, or are extinct because of local overexploitation.

## HABITAT AND ENVIRONMENT

The larger grasscutter (*T. swinderianus*) generally lives in swampy, low-lying areas, especially along river banks and the borders of lakes and streams. Occasionally, it is found on higher ground among bushes and rocks, living where savanna grasses are dense and tangled enough to afford good cover. In Ivory Coast and southern Guinea, for instance, grasscutters are found (and hunted) throughout the savanna zones. And they can occur in close proximity to farmlands and people (for example, in southwest Nigeria).

## BIOLOGY

Although the precise diet in the wild has not been determined,

[4] Information from W. Schröder and S. von Korn.

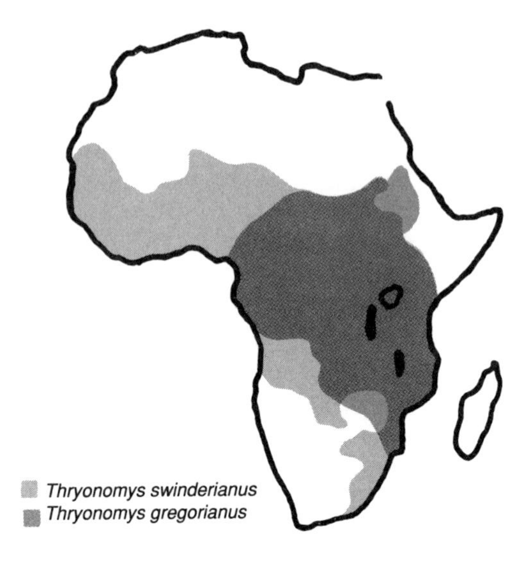


Grasscutters occur across tropical Africa and through the grasslands of East Africa, as far south as the eastern Cape. Throughout this region, where other meat is often expensive and scarce, a grasscutter makes a welcome meal.

grasscutters are vegetarian. They consume nuts, bark, and the soft parts of grasses and shrubs. They particularly favor elephant grass and sweet potatoes. They commonly "raid" cassava and yam plantations, and are considered local pests.

Grasscutters reproduce year-round, although the births seem to peak at certain times of the year, correlated with weather conditions.[5] Probably one male takes several females, and the family group possibly has more than one generation of young. The gestation is about 152 days. Apparently, litters normally contain between 2 and 4 young, but in Benin and Togo some litters of up to 11 or 12 are reported.[6] Newborns are fully developed, their eyes are open, they weigh approximately 80 g, have thick fur, and quickly become accomplished runners.

## BEHAVIOR

Although they commonly forage in groups, grasscutters are generally solitary. They are nocturnal, and they travel at night through trails in

[5] Information from W. Schröder and S. von Korn.
[6] Information from G. Mensah and from W. Schröder and S. von Korn.

reeds and grass, often to water. Most specimens seen in markets are males, possibly because males lead the groups and are thus most prone to being trapped.

When alarmed, these animals stamp their hind feet and give a strange booming grunt. When fleeing, they can run very fast and, given a chance, will take to water. They swim with ease.

For shelter, grasscutters usually weave nests of matted vegetation or scoop out shallow burrows.

## USES

In a broad geographic band across sub-Saharan Africa, cattle raising is severely limited by trypanosomiasis. There, other sources of animal protein, including rodents, are traditionally used. Thus, grasscutter meat constitutes an important food for many Africans. The animals are mostly caught and eaten by families for their own use, but some are sold in markets and especially in roadside stalls. Many families depend exclusively on selling bushmeat, particularly that of grasscutters. In Accra, Ghana, during one year, 73 tons of grasscutter meat were sold in the local market. This represented more than 15,000 animals. In southern Africa, too, people find that these rodents make tasty food, although they may cut off the tail to make the carcass look less ratlike.

The meat is usually eaten smoked, and is so much in demand that grasscutters are hunted in organized drives with spears, dogs, and sometimes fire. It is considered excellent, especially when cooked in soups and stews or barbecued.[7] It has been described as resembling venison in flavor, but it is dark like the meat of wild duck.

## HUSBANDRY

In the savanna area of West Africa, people have traditionally captured wild grasscutters and raised them at home. As an extension of this, organized grasscutter husbandry has been initiated in West Africa. The animals are provided with marshy, tightly fenced areas with plenty of plant cover. The young are harvested from these areas and raised separately.

Ghanaian researcher Emanuel Asibey, a pioneer of this research, reports success at getting such captive stocks to reproduce. To this end, farmers are provided with breeding boxes and foundation grass-

[7] Information from E.S. Ayensu.

cutter colonies. They are taught how to rear and feed the animals for home consumption or for cash income. Basically, the farmers make available large sheds where the animals can move freely. To prevent escape, the walls may be reinforced with cement plaster. The farmers also provide piles of grass, sugarcane, and other foods. A grasscutter reportedly takes about a month to adjust to such confinement. High mortality can occur in this period. The average weight of a mature, home-raised grasscutter is 4–7 kg. The average killing-out (dressed carcass) is 64 percent.[8]

The Wildlife Domestication Unit of Ibadan University in Nigeria, another pioneer of rodent domestication, has also reported the potential of domesticated grasscutter colonies.[9]

Research on grasscutter breeding, husbandry, and feeding is similarly being implemented by the Ministry for Rural Development in Benin and at the Lacena in Ivory Coast (see Research Contacts).

## ADVANTAGES

The demand for grasscutter meat is so large that it is not being met. Markets for it already exist over much of Africa.

## LIMITATIONS

Grasscutters can devastate such crops as rice, sugarcane, soybeans, peanuts, yams, cassava, sweet potatoes, oil-palm seedlings, maize, young rubber, sorghum, and wheat. Therefore, as with most rodents, they should be reared only in areas where they already exist.

In past years, captive animals in Benin have suffered fatal *Clostridium* infections during September and October. In 1986, a broad-spectrum antibiotic was given with outstanding results. During this season, the animals also suffered from ascarid worms, which were also successfully treated with standard drugs.[10]

## RESEARCH AND CONSERVATION NEEDS

Research is needed in the following areas:

- Digestive physiology, feeding habits, feed preferences, feed conversion and growth rate;

[8] Information from E.O.A. Asibey.
[9] Tewe, Ajayi, and Faturoti, 1984.
[10] Information from W. Schröder and S. von Korn.

- Diseases (pathogens and parasites);
- Captive breeding and management (growth rates, space requirements, feed needs, etc);
- Performance under different environments;
- Productivity; and
- Basic biology (for example, chromosome type, reproductive physiology, and social behavior both in its wild state and under controlled conditions).

Moreover, specimens should be gathered from different regions for comparative evaluation. A particular need is to select and breed docile specimens because today, even after several generations in captivity, the animal must still be handled with caution.

Although domestication of the grasscutter is encouraged, wild populations might also be managed to maximize and sustain production through habitat manipulation.

# 20

# Guinea Pig

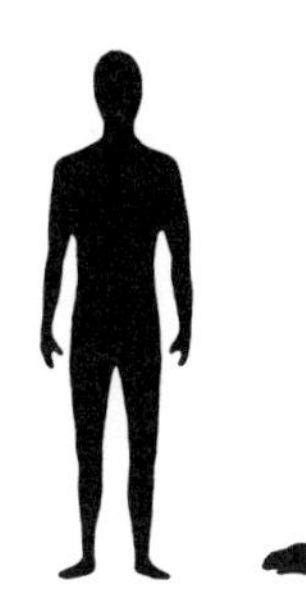

Guinea pigs[1] (*Cavia porcellus*) are promising microlivestock because they require little capital or labor; provide an inexpensive, readily available, palatable meat; have no odor; and are suitable for keeping indoors. In the highlands of the Andes, many Indians raise them to supplement diets based on grains and vegetables. Families eat them mostly on special occasions such as weddings and first communions, or they sell them to restaurants or peddle them in village markets.

The low cost of these small animals makes them available even to many landless peasants. For both the small farmer and apartment dweller, the guinea pig is a possible food reserve. It converts kitchen scraps and marginal wastelands into meat. According to estimates, 20 females and 2 males may produce enough meat year-round to provide an adequate meat diet for a family of 6.[2]

Since husbandry practices are simple and cheap, the guinea pig is an excellent source of supplementary income. An FAO study at Ibarra, Ecuador, showed that on small mountain farms the guinea pig provided more profit than either pigs or dairy cows, partly because its meat fetched high prices.

Although domesticated guinea pigs are mainly a food resource of Latin America, their use has also spread to parts of Africa and Asia. They are raised, for instance, in Nigeria, Cameroon, Ghana, Sierra Leone, Togo, and Zaire. In southern Nigeria, at least 10 percent of all households raise guinea pigs for food, with colonies of up to 30 animals per household. Guinea pigs are also raised in small cages or cardboard boxes by small farmers in the Philippines.[3]

The feeding efficiency is high: studies have shown that it takes between 3.2 and 5.7 kg of forage to produce 1 kg of growth. This makes guinea pigs more efficient than most farm mammals.

[1] See also page 194.
[2] Huss, D.L. and G. Roca, 1982.
[3] Information from L. Fiedler.

The guinea pig is a common household livestock in Peru. They spend their lives around people and become so tame that they can be cradled or carried around. (G.H. Harrison)

The native distribution of the guinea pig's wild ancestor.

Guinea pigs seem especially adapted to the climate and forages of high-altitude zones, but the fact that they are being raised in Central and West Africa indicates that they are also adapted to the lowland tropics.

## AREA OF POTENTIAL USE

Worldwide.

## APPEARANCE AND SIZE

Guinea pigs have stocky bodies, fairly short hind legs, and short, unfurred ears. Adults can weigh up to 2 kg, but an average-sized specimen is about 0.5 kg. They are 20–40 cm long (average 28 cm) and have no tail. In domesticated forms, the pelage may be smooth or coarse, short or long, and in some types the hairs form rosettes.[4] Domesticated types come in colors ranging from white to dark brown, as well as piebald.[5]

[4] In the Andes, animals with curly hair are preferred as meat producers; their bodies are stockier and they are quieter than animals with smooth hair. Information from A. Carpio.

[5] They are polydactyl, and apparently the occurrence of more than five fingers seems to correspond with larger size. Information from A. Carpio.

## DISTRIBUTION

The original home of the wild guinea pig is believed to have been the central highlands of Peru and Bolivia. Its domesticated descendants are important as meat animals mainly in that same area, but, as noted, they are also important in certain African and Asian countries.

A few strains are distributed worldwide as laboratory animals and pets.

## STATUS

Domesticated guinea pigs, as a whole, are in no danger of extinction, although some rare strains are threatened.

## HABITAT AND ENVIRONMENT

These extremely adaptable animals are found in temperate zones and in the highland tropics, but they are usually kept indoors and protected from the extremes of weather. In Lambayeque and other departments of Peru, they are reared at elevations from sea level to more than 4,000 m. In areas where they are raised, daily temperatures fluctuate as much as 30°C. In the Bolivian or Peruvian puna region, for instance, day temperatures can be 22°C, while night temperatures are −7°C. However, they cannot survive freezing temperatures and they may not perform well when exposed to the full tropical heat and sunlight. Many people of the Peruvian highlands keep the animals in darkness (for example, in wood boxes with little or no light).[6]

The animal's original wild habitat is believed to have been an area of grasslands, forest edges, swamps, and rocks.

## BIOLOGY

These herbivores can be raised on kitchen scraps, garden wastes, and weedy vegetation plucked from backyards or roadsides. Andean peasants mainly feed them potato peels, scraps of cabbage, lettuce, carrot, wild grasses, corn stalks, and the foliage of miscellaneous wild plants. Some barley and alfalfa is grown specifically for guinea pigs; it is cut green and sold in small bundles in the markets.

[6] Information from A. Carpio.

The ultimate, user-friendly microlivestock, guinea pigs are quiet, odor-free, and fully at home in houses. They are kept loose in thousands of dwellings in the Andes. A low sill across the doorway—just a few centimeters high—is all that keeps them from straying. (IDRC)

Guinea pigs mate throughout the year except when climate is excessively adverse. Domestic breeds average 2–3 young per litter, although larger litters sometimes occur. The gestation period is 65–70 days with an average of 67. Females come into estrus every 13–24 days, and there is a fertile postpartum estrus.

Females can become pregnant when merely 3 months old, and many produce 4 litters every year from then on. In principle, a farmer starting with 1 male and 10 females could see his herd grow to 3,000 animals in one year.

Newborns are so large that the female's pubic bones must separate for the birth. They emerge fully developed, with fur and open eyes. They look like miniature adults, and they start eating grass and other feedstuffs within hours. (For this reason, babies orphaned at birth have been known to survive.) Weaning may be reached as early as 21 days of age.

The life span in captivity is as long as 8 years, but animals used for breeding usually live only 3.5 years.

## SUPER GUINEA PIGS

Even in their native region, guinea pigs have traditionally received little research attention. However, that began changing in the 1970s with the onset of meat shortages in Peru. (For a time the government restricted beef sales to only 15 days a month.)

For instance, in 1972 Peru began a guinea pig improvement project. Researchers from La Molina National Agrarian University traveled throughout Peru gathering many kinds of guinea pigs—short haired, long haired, black, white, yellow, brown, and even purple. Practically all the guinea pigs eaten in Peru are home grown, and researchers observed that the bigger ones were generally winding up in the stew, leaving the smaller ones for breeding. The people inadvertently were making the

Lilia Zaldivar Abanto leads research on guinea pigs in Peru. Here she shows some of the "super" guinea pigs. These animals are used to improve the Peruvian stock through an ingenious government program: every time a family eats one of their unimproved males, they are given a fresh male of the giant stock. In time, therefore, the genes for large size and quick growth will come to pervade Peru's guinea pig flocks. (R. Kyle)

animals smaller. (This is a common phenomenon for many animals.)

To overcome it, the university research workers compared the mature size and growth rates of all the different guinea pigs. They selected and cross-bred the biggest, meatiest, and fastest-growing ones. This program, later taken over by Instituto Nacional de Investigación y Promoción Agropecuaria (INIPA), produced remarkable results. The starting animals averaged little more than 0.5 kg, the resulting ones averaged almost 2 kg.

Peru's "super guinea pigs" are now getting international recognition. They have been introduced into the highlands of Honduras, where the animal is also part of the Indian cuisine. The FAO has shipped some to the Dominican Republic. In addition, Bolivia, Ecuador, and Colombia have all begun their own guinea pig improvement programs.

Within Peru the government has established 11 breeding stations to encourage the farming of guinea pigs for food. The goal is to provide better stud males to the people so that future animals will grow more quickly and reach a greater weight.

## BEHAVIOR

Guinea pigs generally congregate in small groups, normally made up of 5–10 adults. In favorable areas, however, such groups may coalesce into large colonies. The animals communicate incessantly among themselves, emitting a variety of squeaks and other noises.

Males, although good-natured with other species, often fight fiercely among themselves.

## USES

Guinea pigs are raised mainly for meat. Peru has about 20 million, which annually provide 16–17,000 tons of meat (only 4,000 tons less than Peru's sheep meat production).

Guinea pigs are used worldwide for studies on disease, nutrition, heredity, and toxicology, as well as for the development of serums and other biomedical research.

## HUSBANDRY

Guinea pigs require so little space that a small cage or pit can house up to 10 females and 1 male. They can be raised in cages with wire floors of small mesh as well. The labor required is low. A colony of 1,000 females reportedly can be properly cared for by one person. A layer of wood shavings, shredded paper, straw, and dried corncobs is usually recommended for bedding. The droppings are odorless, so the bedding does not need changing as often as with other animals. When the diet mainly consists of greens, much urine is produced, and then the beds have to be changed frequently.

When grown for meat, the young are weaned at 3–4 weeks and are ready for market in a matter of 10–13 weeks. Weight gain is rapid for the first 4–6 weeks, and then decreases. The carcasses normally dress out at about 65 percent, including the skin and legs. The meat's protein content is approximately 21 percent.[7]

In a few regions of Peru, guinea pigs are "herded" on the open range and retired at night into small adobe coops.

## ADVANTAGES

This small, inoffensive animal rarely bites, is easy to manage, and has no smell. It is an excellent supplemental meat supply. The improved breeds cannot climb or jump so that they are easy to contain. (Primitive "criollo" types, however, can jump.) If kept dry and given green vegetation, grain, and water, it survives in many environments.

## LIMITATIONS

A major constraint is consumer reluctance. Even in Latin America, attempts to promote guinea pig consumption outside the Indian communities have failed.

When raised in a clean environment and under normal feeding conditions, guinea pigs thrive and reproduce and do not need routine vaccinations or antibiotics that cattle, sheep, and pigs often require. However, guinea pigs can be carriers of Chagas' disease and salmonella. Further, they are susceptible to pneumonia if temperatures change abruptly when conditions are wet. Coccidiosis and internal and external parasites are also common.

Green forages and surplus fruits or by-products are critical to provide vitamin C, which the animal is unable to synthesize for itself.

[7] Information from S.D. Lukefahr.

## RESEARCH AND CONSERVATION NEEDS

Since the greatest concentration of guinea pigs is found in the Andes, particular efforts should be directed towards this region. Already, research on guinea pigs has begun in some universities and government research stations in Colombia, Ecuador, Venezuela, Peru, and Bolivia, but more work is needed on matters such as:

- Breeding "elite" stock for distribution;
- Feeding and nutritional-requirement trials, especially for creating alternate feeds that peasants and commercial producers can use during seasons when conventional feeds are hard to get;
- Diseases and parasites that may limit production in small farming systems;
- Management practices concerning reproduction, housing, herd size, and feeding; and
- The genetic basis for weight gain and productivity.

Some research should be directed towards developing rations or introducing drought-resistant forages for the dry season because green forage is needed year-round. A range of practical and economical diets needs to be created.

Animal geneticists in Latin American countries should establish "elite" populations that can provide superior stock throughout the world. It can be anticipated that applying modern breeding methods to existing improved strains will result in great advances in a relatively short period and at little cost (see sidebar).

Three species of wild cavies (*Cavia aperea*, *C. fulgida*, and *C. tschudii*), close relatives of the guinea pig, are native to South America and are declining drastically. Research to preserve them is urgently needed. *C. aperea* is a widely used item of food in rural Brazil and other parts of South America.

Hispaniolan Hutia

# 21

# Hutia

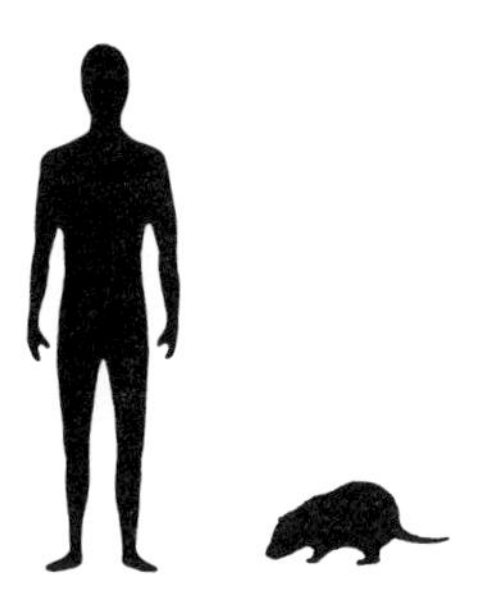

The first meat Christopher Columbus tasted in the New World was probably hutia, a rodent avidly hunted by the Carib Indians. Hutia bones have been unearthed from kitchen middens of pre-Columbian inhabitants of all the Greater Antilles. Indians carried live hutias on voyages, possibly in a semidomestic state, as a source of food. On some islands, hutias were so eagerly sought that their populations were destroyed long before Europeans arrived. Slaves in the cane fields also hunted hutias for food. The surviving species later suffered when forests were cleared and cats, dogs, mongooses, and other predators were introduced. Consequently, the majority of hutia species died out, and today most surviving members of the family (Capromyidae) are facing extinction. Human predation continues in some areas (for instance, in Jamaica) where the tradition of "coney-hunting" still endures in a few regions.

Hutias should be tested as possible microlivestock: success could create the incentive for their complete protection. The animals seem to take well to captivity. The Jamaican hutia is already overproducing in zoos, causing a local glut of animals. And hutias are, or were until recently, kept in barns by some people in Cuba, who fed them on banana and other vegetable waste and ate them regularly.[1]

## POTENTIAL AREA OF USE

The Caribbean.

## APPEARANCE AND SIZE

Hutias are broad-headed, short-legged, robust animals with small

[1] Information from W.L.R. Oliver.

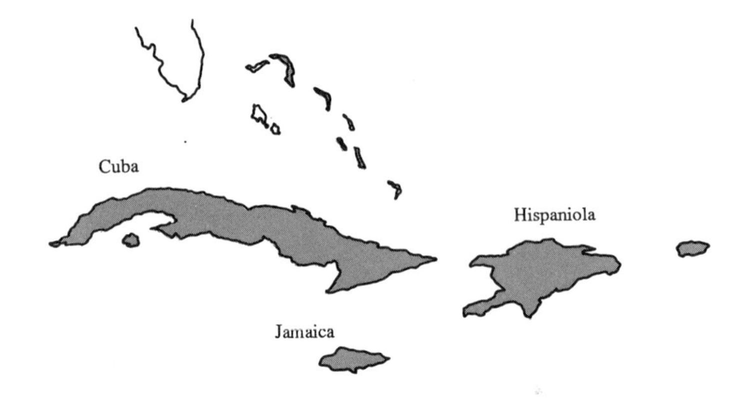


The hutia's native range.

eyes and ears. The various species are from 20 to 60 cm long and weigh from 1 to 9 kg—a size range from that of a guinea pig to that of a small dog. They walk with a slow, waddling motion, but can hop quickly if frightened or pursued. They also climb well.

The 10 living species are all big enough to be candidates for microlivestock. The best known and easiest species to keep in captivity are the Cuban hutia (*Capromys pilorides*) and Jamaican hutia (*Geocapromys brownii*).

The Cuban hutia (also called hutia conga) is about 60 cm long, with coarse fur, a raccoon-shaped body, and a thick tail covered with sparse bristles. A forest dweller, it weighs up to 7 kg.

The short-tailed Jamaican hutia is smaller: it is 33–45 cm long and weighs up to 2.5 kg.

## DISTRIBUTION

Hutias are found only in the Caribbean (Greater Antilles and Bahamas). Most species are confined to a single island, where they represent the only remaining indigenous land mammals. The Cuban hutia is found only in Cuba. The Jamaican hutia is found only in Jamaica, although a close relative occurs on East Plana Cay, Bahamas.

## STATUS

This once widely distributed and plentiful family is now failing. Of the 30 or so known recent taxa, more than half are already extinct, and the remainder all suffer from habitat alteration, predation by introduced animals, and hunting by man. With the exception of the Cuban hutia, all species are included on the list of the world's threatened mammals.

## HABITAT AND ENVIRONMENT

Hutia ranges have been so reduced that these animals survive only in the most inaccessible forests and rocky drylands. Both the Cuban and the Jamaican hutias occur in a variety of habitats from montane cloud forests to arid coastal semideserts.

## BIOLOGY

Most species are terrestrial, but some live in trees. The Cuban and Jamaican species are terrestrial, but they can climb trees if circumstances demand. They maneuver well on trunks and larger branches, descending head first like squirrels.

Hutias are primarily vegetarian, their diets consisting of leaves, bark, fruits, and twigs, as well as incidental catches of small animals such as lizards and invertebrates.

Hutias seem to breed year-round, generally giving birth to litters of 1–4 offspring after a gestation period of 16–20 weeks. The young are well developed at birth, fully haired, open eyed, and capable of most adult movements. After 10 days they begin taking solid food, although they are not fully weaned for at least a month and a half (5 months for the Cuban hutia). Sexual maturity is at 10 months; life expectancy is 8–11 years in captivity.

The Jamaican hutia has one of the highest diploid chromosome numbers ($2n = 88$) of any mammal.

## BEHAVIOR

Most hutias are wary and secretive and are easily displaced by human encroachment. They live like rabbits, hiding among tangled vegetation, in holes, and among rocks—communicating by voice and scent markings. They build shelters mainly in rock crevices, but also

## HISPANIOLAN HUTIA

The Hispaniolan hutia, or zagouti (*Plagiodontia aedium*) as it is known in Haiti, is smaller than the two Capromyids discussed here, weighing just 1.2 kg. It is difficult to breed in captivity and has a lower reproductive rate than either the Jamaican or Cuban hutia. It is therefore less suitable as an economic or food source.

However, there is a significant need for supplemental protein sources in both Haiti and the Dominican Republic. It might be possible to develop a special captive-breeding program for this animal, but it should be done with great care. It is important that a hunting tradition for this animal not be reestablished in rural areas of Haiti or the Dominican Republic, and that local organizations not be misled into believing that there will be a rapid increase in the numbers of this species in captivity.

in the base of thick bushes or in natural cavities in trees. The Cuban hutia is often diurnal, whereas the Jamaican hutia is largely nocturnal.

## USES

Hutia meat is relished, especially in Jamaica. The animals are still hunted, often by using dogs that smell them out and retrieve them from a hole or hold them at bay in treetops.

## HUSBANDRY

Experiences of zoos suggest that the Cuban and Jamaican hutias will thrive in captivity. The animals are generally long-lived and have survived up to 17 years. They are often friendly with their keepers and, when tame, can be held and carried about without any particular danger. However, if angered they can inflict deep bites and should normally be handled with considerable caution.

## ADVANTAGES

These animals are already much in demand. Their meat has an excellent flavor and they are big enough to provide a worthwhile quantity. If husbandry could be developed on a sustainable basis, it could be used as a mechanism for both economic development and for saving the remnant populations.

## LIMITATIONS

Wild populations are threatened. Any captive population must be built up without endangering them.

All hutias are susceptible to predation by domestic cats, mongooses, dogs, and human poachers, so care must be taken to design predator-proof breeding facilities.

These animals are carriers of eastern equine encephalomyelitis, a serious disease of horses.

## RESEARCH AND CONSERVATION NEEDS

Hutias deserve urgent conservation attention. In particular, the following steps should be taken:

- Establish reserves in natural habitats containing breeding populations to ensure the survival of the genetic diversity of these animals.
- Build up breeding populations in suitable zoos and livestock research centers.
- Gather specimens from different regions for comparative evaluation.
- Investigate hutia biology, including chromosome type, reproductive physiology, nutrition, and diseases.
- Assess experiences of zoos.
- Perform captive breeding trials, measuring growth rates, space requirements, food needs, and social behavior (both in its wild state and under controlled conditions).
- Study the social organization and tameability.

Colonies of some species could be established on uninhabited islands, as has been done with the Bahamian hutia *Geocapromys ingrahami*. Even this rare species might eventually be raised to yield meat for local inhabitants, as it is well adapted to dry and barren environments and was a regular food of the pre-Columbian Indians.

# 22

# Mara

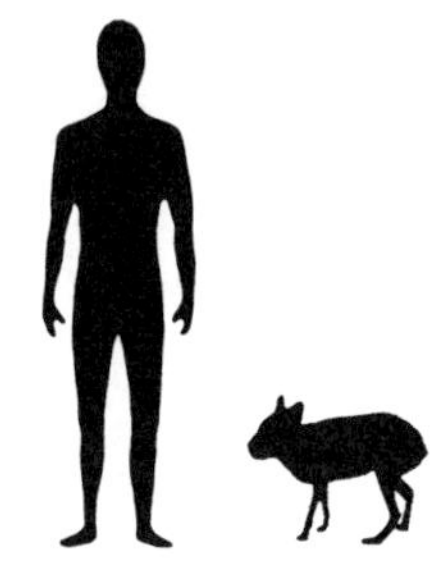

Many wild relatives of the guinea pig are native to South America. Some, such as the guinea pig itself, are small; others, such as the mara, are much bigger. Maras[1] (*Dolichotis patagonum*) are as tall as terriers and, at first glance, look like dwarf antelopes or huge hares. They have large ears and eyes, long legs, and short tails. They generally behave like hares or wild rabbits, but, like deer, they run with a stiff-legged gait when pursued by predators.

These strange-looking creatures are found in temperate regions in the southern half of South America. They are dry-country animals, living on the thorn-scrub, desert plains of Argentina and the stony wastes of Patagonia.

Although exceedingly shy, restless, and watchful, maras tame easily, make good pets, and are much favored by the local populace. They were introduced to France last century, and in Victorian times Europeans sometimes bred them. These big, handsome rodents, grazing in little herds, were considered an attractive addition to the lawns of country estates.

## POTENTIAL AREA OF USE

Maras are best kept in their native region of southern South America. With care, however, they could be used in other areas because they are slow breeders and their population growth is easy to control.

## APPEARANCE AND SIZE

At first sight, these large rodents look like some weird hybrid. They have the long ears of a hare and the tidy body and spindly legs of a

[1] Also known as Patagonian cavy, pampas hare, or giant guinea pig of the pampas.

Mara. (L.T. Blankenship)

small antelope. Although related to guinea pigs, they are long legged. The tail is short; the ears are long and erect.

Average-size maras weigh about 8 kg, but large ones can be 1 m long and weigh up to 16 kg.[2] Females are larger than males. The coat is light in color, with gray upper parts and whitish underparts. The limbs and feet are tinged a yellowish brown. The pelage is dense, with individual hairs standing at nearly right angles to the skin. This gives a harsh texture, even though the hairs are soft and fine.

These animals can hop, walk, gallop, or run. They are extremely swift and can reach 45 km per hour over long distances. They are also accomplished jumpers, often leaping 2 m high from a standing start. The feet are compact and rather hooflike, but with sharp claws. The hind foot has three digits; the front foot has four.

## DISTRIBUTION

The mara's range in the thorn-scrub desert and Patagonian steppe of Argentina extends from about 28°S to 50°S.

[2] Information from A.B. Taber.

## STATUS

Endangered. These animals, once plentiful, are now threatened because of the introduction of the European hare, which is more successful at competing for food. In many of the eastern parts of its distribution (see map) it is now extinct.

## HABITAT AND ENVIRONMENT

Maras inhabit open, dry plains and other treeless semidesert areas of coarse grass or scattered shrubs.

## BIOLOGY

Maras are pure vegetarians. They feed on short grasses and herbs that are sparsely distributed between patches of dry desert scrub. Usually, they are satisfied with a few coarse weeds and the shoots of bushes. However, their overall diet consists of any available vegetation: leaves, roots, fruits, and stems.

Female maras become sexually receptive within a few hours after giving birth. The estrous cycle is 35 days, plus or minus 5 days.[3] The gestation period is 77 days. Each female gives birth to 1–3 young at the mouth of the den; the pups crawl inside to safety. Newborns are well developed, and within a few hours they begin grazing vegetation. They remain in the vicinity of the den for up to 4 months.

Initially, at least, the young are nervous and easily frightened.

## BEHAVIOR

Maras shelter in a burrow that they either construct for themselves or "borrow" from another animal that has abandoned it. They are active during the day and spend considerable time basking in the sun. They are always alert for danger. When alarmed, they flee at high speed. The white rump patch flashes a warning to the others, who then follow this "flag."

A fundamental element of their social system is the monogamous pair bond. Certainly in captivity, and probably in the wild as well, the bond between a pair lasts for life. When breeding, 20 or more pairs may band together temporarily to share a single den for the pups.[4]

[3] Information from L.T. Blankenship.

[4] The combination of monogamy and group life is rare in terrestrial mammals.

The mara's native range.

The animals stand on straight legs, sit on bent haunches with the forepart of the body resting on the fully extended front legs, or recline in a catlike position with the front legs folded under the chest, an unusual position for a rodent. They travel in single file, with the female usually leading. Members of a pair maintain contact by means of a low grumble. Although the long legs can quickly carry it to safety, a mara usually stops every 20–30 m and turns to peer at its pursuer.

The animals clean themselves by licking their sides and apparently by "combing" their fur with their teeth. They wipe their faces as cats do, with the inside of a foreleg.

## USES

Although the light-colored meat is said to be dry and flavorless, it is widely consumed in South America.

## HUSBANDRY

Maras have been successfully raised and bred in many zoos, and, as noted, have been kept as pets.

Adults make little use of any shelter; they seem fond of being out and about in all weather. As long as they have a protected burrow for the use of the pups, mara populations can thrive in severe climates.

In zoos, diets include straw, vegetables, and crushed oats. Drinking water is supplied, although the animals rarely take it if they are feeding on fresh plant materials. They like to have salt blocks, however.

In South America, one mara lived in captivity for almost 14 years; most specimens do not live beyond 10 years.

## ADVANTAGES

Maras are a good size for microlivestock. They have a short gestation period, and they are social and easy to maintain in groups. They can be successfully kept in pens and can be fed relatively low-quality forage. Colonies can grow to be quite large.

## LIMITATIONS

These animals can easily dig under the edges of cages and escape. Extra-deep foundations are needed.

Following heavy rains, care must be taken to keep them from drowning in their subterranean burrows.

If suddenly disturbed, maras can become hysterical, leaping away regardless of anything in the way, and often seriously injuring or even killing themselves as a result. They fear bodily contact.

The mara's monogamous nature in the wild is a likely limitation. But perhaps, like chinchillas, the animal will become polygamous in captivity.

The animals are sensitive to tuberculosis when kept in humid conditions.

## RESEARCH AND CONSERVATION NEEDS

Research needs to increase understanding of the mara include:

- Nutritional trials;
- Husbandry experiments—measurements of growth rates, space requirements, feed needs;
- Productivity tests;
- Grazing-efficiency measurements;
- Exploration of commercial details; and
- Determination of diseases and parasites.

# 23

# Paca

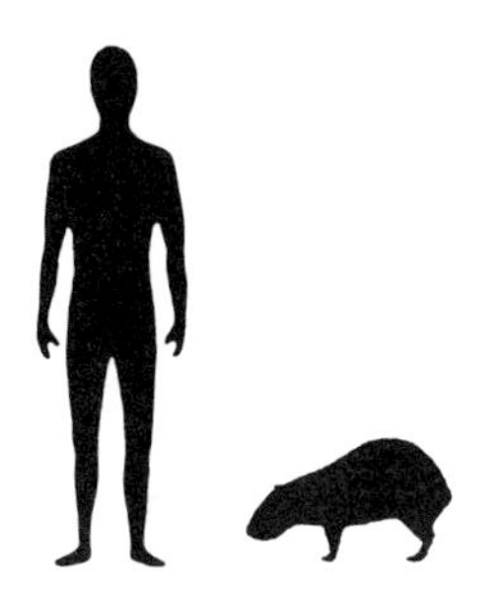

Pacas (*Agouti paca*)[1,2] are large, white-spotted, almost tailless rodents with the potential to become a source of protein for the American tropics. They are found in lowlands from Mexico to northern Argentina. The meat is white and is considered the best of all Latin American game meat. It is common in local markets and restaurants. Tasting like a combination of pork and chicken, it sells at higher prices than beef and is a regular item of diet in some areas. In Costa Rica, paca is served on special occasions such as weddings or baptisms. It has a higher fat content than the lean meat of agoutis, rabbits, and chickens, and has no gamy taste.

Paca has promise as a microlivestock. In several countries, Belize and Mexico for example, people already keep them in cages beside their homes and fatten them on kitchen scraps. In Costa Rica, some are bred on farms, under houses, and even in apartments. Research on raising pacas in captivity is under way at the Universidad Nacional in Heredia, Costa Rica; at the Smithsonian Tropical Research Institute in Balboa, Panama (see page 196); and at the Instituto de Historia Natural in Tuxtla Gutierrez, Mexico. In Turrialba, Costa Rica, an entrepreneur is already breeding and raising paca commercially.

While the paca has potential as a food source, many problems still must be resolved before it can be recommended for mass rearing. If solved, however, this species would become an attractive microlivestock.

[1] The older generic names *Cuniculus* and *Stictomys* are also sometimes still used. There are several common names. A widely used Spanish common name in Central America is "tepezcuinte," while in English-speaking regions of Central America they are called "give not" (because the meat is so good it is not shared with neighbors). In Panama, they are called "conejo pintado," "lapa," "laba," and other names; in Venezuela, its name is "lapa"; in Peru and Ecuador, "guagua."

[2] There is a second species: the mountain paca (*Agouti taczanowskii*). It is a much less likely candidate for microlivestock; it is rare, and has not been studied either in the wild or in captivity. This chapter refers only to the lowland species.

## AREA OF POTENTIAL USE

The paca has potential for use throughout its vast geographical range in Latin America.

## APPEARANCE AND SIZE

In general appearance, pacas are somewhat like giant guinea pigs. The legs are short, the forefeet have four "fingers," and the hindfeet have five small, hooflike "toes." The feet are partially webbed and are adapted both for digging and for swimming. Pacas burrow with all four feet as well as their teeth; even large roots are no obstacle.

Adults weigh 6–14 kg, males being somewhat larger than females. Although they may become bulbously fat, pacas remain "one of the fastest things on four feet." From a standing start, even a fat specimen can jump at least 1 m off the ground. Pacas are also agile. However, their skin's epidermal layer is thin and fragile, and large strips may be ripped off as they rush headlong through spiky undergrowth. However, such wounds heal astonishingly fast—frequently within days.

Pacas are chocolate brown in color. The head is somewhat lighter in shade than the body, and the underparts are whitish or buff colored. There are usually four longitudinal rows of white spots that may merge into stripes along each side of the body. The fur is coarse, spiny, and slippery, and has no underwool. Each hair is stiff, relatively sharp, and very smooth, which makes pacas extremely difficult for predators to hang on to.

Parts of the cheekbones are enlarged, and the cheeks can open to form special pouches. This is more developed in adult males than in females—indeed, adults can be readily sexed by head shape. The pouches are outside of the mouth and are fully haired. The animals use them mainly to create a resonating chamber for their booming bark and noisy tooth grinding. These enlarged cheeks push the large bulging eyes toward the top of the skull. The eyes are suited for nocturnal conditions, the senses of smell and hearing are uncannily acute, and there is an array of long whiskers that is used when maneuvering at night.

## DISTRIBUTION

Lowland pacas are found throughout most of Latin America from east central Mexico to northern Paraguay, Argentina, and Minas Gerais,

The paca's native range.

Brazil. This includes all of Central America and most of Colombia, Ecuador, Peru, Venezuela, and the Guianas. The animal has also been introduced into Cuba.

## STATUS

Burgeoning human populations are severely reducing many of Latin America's native animal resources, and the paca is one of the most persecuted. It has been exterminated within hunting range of virtually all cities, towns, and villages.

Several governments, recognizing the paca's plight, have passed laws prohibiting the hunting and marketing of its meat. Nevertheless, people continue to take it, usually at night, using trail dogs and headlights.

## HABITAT AND ENVIRONMENT

Pacas thrive in a variety of tropical habitats but are most common in forests, swamps, and partly cleared grazing lands. They inhabit most types of forests from deciduous woodland to rainforest. Usually, they stay near streams or rivers, but they often live where there is no permanent water. They are abundant only in little-disturbed forest areas. Although preferring low, dense tree cover, pacas sometimes inhabit open rocky areas and farmland.

## BIOLOGY

These herbivores feed mainly on fruits, young seedlings, and some seeds. However, when fruits are scarce they may switch to browsing leaves and roots. They probably sometimes eat large insects, and, on rare occasion, may perhaps eat small vertebrates. Captive pacas, like many other "frugivores," seem to develop a protein deficiency and will eagerly eat meat scraps on occasion.

The young are usually born singly after a gestation period of 146 days. There are probably 2 births a year. Females have an estrous period that begins shortly after giving birth. If mating does not take place at this time, the female becomes unreceptive until after the 3-month (sometimes 4- to 6-month) lactation is over. The length of the estrous cycle is 30 days.

During daylight hours, pacas seclude themselves in brushy cover, in or under fallen logs, or in extensive underground burrows. The burrows, which may be several meters long, are dug in moist soil or taken over from other animals; they are often in river banks, on slopes, among tree roots, or under rocks. Usually, several exits are provided, often being plugged with leaves as a disguise.

## BEHAVIOR

In the wild, pacas dig large holes and rummage about the forest floor at night, gnawing on fruits. Pairs inhabit a defended area, sometimes living together in the same burrow, sometimes not. Also, they usually travel alone, following paths that lead to feeding grounds and water. Individual home ranges are small (1–3 hectares).

Although pacas are terrestrial, they enter water freely, they swim well, they copulate in water, and, when alarmed, they generally attempt to escape by swimming. They are also lively and playful; however, they can be exceedingly obstinate. Sometimes fighting among themselves becomes very savage. When angered they growl, sometimes noisily, and they can suddenly jump on aggressors, real or imagined, delivering frightful wounds with their chisel-like front teeth thrust forward like a spear.

## USES

As noted, paca meat is tasty and brings high prices in the markets. It is considered a delicacy in fine restaurants and was served to Queen Elizabeth during her October 1985 visit to Belize. In Mexico, pacas, like pigs, are usually boiled unskinned. Even the skin is then edible.

## HUSBANDRY

If treated appropriately when young, pacas become manageable. They undergo "imprinting," a characteristic of most species that have been domesticated. An imprinted paca becomes so tame that it seeks out human company, follows people around like an amiable dog and,

"Domesticated" pacas. (N. Smythe)

if turned out of its cage, returns voluntarily. (To achieve this degree of tameness it is necessary to remove the animal from its mother at an early age.)

Although wild pacas are almost entirely nocturnal, tame pacas are more active during daylight hours.

Young or partly grown pacas are commonly exhibited in zoos. They eat prodigious quantities of almost any vegetation and have been called, "a good substitute for a large garbage pail." Diets can include rolled oats, raw vegetables, bananas, apples, and bread. They probably need additional protein occasionally, and seem to appreciate some fat in their diet.

## ADVANTAGES

If husbandry can be developed, the clamor for paca meat throughout tropical America would be a big economic incentive for farming these animals. The excellence and wide acceptance of the meat is an indication that paca farming would be taken up both in rural and urban areas and by many levels of society.

## LIMITATIONS

Pacas can harbor human diseases, including leishmaniasis and Chagas' disease.

Apart from the project in Panama (see page 194), pacas have bred only sporadically in captivity, with few offspring surviving. However, successes have been recorded in zoos in London, San Diego, and Washington, D.C., and in a research project in Tuxtla Gutierrez, Mexico. In Costa Rica, they are also reportedly breeding well, with a survival rate of 90 percent since 1982, and 80 percent of the females are reproducing.[3]

All adult pacas can be aggressive and dangerous. Their powerful incisors can inflict serious wounds. (They can even rip through planks.) Intraspecific aggression is one of the most serious impediments to captive breeding.

Unlike the capybara, the paca not only has a long gestation but usually bears a single young. Thus, the output of a single breeding female may be, at best, two offspring per year (at least this is the expected production in the wild). This "slow" breeding is a limitation.

[3] Information from Y. Matamoras H.

In captivity, however, there is a possibility that it can be speeded up.[4]

The fact that pacas bond together in pairs is a limitation. If every female has to be accompanied by a male, then many (otherwise unnecessary) males have to be fed and maintained.

Male pacas are considered difficult to keep as household pets because they spray females (or human substitutes) with a mixture of urine and glandular secretions. This can occur several times a day. In addition, they have anal glands that produce a musky odor that some people find objectionable.

## RESEARCH AND CONSERVATION NEEDS

Although pacas are common in some areas over the vast region from Mexico to Argentina, they are little understood, even by zoologists. In fact, most data concerning this animal have come from interviews with local hunters. Intensive field work is needed to develop an understanding of the paca's biology, status, and habitat requirements.

The popularity of paca meat makes it urgent to start this work as well as to begin breeding pacas on an organized basis. Such projects would lay the groundwork for preserving the species.

Particular research needs concern the following:

- Age structure and reproductive performance;
- Growth rates and feeding habits;
- Behavioral patterns in captivity;
- Nutritional requirements;
- Meat quality;
- Helminth and arthropod parasites;
- Role in transmitting or perpetuating diseases;
- Reproduction (such as external manifestations of estrus in females); and
- Genetic variations that would allow the selection of animals adapted to captivity and females that produce multiple offspring—twins, triplets, or more.

Ways must be found to introduce more than one female to each male without inciting aggression.

[4] Females show considerable individual variation: some breed readily, whereas others remain stubbornly unreceptive to the males. Imprinted females seem to be less receptive than wild ones, but they also resist less aggressively. Thus, tame pacas are manageable but difficult to breed; wild pacas are unmanageable but easier to breed. Information from N. Smythe.

# 24

# Vizcacha

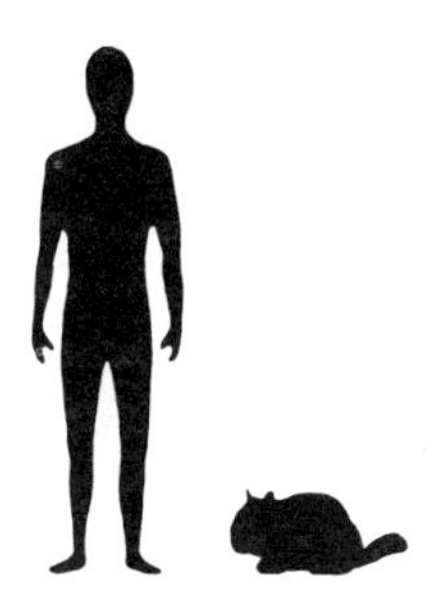

Vizcachas[1] (*Lagostomus maximus*) are soft-furred South American rodents that look like long-tailed guinea pigs. They can weigh as much as 8 kg and are resilient animals, inhabiting dry pampas and shrub lands in northern Argentina and neighboring countries. They seem to have promise for producing meat and hides in marginal zones within their native habitat.

Like chinchillas (page 277), these rodents provide a prized furry pelt. They also provide meat that reportedly tastes "as good as hare," which in Europe is considered the epitome of dining. They are easily trapped alive in cheap, homemade, multiple-catch, funnel traps. And they are thought to be suitable for farming on a large scale.

On the other hand, vizcachas are currently considered pests because they take grazing from cattle and sheep and because they build large burrows that undermine the land. Government campaigns have eradicated them in the richer agricultural areas of Argentina, but the animals are still common in marginal zones. There is evidence that they become more abundant when domestic livestock overgraze the land. In improverished marginal sites, where other livestock enterprises are unsuitable, the potential exists for game-ranching vizcachas.

## AREA OF POTENTIAL USE

Because of the potential hazard to new areas, vizcachas can be used only in the pampas regions of southern South America where they are already widespread.

## APPEARANCE AND SIZE

Vizcachas have short front legs, long, muscular hind legs, and round

[1] The common name is also spelled viscacha.

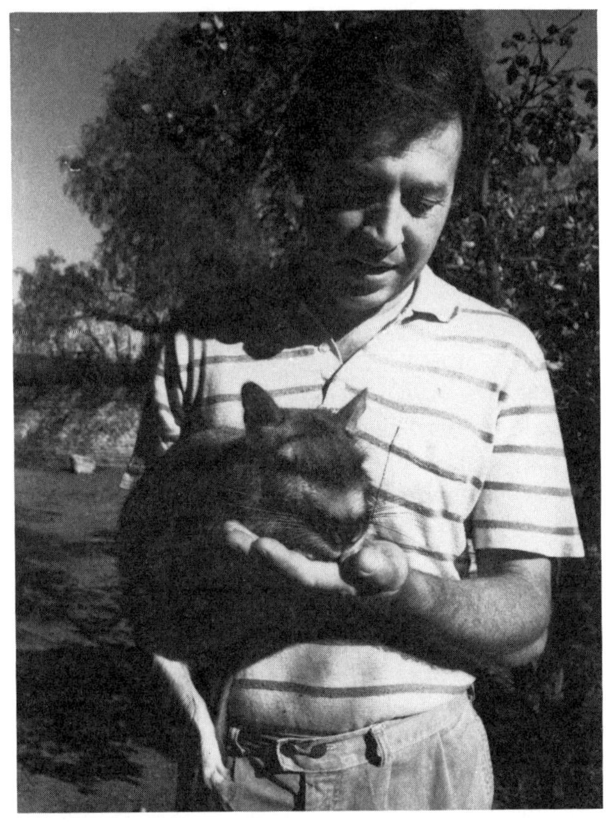

Rio Cuarto, Argentina. In this area of central Argentina, vizcacha are common. One slaughterhouse alone handles 10 tons of the animals weekly, and the meat can be bought in any restaurant. (E.L. Marmillon)

eyes and ears. Their heads seem oversized in proportion to their bodies. Males weigh 5–8 kg; females 2–4.5 kg.

Members of the same rodent subfamily as the chinchilla, they have a thick, soft, valuable fur that is gray or brown above, whitish or grayish below. They are, however, much larger than chinchillas.

Although basically running animals, vizcachas often jump bipedally (like kangaroos), and they sit erect while eating or grooming. The forefeet have four long flexible digits used to grasp food. Their soles and palms are naked and have fleshy pads (pallipes).

## DISTRIBUTION

Vizcachas once swarmed widely over the savannas of southern Paraguay, Bolivia, and Argentina, but they are being systematically exterminated. Today they inhabit isolated areas of north, central, and western Argentina and southern Paraguay.

## STATUS

Since 1907, these animals have been mercilessly hunted. The governments of the Argentine provinces where they are mostly found, reward hunters for killing this "pest." However, the numbers are so reduced that now there is no need for a bounty system.

## HABITAT AND ENVIRONMENT

Vizcachas live in flat, dry, steppelike plains; in dry woodland (Chaco); and in low mountains.

## BIOLOGY

The fact that these rodents eat their own droppings (coprophagy) augments their ability to utilize natural forages, and allows them to abound in degraded zones. They feed on any plant materials they can find near their colonies, particularly grasses. In feeding trials, their daily dry matter intake was 2–5 percent of the body weight. The metabolic efficiency (dry matter per kg) was 33–56 percent; the digestive efficiency was 50–60 percent.[2]

Male vizcachas become sexually mature at about seven months of age and remain fertile throughout the year. The gestation period is long: 154 days. Litters contain one or two young. Newborns are well developed, fully furred, and open eyed, although they cannot fend for themselves for at least three weeks. In the wild, one or two litters are reared each year.

## BEHAVIOR

Vizcachas are nocturnal and are active year-round. They inhabit underground burrows, living in colonies often containing many individuals. They collect a variety of materials (for example, bones, sticks, and stones), and heap them in piles above the entrances to their burrows.

Their hearing and sense of smell are acute.

[2] Information from J.E. Jackson.

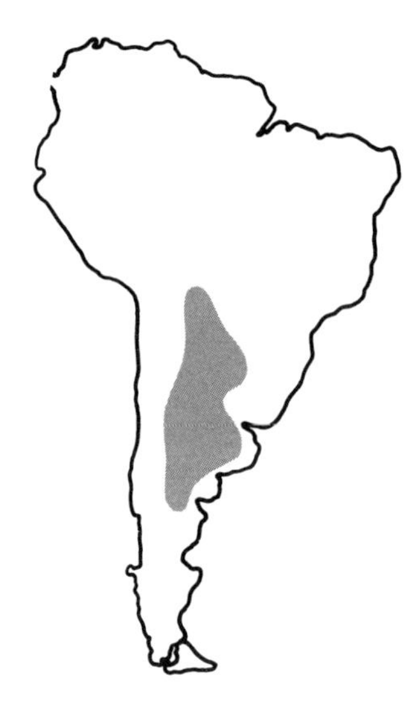

Vizcacha's native range.

## USES

Vizcachas have long been hunted for food as well as for their fur. Their meat is often consumed in pickled form in southern South America.

The skins are fabricated into table runners, rugs, bedspreads, slippers, and belts. The skins are also popular for overcoats.

Vizcachas can be kept in captivity without major difficulty.

## HUSBANDRY

In zoos, vizcachas are fed the typical diets furnished for vegetarian rodents: rolled oats, green vegetables, bananas, apples, and bread. They are usually kept indoors in wire-fronted cages, about 1 x 2 m in size, and provided with sleeping boxes.

Little is known about vizcacha husbandry, but in one trial, weight rose most rapidly in males until age 18 months (the average size was then 5.3 kg) and subsequently slowed. The heaviest male was 7.3 kg at 30–32 months. The female's weight gain was greatest until 16–18 months (average size 3.3 kg).[3]

[3] Information from J.E. Jackson.

## ADVANTAGES

The animals seem well adapted to harsh sites where the climate and forage make raising conventional livestock difficult.

The meat is white and has a good nutritive value because of its high digestibility, low levels of saturated fats, and high levels of proteins.

In marginal zones of the pampas, these rodents appear far more productive than traditional livestock.[4]

## LIMITATIONS

Vizcachas may do considerable damage by foraging in cultivated crops. As noted, ranchers claim that they take grazing away from domestic animals, 10 vizcachas eating as much as a sheep. In addition, they claim, vizcachas destroy pasture with their acidic urine. And the large burrow systems sometimes create a hazard.

Because their reproductive rate is low and their growth rate is only moderate, their commercial breeding might not be profitable except in well-designed projects with clear markets.

Vizcachas require sturdy pens, which implies a high initial cost for materials such as concrete and brick.

Vizcachas can be aggressive to one another, especially in captivity.

## RESEARCH AND CONSERVATION NEEDS

There are several possible research projects, including:

- Gathering specimens from different regions for comparative evaluation of characters such as biology, chromosome type, reproductive physiology, social behavior (both in its wild state and under controlled conditions);
- Attempting captive rearing and small-scale husbandry;
- Assessing performance under various environments; and
- Quantifying productivity and population dynamics in relation to rangeland use and improvement practices.

A rational cropping program based on wild stocks is perhaps more viable than captive breeding. This could be organized so as to keep vizcacha numbers in check while sustaining a small chain of processing plants by providing a constant supply of meat and skins.

[4] Information from J.E. Jackson.

Rock Cavy

# 25

# Other Rodents

The 10 previous chapters have described some rodent species that show promise as microlivestock. Rodentia, however, is one of the largest families of mammals, and the species highlighted by no means exhaust the possibilities. In this chapter we briefly mention others that deserve consideration and exploratory research. These might prove to be potential resources, at least in localized situations. Several are fast nearing extinction and they deserve protection and immediate attention from animal scientists.

## CHINCHILLAS

In the high Andes of South America are found the short-tailed chinchilla (*Chinchilla brevicaudata*) and the long-tailed chinchilla (*Chinchilla lanigera*). These plump little rodents have fur that is possibly the thickest, softest, and warmest of any animal's.

On the surface they would seem to be ideal candidates for microlivestock in Third World regions. Indeed, in recent years chinchilla-rearing has been successfully carried out in northern India. However, raising these animals for commercial markets is a highly specialized and costly business because only rare and expensive breeding stock produces top-quality pelts, and all other pelts are worthless in today's marketplace.

These guinea-pig-sized animals have round ears, a bushy tail, and range from 25 to 50 cm long. Adult males rarely weigh more than 500 g, but females may weigh up to 800 g.

## PACARANA

The pacarana (*Dinomys branickii*) is the third-largest living rodent—only capybaras and some beavers are larger. But little is known about this seldom-encountered, forest-dwelling species. Nevertheless, pacaranas appear to be likely candidates for domestication. They are amazingly even-tempered and peaceful, and become surprisingly tame.

Because these animals are endangered they are unlikely candidates for microlivestock in the short run, but their large size and good meat could be the stimulus for an international effort to study, protect, and rear them in large numbers before it is too late.

Pacaranas are found along the eastern foothills of the Andes, including parts of Bolivia, Brazil, Colombia, Ecuador, and Peru. Their area of potential use is in Latin America and the Caribbean.

It would be advantageous for livestock scientists to investigate this species in captivity. Studies are needed of the animal's general biology, including its nutritional requirements, reproductive capacity, behavior, and physiology. For wildlife specialists, there is a compelling need to protect this species.

## SPRINGHARE

Scattered through the dry lands of eastern and southern Africa—on numerous grasslands, plains, fossil lake beds, hill slopes, and floodplains—is the springhare (or springhaas). With its powerful hind legs, tiny forelegs, upright stance, and hopping gait, it looks something like a tiny kangaroo. It is commonly seen at night, eyes glowing a characteristic red in the headlights.

This animal is an important source of food and skins for rural peoples throughout southern Africa. In Botswana, it is the principal bushmeat in the human diet.

Springhares are 35–40 cm long and weigh 3–4 kg. There is one species (*Pedetes capensis*) and two subspecies: the East African springhare and the Cape springhare (formerly *Pedetes cafer*).

## ROCK CAVY

The rock cavy (*Kerodon rupestris*) is closely related to the guinea pig and occurs in the impoverished semiarid region of northeastern Brazil. It is large and lean and has a face somewhat resembling a dog's. It is hunted extensively and is an important source of meat for country people, who consider it a delicacy.

This creature might be suitable as a microlivestock. It consumes leaves and bark, and breeds well in captivity. Famine is a serious periodic problem in the often-drought-stricken area, and protein deficiency is common. The rock cavy, like the guinea pig, may be amenable to domestication and may be able to provide the people with better nutrition. However, it is difficult to keep in a cage because it moves fast, climbs well, and easily slips out.

Because these rodents occur in rare and patchily distributed habitats, they are in desperate need of protection, whether or not they prove to have any long-term utility.

## SALT-DESERT CAVY

A smaller relative of the mara (page 256), the salt-desert cavy (*Dolichotis salinicola*) is rabbitlike in appearance and behavior. It has large ears and eyes, long legs, and a short tail. It lacks the mara's white rump patch.

This animal inhabits dry, salty areas of the Chaco desert, particularly areas of dry, woody brush. Specifically, it is found in the saline western Chaco of Paraguay and northwestern Argentina, as well as in the extreme south of Bolivia. It is about 45 cm long and weighs up to 4 kg.

Although it might make a useful microlivestock in its native habitat, the salt-desert cavy breeds rapidly and can cause much devastation; it should never be introduced to new regions.

The salt-desert cavy's life in the wild is largely unknown; however, some have been successfully raised and bred in zoos. Exploratory research on keeping and managing these little creatures is warranted.

## OTHERS

As noted earlier (page 193), it has been estimated that 42 peoples of various cultures eat rodents. Most of these eat locally available species, some of which are listed below. Whether any have long-term usefulness is uncertain, but study of them in the wild and in captivity could result in some interesting and valuable scientific discoveries.

### Solomon Islands Rodents

The Solomon Islands in the southwest Pacific are home to a collection of six rare giant rodents that live in the rainforest canopy. At one time they were important food items. Archeologists have dug up tens of thousands of rodent bones on sites where people lived as long as 30,000 years ago. These mysterious animals were classified last century, but most have not been seen by biologists for decades. However, one, the thinking rat (*Solomys sapientis*), was rediscovered in 1987—the first time it had been reported seen since 1901. The others may also be inhabiting the dense and undisturbed forests.[1]

The thinking rat was located on the island of Santa Isabel. It proved to be gentle, unafraid, and friendly. It lives in the forest canopy, weighs up to 1 kg, and feeds on nuts and fruits. Efforts to build up the population are urgently needed. The other species should also be sought.

Such animals could be a resource for rainforest production (perhaps in the manner of butterfly farming in Papua New Guinea[2]). They may never be plentiful enough to be food sources again, but they could nonetheless become valuable. Zoos the world over are likely first customers for these scientific curiosities if production can be boosted and the populations secured.

## Giant New Guinea Rat

The giant New Guinea rat (*Mallomys rothschildi*) is often eaten in New Guinea. It is little known to science but is odorless and easily tamed. It grows so fast that it becomes as big as a guinea pig even before it is weaned. It feeds on a wide variety of tubers and vegetable products. An attractive species native to mountain forests, it has a long heavy tail and black fur with white ticking.

## Porcupines

Porcupines, brush-tailed porcupines, and their relatives are distantly related to guinea pigs and are widely consumed as food in tropical regions. Examples are:

- Indian porcupine (*Hystrix indica*) and other species of South Asia.
- Cape porcupine (*H. africaeaustralis*) of southern Africa.
- Prehensile-tailed porcupine (*Coendou prehensilis*), which inhabits Central and South America. These nocturnal creatures live in the thick, leafy crowns of trees or in hollow trees or holes in the ground. Although often belligerent among themselves, they can be very friendly and tame toward humans.

[1] The other species are *Uromys imperator* (last collected 1888), *U. rex* (1888), *U. salebrosus* (1936), *U. salamonsis* (1883), and *Solomys porculus* (1888). Information from Tanya Leary.

[2] See companion report, *Butterfly Farming in Papua New Guinea*. 1983. Washington, D.C. National Academy Press.

## Kiore

The kiore (*Rattus exulans*) was formerly an important component of the diet of most Polynesians, including the Maoris of New Zealand. It has been successfully reared in captivity in recent times. Unlike many other rodents, this animal is normally not a scavenger; it is a clean, even fastidious feeder that is basically a vegetarian (flowers, berries, nuts, and seeds) and is reckoned remarkably good eating.

## Soft-Furred Rat

The soft-furred rat (*Praomys*) is relatively large and slender and is found in the tropical forests of Africa from sea level to more than 3,000 m. Among the most common rodents of the African jungle, it is trapped almost everywhere. It feeds largely on plants, but eats large quantities of ants and other insects. It has been raised successfully in captivity and is eaten by villagers in Malawi.

## Giant Squirrels

The giant squirrel (*Ratufa bicolor*) occurs throughout Indonesia, Malaysia, and Sri Lanka. It is the largest squirrel on earth, almost the size of a cat. A related species, the palm squirrel (*Funambulus*), also gets very large. It can crop nuts in the very high treetops that are inaccessible to people. It is widely used as food.

## Squirrels

Squirrels of the genus *Callosciurus* (notably *C. notatus* and *C. prevostii*) are significant pests on cocoa, oil palm, and mixed fruit plantations in Southeast Asia. They can be reared and bred on a diet of most types of fruit as long as a little protein, in the form of insects or cooked wheat, is available. The systematic use of this animal may offer a chance to turn a pest to advantage.

## Cloud Rat

Two species of slender-tailed cloud rats (*Phloeomys* spp.) are found only in southeastern Asia. They live in tree cavities, climb well, and are well adapted to tree life.

One of these (*P. cumingi*) lives in the northeastern part of Luzon in the Philippines, where it appears to be thriving. It has a large body and long tail and is the largest member of the mouse subfamily.

The other (*Crateromys schadenbergi*) has long, thick hair and a thick, bushy tail. It lives in the mountainous areas of northern Luzon. A nocturnal animal, it feeds on buds, bark, and fruits.

Both are attractive, even fascinating, creatures that are relentlessly hunted for food. They might make useful livestock in forest situations.

## Spiny Rat

The Cayenne spiny rat (*Proechimys guyannensis*) is found throughout most of South America. It is tasty, easily kept in captivity, and is popularly used in Colombia for food. A nocturnal animal, it is one of the most common mammals in many areas. It has been raised in captivity on bananas, sweet corn, coconut, grain, and various seeds.[3]

## Bamboo Rat

The bamboo rat (*Rhyzomys* spp.) is the largest rodent on the island of Sumatra in Indonesia; it weighs 2–4 kg and the body can be as long as 45 cm. It prefers to live in bamboo thickets and is hunted and eaten by many local peoples.

[3] Information from Lynwood Fiedler.

*There is a need—in considering any sort of game farming—to relate the animal's "preferred" conditions (range, feed, temperature, etc.) to the varying land and climate types not being effectively utilized. In other words, rather than looking at animals that might be farmed, it might be necessary to consider the terrain and climate and then seek animals that would "do" well under those conditions. One major fault with traditional farming has been the tendency to force traditional livestock onto unsuitable land. This has given rise to numerous serious problems. In many cases the most suitable use of an area is provided by several species grazing together to mutual benefit (e.g., goats and sheep on New Zealand hill country improve grazing for each other if ratios are right).*

David Yerex

# Part V

# Deer and Antelope

Several types of tropical deer[1] and antelope are no bigger than an average-size dog. These "microdeer" and "microantelope"[2] are the smallest of all ruminants. Although there is considerable experience with rearing and utilizing the larger species, little is known about these miniature ones.

Given research, mouse deer, muntjac, musk deer, pudu, brocket, huemul, and water deer, as well as half a dozen small antelope, might prove to have considerable potential. Collectively, they come from diverse habitats, ranging from equatorial to subarctic and from moist rainforest to arid savanna. They are adapted to some environmental conditions that are only marginal for production of conventional livestock because of drought, heat, diseases, altitude, or other constraints.

## DEER FARMING

Deer appear to be unlikely candidates for livestock, but reindeer were probably among the first domesticated animals and have been draft animals for perhaps 20,000 years. Even today, tens of thousands of reindeer pull sleighs in the European arctic. On military expeditions, the ancient Romans took along herds of fallow deer as a source of

[1] This section includes both mouse deer and musk deer, which are not true deer species. True deer belong to the Family Cervidae; mouse deer to the Family Tragulidae; musk deer to the Family Moschidae. Their common names are technical misnomers but, for convenience, we include them here.

[2] We are using these names for animals whose mature body weight is less than about 20 kg.

meat, and more than 1,000 years ago deer were annually herded off the Scottish Highlands for winter meat supplies.

In recent years, there have been breakthroughs in the "domestication" of deer. Species already being farmed are: red deer (New Zealand, Australia, Taiwan, Korea, Russia, China, Scotland, the United States), elk (New Zealand, Canada, the United States), fallow deer (New Zealand, Australia, England, Denmark, Sweden, Switzerland, Germany, the United States), rusa deer (Australia, Mauritius, New Zealand, Papua New Guinea), sika deer (Taiwan, New Zealand), musk deer (China, India), and Père David's deer (New Zealand). Although not truly domesticated, even the moose has been tamed in Scandinavia and the Soviet Union, the calves being bottle-raised from three days of age.

New Zealand has made particular progress in domesticating large deer (see sidebar). It seems probable, therefore, that similar success with small deer could be achieved. For those seeking interesting, pioneering research, microdeer are good candidates.

## ANTELOPE RANCHING AND FARMING

The worldwide experiences in domesticating various deer species suggest that the organized production of small antelope should also be considered. Several large species have already been studied and are used in game farming in eastern and southern Africa. Similar research on the smaller species, which so far have received little or no attention, is one of the more speculative ideas in this report. We put it forward only for consideration by researchers, but if exploratory studies prove successful, this is a topic deserving international support.

In some parts of Africa there are large expanses of uninhabited lands, and producing any sort of livestock there is limited by aridity and by the presence of tsetse flies. But in this habitat live tiny antelope such as dikdik, suni, and klipspringer. In the rainforests and secondary forests are found duikers and the royal antelope. All these creatures have advantages that justify their consideration as microlivestock: they have a more rapid turnover than the big species, and they produce a high yield of quality meat. In addition, compared with cattle, these native ungulates make better use of the habitat. Cattle select a limited number of grass species; antelopes choose a wider range, and also include forbs, bushes, and trees.

More important perhaps is their resistance to many diseases. Most, if not all, are resistant to trypanosomiasis, the disease carried by the tsetse fly. They are not immune to this and other tropical diseases, but they are less susceptible than cattle. However, part of this may be owing to their ability to roam widely; if confined and treated like domestic animals, they may also require some protection against parasites and diseases.

Antelopes are also more productive than cattle; that is, they produce a given quantity of meat more quickly because they breed better in the African hinterlands and grow more rapidly on its existing forages. On the other hand, they generally require a richer diet than cattle.

Finally, in their favor, antelopes affect the habitat less than the same density of cattle does; they spread out more while feeding and thereby cause less erosion.

There are two ways of exploiting this potential. One is by "cropping"—taking a controlled offtake from free-ranging populations without depressing the overall population. Several methods for producing meat this way from large antelope have been attempted in countries such as Kenya, Tanzania, and Mozambique. Few have persisted. Often this has been due to opposition from the vested interests of the cattle industry and from stringent veterinary requirements. Nonetheless, game ranching offers a means by which marginal lands could produce food of a high nutritional quality on a sustained basis.

The other method is by farming—that is, domesticating or partially domesticating the animals, keeping them in pens or herding them like cattle. Experiments in farming antelopes have been less common than game ranching, but one of the most interesting is that conducted on the Galana Ranch in Kenya. Three wild species—buffalo, eland, and oryx—were selected for comparison with cattle. Half-grown animals were preferred for capture, and it was found that if they were kept in the dark for the first week after capture, and then gradually provided with more and more space while they became familiar with people, after about six weeks they could be released into the open and herded from place to place. Grazing during the day under the eye of a herdsman, they allowed themselves to be herded back to a pen at night in the traditional African manner, where they would sleep around the campfire. This was a promising advance in behavior modification leading toward domestication. The oryx, for example, gained weight on grazing that would not even sustain cattle, and it required only a quarter of the amount of water.

At bottom, the question is not what contribution antelopes can make to the African larder, for they already make a significant contribution through (largely illegal) hunting. The question is whether farming could make them a sustainable asset rather than their being senselessly squandered, as is the case at present. Although its potential has yet to be realized, antelope farming is not a panacea for Africa's food problems, and certainly not the world's, but it might pave the way to a new and more gentle way to make savannas useful.

## TAME GAME

Since the 1970s, deer have taken the place of sheep on many New Zealand pastures, and today the country has more than 5,000 deer farms carrying over one million head, mainly red deer. It is now common along country roads to see tall fences surrounding graceful deer quietly grazing ryegrass and clover. And there are all the appurtenances for deer that exist for cattle and sheep. Auctions and shows are held regularly. Deer farmers have a professional association and produce their own glossy magazine. Government scientists publish pamphlets on the care and management of deer. There are recognized stud stags, computerized recording schemes for breed improvement, and even veterinary services specifically for deer. Hybridization between wapiti and red deer, and Père David's deer and red deer, is accepted practice. The animals are moved by use of dogs (which command by mere presence rather than by bark or bite), and herds of up to 80 are shifted by truck. Slaughter facilities specifically for deer are in operation throughout the country.

This transformation of a nervous, jumpy, and retiring wild species into a farm animal is a remarkable achievement. Once accustomed to people, many specimens become gentle, even affectionate, and will come at a farmer's call. Males are generally as easy to handle as females, except during the rutting season when they become aggressive and cannot be handled at all.

However, even at the best of times the farmed deer must be handled gingerly. If the causes of stress are not quickly suppressed, hysteria can erupt throughout a herd; in an instant, quiet animals can be leaping suicidally in all directions, disoriented, diving head-on into fences, charging gates. Chronic stress, the causes of which are not always obvious, can result in illness or death, although this trait diminishes in subsequent generations of farm-born stock.

A deer farm has to be laid out to certain special specifications. To prevent escapes, the boundary fences must be 2 m high with netting of 15- or 30-cm mesh. Inner fences need only be 1.5 m high. Water troughs are placed in the middle of the fields, and nothing is allowed to jut inwards from the fences because the animals tend to walk fencelines and take comfort from the illusion of openness. Because deer like to wallow in hot weather, some farmers also provide shallow waterholes.

Deer yards can be of any design, but the sides of the passageways and holding pens should be solid, as deer do not see fences very well, particularly when under pressure, and may injure themselves in a leap towards what appear to

Amy Roydhouse, age 5, on her family's deer farm near Napier, New Zealand. (Fraser Duncan, *The Napier Daily Telegraph*)

be wide open space. (New Zealanders usually make the sides of plywood.) Also, the holding pens should be roofed, as semidarkness has a calming effect. Animals that in the sunlight become hysterical on seeing a person in the distance, can, in the relative darkness of a roofed shed, be touched and even given injections.

Despite the special facilities, however, handling deer takes time and care and experience. The most successful farmers spend much time among the deer so that the animals become accustomed to human presence. This helps to make yard work easier. Also, new arrivals are allowed to wander through the yards on their own to become familiar with them. In addition, special tame deer are used as leads or decoys to encourage the rest of the herd to follow. Using such simple techniques, a formerly intractable species has become almost fully domesticated.

Lesser Malayan Mouse Deer

# 26

# Mouse Deer

Mouse deer[1] (*Tragulus* spp. and *Hyemoschus aquaticus*) are among the smallest ruminants known. The lesser mouse deer of Southeast Asia is probably *the* smallest; an adult stands only 20 cm high and weighs a mere 1–2.5 kg.

Although they look vaguely like tiny deer, mouse deer differ in several particulars. The stomach is simpler and (like the camel's) has three instead of four effective compartments. Rumination occurs, but mouse deer are the most primitive of all ruminants. Indeed, they share a number of characteristics with nonruminants, including lack of horns or antlers; continually growing, tusklike upper canines in males; sharp-crowned premolars; and four fully developed toes.

Virtually unchanged in 25 million years of evolution, these are solitary, nocturnal, retiring animals that have seldom received detailed research. Whether they might make suitable microlivestock is unknown. However, they seem to be tractable, and people in Southeast Asia (Sarawak, for instance) have traditionally kept at least one of the species as backyard pets. Moreover, mouse deer are indigenous to tropical lowland regions and withstand the heat and humidity that are stressful to most conventional livestock species. They probably also are resistant to many diseases of those torrid regions.

In the United States, mouse deer are being raised as laboratory animals for basic research on ungulates. This is because the animals are easier to handle than large deer or goats.[2]

## APPEARANCE AND SIZE

Mouse deer are graceful, lithe, and look somewhat like large rodents.

[1] Also known as chevrotains. These animals are not true deer; their common name is a misnomer. Although they resemble deer, they also have some features of pigs. Zoologists classify them as a separate family, the Tragulidae. There are three Asian species: the lesser Malayan mouse deer *Tragulus javanicus*, the Indian mouse deer *T. meminna*, and the larger Malayan mouse deer *T. napu*. There is one African species, the water chevrotain *Hyemoschus aquaticus*.

[2] Information from I. Muul.

The Asian species are the shape and size of an agouti (see page 198); the African species is more like the paca (see page 262). All have short legs, a small head, and a pointed snout. Adults weigh from 1 to 5 kg, depending on species. The head and body are only 0.4–1 m long, and the shoulder height is merely 20–36 cm. Males are generally smaller than females.

In most species the body is a rich brown with white spots and stripes. The belly is usually white. The animals stand on the middle toes, so that the lateral ones do not touch the ground. Neither sex bears antlers. In males the upper canines form long tusks that may extend outside the lips and even to below the line of the jaw.

## DISTRIBUTION

Twenty-five million years ago, early forms of mouse deer existed throughout Asia, Africa, and Europe. Today's species are restricted to tropical forests and mangrove thickets of Southeast Asia and Central Africa.

Of the three Asian species, the Indian mouse deer occurs in southern India and Sri Lanka; the larger Malayan mouse deer occurs on the mainland of Southeast Asia and the lesser Malayan mouse deer occurs on Java as well. The water chevrotain, a related African animal, is found from eastern Zaire to the Atlantic coast.

Range of Asian mouse deer: (1) Indian mouse deer; (2) larger and lesser Malayan mouse deer.

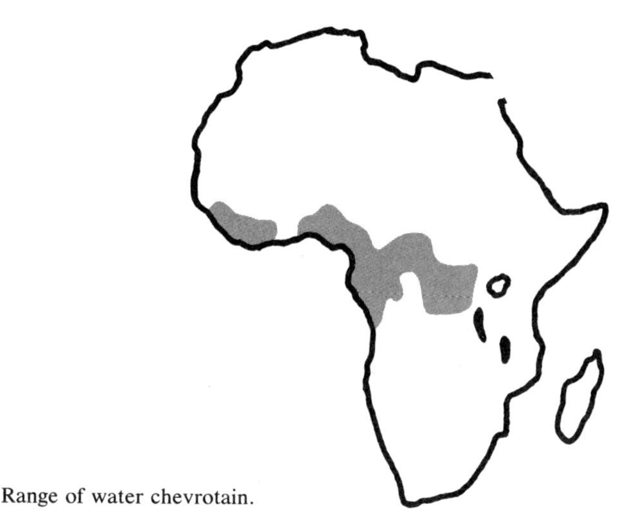

Range of water chevrotain.

## STATUS

In recent years, human encroachment into the forest has caused the destruction of the mouse deer habitats and has put various mouse deer species under a pressure that is causing their populations to decline.

## HABITAT AND ENVIRONMENT

These animals inhabit equatorial forests and mixed secondary tropical forest. They generally live among undergrowth on the edges of dense lowland rainforests. They especially haunt rivers and swampy bush areas, seeking escape by water when in danger.

## BIOLOGY

Little about these animals is recorded. Essentially vegetarians, they feed chiefly on fruits, supplemented by leaves. They also eat insects (for instance, ants), if available. They do not seem to eat grass.[3]

The premolars of the mouse deer are designed for piercing and chopping food rather than for chewing. As noted, the stomach consists of three functional compartments: the rumen, the reticulum, and the abomasum. (The omasum of ruminants is represented only by a rudimentary area.) The stomach occupies almost the whole of the

[3] Information from G. Dubost.

abdominal cavity, extending from the diaphragm to the pelvic inlet, which provides this small animal with large food-storage capacity.

The blood has a very high erythrocyte (red blood corpuscle) count as well as the smallest erythrocyte size of any mammal. The flesh is "white" and the muscles contain little myoglobin.

Mature females reproduce almost continuously, and usually regardless of season. In the female larger Malayan mouse deer, mating occurs within 2 days of giving birth. In the African species, many births are synchronized with the rainy seasons, when fruits are plentiful. The gestation period is about 5–6 months, depending on the species. There is only one young per birth. Weaning normally occurs at 2–3 months, but can occur as early as 3 weeks, with sexual maturity achieved at 4–5 months (Asian species) and 10 months (African species). The young stay alone, hidden in vegetation during the first month or two.

## BEHAVIOR

Mouse deer are shy, keeping to dense jungle and depending on concealment for protection. Although often present in large numbers, they are seldom seen. Preferring to be near lakes, rivers, or streams, they can nevertheless wander 1 km or more from water. They feed mostly at dusk or at night, sheltering in undisturbed areas or under shady bushes during the day. They utter weak, bleating sounds, and when frightened, jump a meter or more in the air.

Communication is by scent and calls. The African species possess anal and preputial glands, with which, along with urine and feces, they mark their home ranges. Males of both Asian and African species possess a chin gland to mark either the vegetation or their mates.

Mouse deer are among the most excitable, nervous, and jumpy animals. One must tread softly in their presence for fear of causing absolute pandemonium and mishap.

## USES

Mouse deer are widely sought by native people for food, and their meat is highly regarded. Dressed carcasses have a high proportion of muscle (84 percent in Asian species), low proportion of bone (15

[4] Information from G. Dubost.

Opposite: Water chevrotain. (A.R. Devez, CNRS, Mission Biologique au Gabon)

percent), and an insignificant amount of fat. The ratio of muscle to bone is large—5.6:1. The mean dressing percentage of 62.1 percent is greater than that reported for cattle, water buffalo, or goat.[5]

## HUSBANDRY

Adults are wild, generally intractable, and "flighty," but young animals (at least of the Asian species) tame readily and make good pets. Nevertheless, these are delicate creatures and must always be handled gently. Individuals caught in the wild tend to bash against the sides of cages.

Despite an unpromising temperament, Asian mouse deer are regularly bred in zoos, including those in Amsterdam, New York, and Zurich. They also have been reared successfully in small enclosures at several research institutes, such as the I.R.E.T Institute, Makokou, Gabon; the Institute of Medical Research in Kuala Lumpur, Malaysia; and at Fort Detrick in the United States.

Perhaps the best way to breed this animal is by using the battery system of small units comprising one male and two females per cage. The costs are mostly for obtaining suitable enclosures and for feeding and watering troughs.[6] The cages must be covered with mesh because the mouse deer can jump. However, the covering must be sufficiently high to allow the male to stand with its body vertical during copulation.[7] They can be fed a variety of foods and grow well on stems of bean plants.

## ADVANTAGES

As noted, these are small, seemingly tractable creatures that are at home in the heat, humidity, and diseases of tropical lowlands. They might play a particularly important role as livestock for tropical rainforests; the forests could be left standing while the animal still produces meat. Today, in a widely condemned process, tropical rainforests are being felled in order to raise cattle for meat.

[5] Information from M.K. Vidyadran.
[6] Information from Roy A. Sirimanne.
[7] Information from I. Muul.

## LIMITATIONS

Small size makes mouse deer easy prey for various predators. In the wild, snakes, crocodiles, eagles, and forest cats feed on them.

Mouse deer are among the most excitable, nervous, and jumpy animals. One must tread softly in their presence for fear of causing absolute pandemonium and mishap.

The different species are solitary, and it is difficult to keep many individuals (especially males) in a restricted space. They must be kept in a quiet enclosure, with cover or good shelters.

## RESEARCH AND CONSERVATION NEEDS

The survival of these four "living fossils" depends on conserving their rainforest habitat and restricting hunting, especially night hunting. But studies of their propagation and management are also imperative.

In particular, research is warranted on various aspects of their husbandry, such as enclosure design, space requirements, and health. A special research need is to understand the animal's nutritional requirements and to develop diets for use in captivity.

# 27

# Muntjac

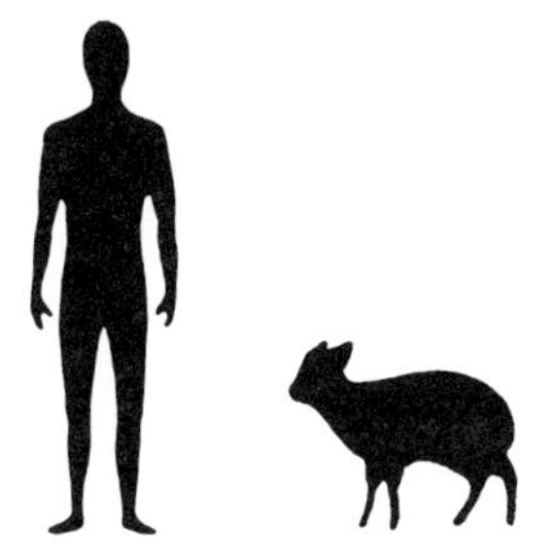

Muntjac, or barking deer (*Muntiacus* species), are among the most widespread but least known of all Asian animals.[1] They are almost the size of an average dog and they bark, but they are true deer. These little animals adapt well to captivity and have been introduced to zoos and wildlife collections throughout the world.

In recent years one species, the Reeves' muntjac, has become established in England, and a few specimens have settled into semi-captivity—staying behind fences, accepting human presence, and even eating out of people's hands. In Sarawak, villagers have also been known to keep muntjac, feeding them a diet of rice with some leafy matter occasionally added.[2] Such experiences show that these small, shy deer can be calm and adaptable. It also shows that they can be raised on practical, artificial diets and that they are not strictly browsers. This creates the possibility (admittedly highly speculative) that they might make future microlivestock.

Muntjac produce lean, palatable venison and perhaps could be farmed on an organized basis. They are native to severe environments where heat, humidity, and endemic diseases make raising conventional livestock difficult. In future, given research, muntjac might become widespread contributors to Asian economic development. The lessons learned in captive breeding could also be important for conserving endangered muntjac species.

One species, the Indian muntjac, has an incredibly low chromosome number (2n = 7 in males; 2n = 6 in females), which makes them partic-

[1] There are five species: *Muntiacus muntjak* (red muntjac, Javan muntjac, Indian muntjac), *M. crinifrons* (black muntjac), *M. reevesi* (Reeves' or Chinese muntjac), *M. feae* (Fea's muntjac), and *M. rooseveltorum* (Roosevelt's muntjac). This chapter is based almost entirely on Reeves' muntjac, the only species that is scientifically well known. The taxonomy of muntjacs is currently under revision.

[2] Information from R. Basiuk.

ularly promising candidates for mammalian genetic studies. This species has regularly bred well in both zoos and research institutions.

## APPEARANCE AND SIZE

Muntjac are small and slender. Reeves' muntjac, the smallest, has a shoulder height of 45–60 cm. Fully grown, it weighs less than 20 kg; commonly it is merely 10–12 kg. The Java subspecies of the Indian muntjac is the biggest, with a shoulder height of 58 cm and a mature weight of 43 kg.

Antlers on the males usually include a main prong as well as a much shorter brow tine. Even the main prong is no larger than a finger, but its pointed tip is hooked and must be treated with caution. There are also two tusklike canine teeth that protrude from the mouth. These have sharp points and a knifelike posterior edge, capable of cutting to the bone a person's finger or another muntjac's rump.

Coloration varies from deep brown to yellowish or grayish brown with cream or whitish markings, depending on the species.

## DISTRIBUTION

Muntjac are native to a vast region from eastern China to Nepal, India, Sri Lanka, and Indonesia.

Almost a century ago, Indian and Reeves' muntjac were introduced to a deer park in southeast England. Some escaped, and (as noted) the Reeves' muntjac has adapted, spread, and settled down to life in the countryside.

## STATUS

Of the five muntjac species, the Reeves' and the Indian are well known and in no danger. For instance, recent estimates of annual game production have shown that there are about 650,000 Reeves' muntjac in China. The other three are threatened with extinction. Black, Roosevelt's, and Fea's muntjacs are virtually untried in captivity, but the success of raising Reeves' muntjac in English country gardens suggests that perhaps their populations could be saved through captive breeding.

## HABITAT AND ENVIRONMENT

In their native habitats, these small deer are usually found in dense vegetation on hilly ranges at elevations up to at least 3,000 m.[3]

[3] Information from H.A. Jacobson. The species was the Indian muntjac at Langtang National Park in Nepal.

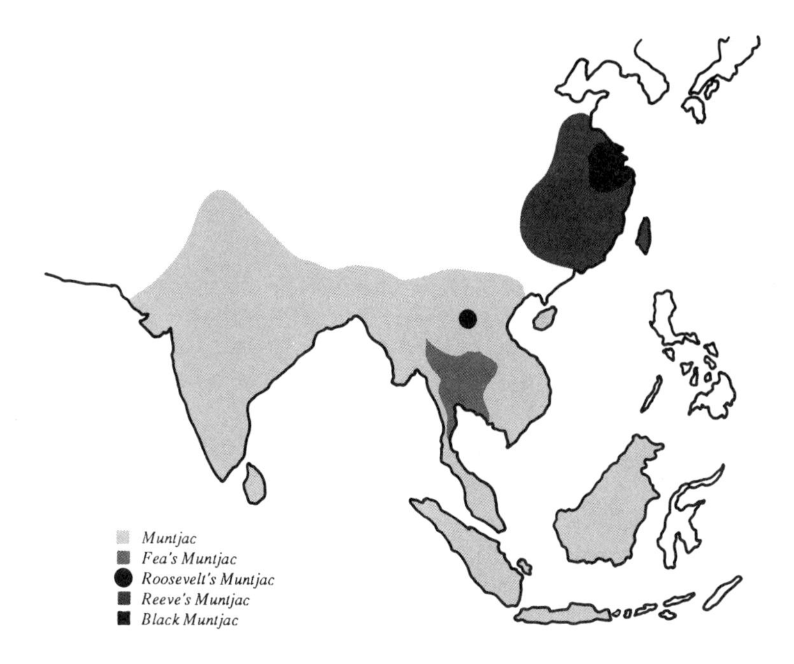


The native ranges of the various muntjacs.

## BIOLOGY

Muntjac seem to be primarily browsers. However, in captivity they eat fresh grass, alfalfa hay, and feed pellets. They also readily eat root crops such as potatoes, carrots, and parsnips. They are "concentrate selectors," preferring foods low in fiber and rich in protein and nutrients. Captive specimens reportedly need a supply of cut browse.

Breeding may occur year-round, but in practice it is synchronized with certain seasons. The first conception can occur as early as 6 months of age. One, occasionally two, young are born after a gestation of about 200 days. They weigh around 1 kg at birth, and the fawns usually remain hidden until they can move about with the mother. The females mate within a few days of giving birth. Life spans up to 16 years have been recorded.

As noted, the Indian muntjac is a species of great cytogenetic interest. It has the lowest diploid chromosome numbers yet found in a mammal.[4] The large, easily distinguishable chromosomes are a great

[4] Diploid chromosome numbers for other species are Reeves': male 46, female 46; Fea's: female 13; black: male 9, female 8; Roosevelt's from Laos: female 8. Information from K. Benirschke, H. Soma, S. Liming, and D. Wurster.

advantage in tissue culture, and many laboratories now have muntjac cell lines. The karyotypes of the different species' chromosomes are very different; the Indian and Reeves' muntjac can hybridize, but the offspring are infertile.

## BEHAVIOR

Muntjac are dainty and have a captivating charm. Always on the alert, they are active both day and night. Often they will bark for an hour or more, but typically they bark for only a minute or two. When panicked, captive muntjac may rush into fences or walls. They can easily leap barriers 1.5 m high.

Males are highly territorial and defend their territories vigorously. Adult females also inhabit a specific territory, which they defend against strange individuals.

Males mark the ground at intervals by lowering the head and rubbing the frontal glands on the ground and by scraping their hooves against the ground. They mark trees by scraping the bark with the lower incisors and rubbing the base of their antlers.

## USES

In Asia, muntjac are hunted for meat, skins, and antlers.

## HUSBANDRY

Muntjac thrive and, at least under ideal conditions, breed freely in captivity. However, they may stop breeding if they are crowded.

In England, zoos and private collections keep as many as six muntjac in an area of 40 x 20 m. Plenty of cover is provided for the animals to hide in. Fences almost 2 m high are used. (Lower fences are reportedly adequate where there is no risk of the deer being panicked by dogs or people.)

## ADVANTAGES

Like other deer, muntjac produce extremely lean meat.

They seem to be healthy animals. In Britain, wild and captive muntjac have few gastrointestinal worms, and ectoparasites such as ticks and lice are not a problem.

Mature Reeves's muntjac buck in an English woodland. (R.D. Harris and K.R. Duff).

## MUNTJAC

In England, far from Asia's forests and mountains, the Reeves' muntjac is the subject of a peculiar chapter in the otherwise unhappy history of introduced species. Charmed by the deer's odd characteristics, the British have welcomed its invasion into gardens around the country. Now, there are tens of thousands of muntjacs on the loose, and some people have even adopted them as pets.

Although other alien species have wreaked havoc on native plants and animals, muntjacs appear to be a merciful exception. At least over the short term, muntjacs "have proven to be an almost innocuous asset to the countryside. They give pleasure to thousands and pain to few. Eating mostly ivy, grass, leaves, and prickly bushes, they rarely feed in one place long enough to do much damage—except to an occasional suburban garden."

Oliver Dansie and one of his trusting charges from the woods behind his backyard in Welwyn, Hertfordshire. (G.H. Harrison)

Life for many muntjacs is made soft by homeowners who find the deer's large, dewy eyes and tiny antlers irresistible. The kind-hearted suburbanites put out salt licks, water, and kitchen scraps, and they built snug little shelters against the cold north wind. "All our adult deer will take food from the hand," says Walter Buckingham, who has kept muntjacs for five years in his garden in the county of Hertfordshire, just north of London.

The adaptable immigrant is colonizing new areas so rapidly, say biologists, that soon there may be more in England than in Asia. "In time," predicts muntjac-researcher Oliver Dansie (pictured opposite), "it may eventually establish itself as our most widely distributed deer species."

## LIMITATIONS

In the wild these animals are not gregarious and are generally found alone or in pairs. Because of their strong territorial instincts, large males may not be able to be kept together without fighting. The upper canine teeth can inflict serious wounds. Some females are intolerant of each other as well.

Muntjac are fragile; they cannot be held by the legs, for example.

Some infectious diseases may prove to be of epizootic importance: foot-and-mouth disease, mucosal disease, epizootic hemorrhagic disease, rinderpest, and tuberculosis (all three types). This may be a problem, especially when people are raising the animals by hand.

## RESEARCH AND CONSERVATION NEEDS

To ensure a better understanding of their potential as microlivestock, muntjac deserve research and recognition from animal scientists and conservationists from Nepal to China. The English experience shows how populations of the endangered species might be built up. At present, however, none are receiving any husbandry research; only the two most common species can generally be found in zoos.

These animals deserve investigation into physiology, reproductive requirements, fertility, nutrition (for example, food preferences, feeding strategies, and food utilization), growth, adaptability and environmental tolerance, diseases, management, social structure, and selection for calm temperament.

# 28

# Musk Deer

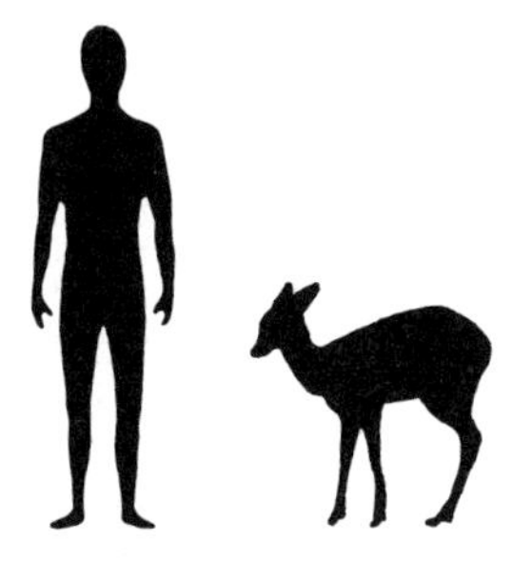

The musk deer (*Moschus* species)[1] is so small that (like other ungulates in this section) it is only as large as an average-size dog. A gland in males produces musk, a thick, oily secretion that is one of the most valuable substances in the animal kingdom. Musk is used in Oriental medicines as well as in European perfumes, and in recent years it has sometimes sold for as much as three times the price of gold.

Musk is traditionally obtained by killing these deer and removing their glands. The dried glands, called pods, contain a reddish brown musk powder that has been a commodity in international commerce for more than 1,000 years. Despite bans in India (1972) and Nepal (1973), musk continues to be illegally exported, mainly via Hong Kong, for use in Japan and Europe. In Japan, for example, it is an ingredient in more than 200 different medicines.[2] In Europe, musk goes into some of the most famous perfumes.

The international trade in Himalayan musk, originating from both northern and southern sides of the Himalayan divide, amounts to 200 kg per year, representing an annual slaughter of 20,000–32,000 male deer.

The commercial value of the animal makes it highly attractive for development as a livestock species. The economic force causing its slide toward extinction could be employed to protect and restore both the species and its habitat. Ranching these deer might put musk production on a sustainable footing. It might also encourage habitat protection, because in the harsh climate of the high Himalayas, rearing

[1] This is not a true deer. Although superficially similar to living deer, it is currently placed in its own family, the Moschidae. Traditionally, only one species, *M. moschiferus*, was recognized. However, the taxonomy of the genus is currently under revision by P. Grubb, Wang Yingziang, and C.P. Groves, who are proposing several species. In their classification, the common species in western China and most of the Himalayas is *M. chrysogaster*.

[2] Information from S.K. Dhungel.

musk deer could be much more profitable to villagers than raising crops or cattle.

Musk deer are already being farmed under primitive conditions in China, where techniques for extracting musk without killing the animal have been developed. In India, small collections of musk deer have been established by the forest departments of Himachal Pradesh and Uttar Pradesh. For several years authorities in Nepal have successfully extracted musk from an adult male at the Kathmandu Zoo, without apparent harm to the animal.[3] These experiences suggest that musk might become a farmed product. So far, however, success has been limited. The Chinese animals, for example, have a high mortality rate and the musk is said to be of poor quality. Nonetheless, these examples are valuable pioneering case studies that deserve recognition, support, and further development.[4]

## APPEARANCE AND SIZE

With their long ears, arched back, and bounding gait, these diminutive deer remind one of large hares. The pointed face and large ears make their heads strikingly reminiscent of kangaroos. The coarse hair gives them a stocky appearance. The color varies according to species (and subspecies) from rich reddish brown to dark gray or black. The peculiarly brittle and wavy hair probably has good insulating properties, as it consists of air-filled cells arranged like a honeycomb.

Musk deer have an average mature weight of about 6–11 kg and a body length of 50–90 cm. They stand 50–60 cm high at the shoulder and 5 cm higher at the rump because the hind legs are longer than the forelegs. Some dwarf types are only 40–46 cm high. The tail is short, and in males it is naked, except for a terminal tuft of hair, because they mark their territories by constantly rubbing the caudal gland, which is located near the tail, onto objects.

All four toes are flexible, which, compared to the rigid hoof of other ungulates, gives a firmer grip on precipitous slopes. The dew claws are enlarged and, together with the central digits, splay out, to minimize sinking in soft snow.

Neither sex possesses antlers, but males have long upper canine teeth that project well below the lips. The lower front teeth have a spatulate form that probably helps the animals scrape lichens from the surfaces of rocks and trees in winter, when most vegetation is snow covered.

[3] Information from B. Kattel.

[4] In France, at least one perfume factory is keenly interested in the domestication of the musk deer because it needs musk for its perfumes and wants to regularize the trade.

The musk deer's native range.

## DISTRIBUTION

The genus is distributed patchily throughout the forested mountainous parts of most of Asia. One population extends from just north of the Arctic Circle southward to the northern edge of Mongolia and Korea. Others occur in China, northern Vietnam, and the Himalayan region including Bhutan, Assam, Tibet, the Indian Himalayas, Nepal, and northern Pakistan.

## STATUS

During this century, the musk deer has rapidly declined throughout its former regions. In many parts of Afghanistan, Pakistan, India, Nepal, and probably Tibet, it is already regarded as rare, with a distribution that is becoming increasingly localized. Possible exceptions are Bhutan and most of China, where its population is thought to be stable. In southern China, a recent estimate puts the musk deer population at 100,000 head. In western and northwestern China, the population is estimated at 200,000–300,000 head.[5]

It is the uncontrolled hunt for musk that in most places is driving this animal toward extinction, but its habitat is also being increasingly destroyed by livestock and woodcutters. Part of the loss to hunters is

[5] Information from Sheng Helin and Lu Houji.

owing to the mindless way in which the animals are caught. Most are snared in traps or nets or killed by poisoned stakes set on trails. This kills all the animals indiscriminately, even females and fawns, which produce no musk. This waste of reproductive animals is extremely destructive to the populations and is senselessly hastening the musk deer's extinction.

## HABITAT AND ENVIRONMENT

Musk deer mainly occur in upland woodland and scrub areas. They prefer remote, dense vegetation, especially birch-rhododendron forests in mountainous terrain. They are seldom found in treeless regions or areas thickly populated by people. In the Himalayas, the upper limit coincides approximately with the tree line, which is as high as 4,600 m at the eastern end.[6]

## BIOLOGY

Despite its economic importance and wide distribution, little is known of the musk deer's biology. Nonetheless, it is known that musk deer have a gall bladder, a bovid feature that distinguishes it from the true deer.

The animals are browsers, relying on young leaves, buds, fruits, and flowers. During the winter, as snow deepens, they depend more on lichens growing above the snow on rocks or tree bark, although in shallow snow they scrape for vegetation with their hooves.

The male's musk sac is unique among deer. Situated between the umbilicus and penis, it contains the gelatinous, odoriferous oil. The amount varies with the season and the age of the animal, but pods of adult males usually weigh about 30 g; occasionally up to 45 g.

Males also have a caudal gland under the tail, which secretes a viscous yellow substance with a goaty smell. They mark vegetation with this secretion by rubbing their hindquarters against stems and branches.

Males become sexually mature at about 18 months, but females seem capable of reproducing in their first year. The estrous cycle is 18–25 days; the receptive period lasts 36–60 hours. Gestation varies from 178 to 192 days. Each female usually bears one or two fawns, rarely three. The fawn is precocious—able to stand and move within 15 minutes.

[6] Information from M. Green.

## BEHAVIOR

Musk deer are shy, furtive animals with keen senses of hearing, smell, and sight. They are normally solitary and are most active at dawn and dusk. Only under cover of darkness do they frequent the more open spaces. Except in the rutting season, more than two animals are seldom seen together; groups usually consist only of a female and her young. The alarm call, a loud hiss, is often accompanied by a high-stepping, springy gait.

In the wild, musk deer lead "orderly" lives. They use well-established trails connecting well-established feeding places, resting places, and "latrines" where they deposit their droppings. Migration is uncommon.

Remarkably sure footed, musk deer climb cliffs and even the trunks of leaning trees. Being small and solitary, they rely on camouflage to avoid predators but flee through established escape routes when disturbed. If cornered, males defend themselves by slashing with their tusks, often inflicting deep cuts and severe injuries.

## USES

The strong-smelling, reddish-brown musk is obtainable only from this animal.[7] It is used as a fixative in expensive perfumes to increase the retention of the fragrance on the skin. In Oriental medicines it is used in stimulants, sedatives, and other products. Some medicinal properties appear to be genuine.[8] Recently, highly purified musk has been selling for as much as $27 a gram.

Trade in the Himalayan musk deer or its products is banned by all countries that are parties to the Convention on International Trade in Endangered Species of Wild Fauna and Flora (CITES). However, products from musk deer in the Soviet Union and China can be traded legally under license. It seems likely that if formal, self-sufficient, musk-deer farming projects can be established elsewhere, with safeguards to minimize poaching, a wider trade would be officially sanctioned.

## HUSBANDRY

Since at least 1919, Chinese scientists have been experimenting with the extraction of musk without killing the males. When the sexual

[7] Substances similar to musk are also found in some small solitary bovids such as the steinbok, but are not used commercially.

[8] Information from M. Green.

activity is at its peak, the males are caught and musk removed from the pod with a runcible spoon (curved fork) inserted into the sac's aperture. The procedure takes only minutes. There are records of up to 9 g being recovered at a time.

As noted, China established formal musk deer farms in 1958. They are clustered in the Maerkang, Miyalo, and Manchuan areas of Sichuan Province; the Zhenping county of Shanxi Province; and the Fuziling area of Anhui Province. Despite heavy initial losses of animals, mainly during transportation and acclimatization, the Chinese now breed musk deer in considerable numbers. However, juvenile mortality is still high and longevity relatively short. Zoos in other parts of the world have also had difficulties maintaining the animals.[9]

In captivity, musk deer readily accept many foods: lettuce, carrots, potatoes, apples, rolled oats, hay, alfalfa, bananas, some grass, bamboo leaves, and pumpkins, for example.

## LIMITATIONS

The musk deer's social system may represent an impediment to its successful reproduction in captivity. It is irascible and scares easily. In close confinement, males fight and may have to be isolated from each other.

There is an inherent danger in any captive-breeding scheme in the Himalayas: resumption of legal trade in natural musk could damage the remaining populations by stimulating the market and providing an outlet for illicit musk from poached animals. Some biologists (notably in India) consider that a total ban on the trade in natural musk from all sources is essential.

So far, removing the musk without damaging the animals has not proved commercially successful because of market resistance. Most purchasers require the entire pod in order to be certain that they are receiving the genuine product. Given a regularized farming program, however, it seems likely that mutual trust would circumvent this lack of confidence.

## RESEARCH AND CONSERVATION NEEDS

To protect this species will not be easy. It occurs in vast, remote areas that are difficult to patrol. The local people are poor and traditionally have used it as a source of income and food. The value

[9] Information from M. Green.

of musk is so high that smuggling is already well organized and will be difficult to eradicate.

It is imperative that the status of existing musk deer populations be established. This is especially important in Nepal. The total population may be not more than 500 in the wild.

No matter what is achieved in farming, pressure on populations will only be reduced by protecting the wild specimens. Thus, the dietary requirements and behavior of musk deer should also be evaluated with a view to building up the wild populations directly. If the techniques of breeding and handling can be improved, farming the animal may also indirectly help wild populations by reducing the pressure to harvest them.

An alternative to captive breeding might be ranching the wild animals. In this process, males would be caught periodically and the musk extracted before releasing them. The organized, sustainable harvesting is particularly attractive if developed at the rural level with revenue going directly to local people; it would provide them incentive to conserve not only "their" musk deer but also its habitat. For this purpose, today's musk deer hunters could be trained to extract musk from live animals, releasing and recapturing them on a controlled basis.

Another alternative could be controlled culling at a sustainable level, as is now done in the Soviet Union, where about 5 percent of the population is harvested each year. However, elsewhere annual culls at any level would not be feasible for at least 10 years because the populations are now so low.

Pudu

# 29

# South America's Microdeer

South America contains three types of tiny, indigenous deer. None are well known to science, yet they are of microlivestock size, and if given research attention at least two might respond to rearing in captivity.

## PUDU

The pudu (pronounced "*poo*-doo") is native to temperate forests of the Andean region. It is among the smallest of all true deer, adult males being merely the size of small terriers and the females being smaller still. It is very shy and retiring and is endangered.

All things considered, this animal would appear to be an unlikely candidate for microlivestock. But wherever it is found, the pudu is mercilessly hunted, and captive rearing might be the only way to save its populations from extinction. Indeed, it is already being raised in experimental herds in Chile and Argentina.

Pudus (also called the Andean dwarf deer) once ranged widely through the foothills, valleys, and lowlands of the Andes. They prefer the dark, dank underbrush of the cool rainforest, particularly thick bamboo stands. There are two species: *Pudu pudu* is distributed in parts of southern Bolivia and throughout much of southern Chile nearly to the Straits of Magellan. It is also found on islands off the Chilean coast. *Pudu mephistophiles* is distributed throughout the highlands of Ecuador, where it occurs only in cool areas at great height.

With their short legs, stocky bodies, and compact heads, pudus do not look much like deer—more like small antelopes with foxlike faces and spiky antlers. Full grown, they are only 40 cm tall and weigh less

than 12 kg. They have thick fur ranging from reddish brown to pale gray.

Because of the pudu's small size, shy and secretive nature, and forbidding habitat, few people have ever even seen one. Nonetheless, these animals tame easily and reportedly were once kept by South American Indians. Several generations were also once bred in a Paris apartment and were treated exactly like domestic dogs, which most people who saw them for the first time thought they were.[1]

In recent years, habitat destruction has greatly reduced the range and numbers of these attractive and fascinating little creatures. The International Union for the Conservation of Nature already lists them as vulnerable to extinction. Pudu studies are highly recommended, and raising pudus promises to be an interesting and valuable activity that may one day lead to one of the most intriguing microlivestock of all.

## BROCKET

Brockets (*Mazama* spp.) are small deer that occupy the place in South America's environment that duikers occupy in Africa (see page 326). They typically reside in thick brush. They occur widely throughout South America and are found in every country except Chile and Uruguay. They also occur in Central America, the West Indies, and Mexico.

There are four species:

- Red brocket (*Mazama americana*), the most common and widespread, is found from Mexico to Argentina. It is also the largest species, with a mature weight of about 20 kg.
- Gray (brown) brocket (*Mazama gouazoubira*) is also found throughout Latin America. It is slightly smaller, weighing about 17 kg.
- Lesser brocket (*Mazama rufina*)[2] resides in small and scattered locations in Venezuela, Ecuador, Peru, Bolivia, Brazil, Argentina, and Paraguay. It weighs 10–20 kg.
- Dwarf brocket (*Mazama chunyi*) is found only in pockets of forest and brush on certain mountainsides in Venezuela, Colombia, Ecuador, and northern Peru. The smallest brocket, it weighs only 8–12 kg.

[1] I.I. Sanderson, *Living Mammals of the World,* Doubleday and Co., Inc., New York. ND.

[2] This includes the formerly recognized species *Mazama nana* and *Mazama bricenii.*

Opposite: Little is known about the shy, secretive pudu—not even how many there are in the wild. However, a handful of researchers, including Mark McNamara (shown here) have started to study them. (New York Zoological Society Photo)

Except for their size and color, all brockets look alike. The head, neck, and tail are short; the ears are wide. The lumbar region is higher than the shoulders, and this, together with an arched back, gives them a hunched appearance not unlike a duiker's. The antlers are simple spikes, never longer than a person's hand.

The different species are similar in behavior, too.[3] They generally wander around singly or in pairs. Although frequenting dense cover during the day, they emerge at night to feed in open areas. Little is known of their food preferences. But farmers know only too well how fond they are of melons, beans, peppers, and corn. Doubtless, wild forest fruits dominate their native diet.

Although extensively hunted, brockets are so adept at dodging into dense brush that relatively few get caught. However, small size makes them vulnerable to many other predators: puma, jaguar, ocelot, and eagles and other large birds of prey. Near villages the domestic dog is probably their worst enemy. (Infuriated vegetable growers commonly set their dogs on them.)

Although at first sight these retiring, nervous, and agile creatures seem unlikely to be even potential microlivestock, young brockets are sometimes caught and raised by people. It is not uncommon to find them as pets on farms and in gardens. They seem to become very tame and might therefore make useful livestock at some future time. At least one species, the gray brocket, adjusts particularly well to life in and around human settlements.

## HUEMUL

The third type of South American microdeer, the huemul,[4] is a much less likely candidate. It is very rare, very shy, and has so far shown little likelihood of settling into captivity. However, huemul conservation is critical: without urgent attention, the animal will become extinct. Although totally protected by law, it is declining owing to poaching, farm dogs, habitat loss and diseases transmitted by cattle and other livestock.

There are two species:

- The Chilean huemul (*Hippocamelus bisulcus*)[5] occurs in high-

[3] The dwarf brocket was first described in 1908. The habits of this species are virtually unknown. It is so like the pudu (see above) that it was formerly classified as *Pudu mephistophiles*.

[4] Sometimes called "guemal."

[5] Their strange generic name, *Hippocamelus*, arose because initially these animals were not recognized as deer. No one knew how to classify them, and they have been given more than 20 different names. At least one taxonomist considered them to be "horse-camels."

altitude forests, thickets, and grasslands in the Andes of southern Chile and Argentina.

• The Peruvian huemul, or taruca (*H. antisensis*), occurs in parts of the Andes of southern Peru and Bolivia as well as of northern Chile and Argentina.

Both species live in small herds above the tree line. They are very shy, and even though the Chilean national seal bears the depiction of a huemul, almost no Chilean (or anyone else, for that matter) has ever seen a live one.

At less than 85 cm tall and probably weighing under 15 kg, huemuls are sized to be microlivestock. However, previous attempts at rearing them in Chile have met with little success. Nevertheless, huemuls have been kept in zoos in Germany, and such experiences—together with the increasing knowledge of how to raise red deer and other species—may eventually provide the keys to their continued existence.

# 30

# Water Deer

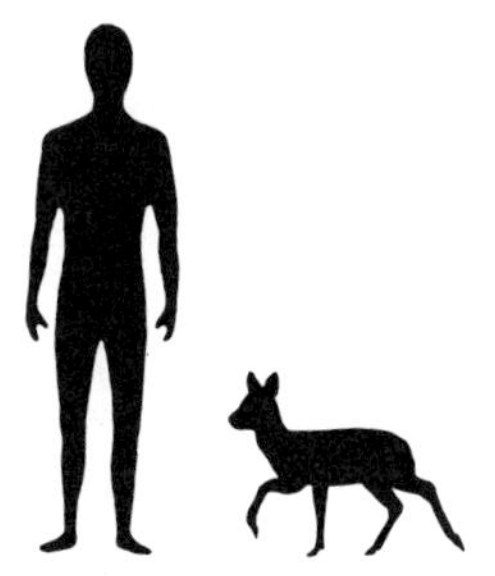

Little is known about the water deer[1] (*Hydropotes inermis*), but it should be considered along with mouse deer, muntjacs, musk deer, and others as a possible species for microlivestock development. It is comparable in size, and it is unusual among deer for producing large litters; births of triplets or more are common.

As with the other species in this section, this is a highly speculative notion; however, there is some justification for it. The Zoological Society of London has successfully established breeding colonies, and other zoos have also bred the animal in captivity. The water deer has the advantage of rapid growth, early maturity, and high fecundity. Indeed, given protection, its populations have been known to increase rapidly.

## APPEARANCE AND SIZE

The animal has a graceful and delicate appearance, its best known characteristic being the male's long tusks. Both sexes are about the same size, standing 45–55 cm at the shoulder and weighing up to 19 kg. Body length is up to about 1 m, and the tail is so tiny that it is barely noticeable. The round-tipped ears are characteristically held erect above the head.

Water deer are somewhat like muntjacs, but they are longer in the leg and lighter in build. Their forelegs are shorter than their hindlegs so that they stand slightly higher at the haunches than at the shoulders. This gives them a hunched appearance.

The hair is generally thick and coarse, longest on the flanks and rump. The backs and sides are usually yellowish brown, finely stippled with black. In winter, the coat is thick and variable in color; pale fawn

[1] They are also known as Chinese water deer, Chinese river deer, or the Yangtze River deer.

and peppery gray-brown are common shades. In summer, the coat is sleek and reddish. Fawns are white spotted at birth, but this dappling soon fades.

Neither sex has antlers. The upper canine teeth, especially in the males, are enlarged, forming curved tusks 7 cm long. These are much bigger than those in muntjacs and can protrude well below the jawline.

In both sexes there is a small inguinal gland present between the hind legs, the only instance among deer.

## DISTRIBUTION

In China the water deer has a wide distribution range. It is mainly found in the provinces of Jiangsu, Zhejiang, Anhui, Jiangxi, Hunan, Hubei, and Fujian, the mid and lower Yangtze River Basin, and coastal areas and islands in central and eastern China. It is also found in Guilin, southern Sichuan Province, Guangxi Province, and Guangdong Province in the south.[2]

In Korea, the animal lives along the lower reaches of most rivers, except those in the extreme northeast. Its northern limit of distribution is probably about latitude 43°N.

At Woburn Park in England, a few escaped from a herd early this century and have increased and become established in a number of counties, particularly Bedfordshire, Hertfordshire, Cambridgeshire, and Norfolk.

## STATUS

In China, owing to increasing reclamation and cultivation of wetlands, the habitat of the water deer is gradually shrinking. At present the animal is protected by the government, which designates appropriate hunting seasons. It is estimated that about 10,000 are killed each year by hunters. In Korea the water deer is still plentiful. As noted, in England it is maintaining itself and is thriving in some protected parks. It is also reported to be present in France.

## HABITAT AND ENVIRONMENT

In its home range, the water deer is usually found among reeds and rushes in swampy areas. It also frequents the tall grasses and sparse shrubs of mountainsides and cultivated fields. In England it has adapted to a variety of habitats, including woodlands.

[2] Information from Lu Houji.

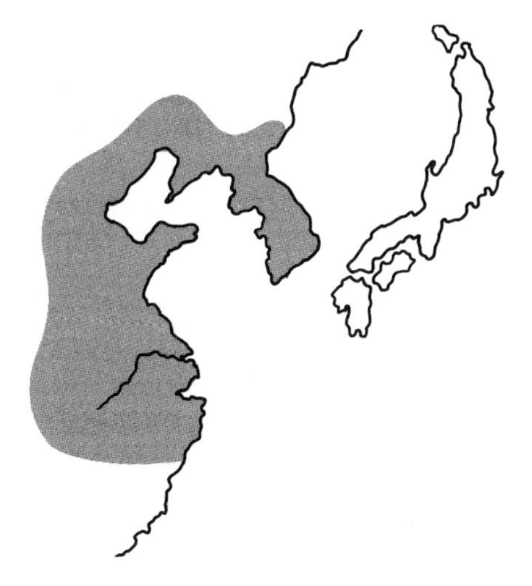

Native range of the water deer.

## BIOLOGY

The water deer is chiefly a grazer. It feeds mainly on reeds, coarse grasses, and some tree leaves.

As previously mentioned, this is the most prolific species of deer. Up to seven fetuses have been observed in a single pregnant female, although the normal litters are two or three. The gestation period is about 176 days. Fawns are born in late spring or early summer, and weigh only about 0.5 kg at birth. Within 4 days the newborns can live almost exclusively on grasses. Usually, however, they are fully weaned after 4–8 weeks, but remain socially attached to the mother. They appear to become independent after about 4 months. Males become sexually mature at about 5–6 months; females at about 7–8 months.[3]

## BEHAVIOR

Water deer are generally seen alone. Even where abundant, they seldom congregate in herds. Females are sometimes intolerant of each other, as are adult males or young males. However, in captivity, several females can graze and rest in loose aggregations. The peak period of grazing activity occurs around dusk. Feeding sessions are interspersed with periods of passive cud-chewing.

Water deer "bark" at intruders. During the rut, males are especially noisy and aggressive, and defend their territories with vigor. Fighting involves striking their tusks into the shoulder or back of their opponents.

[3] Information from Lu Houji.

Mature buck, showing his long tusks and winter coat. (R. Harris and K.R. Duff)

These are extremely excitable little animals. When upset, they often "hump" their backs and bound away like rabbits. They are also good swimmers.

## USES

In China the water deer are hunted for their meat and skins, and newly born fawns are killed to obtain the mother's colostrum for medicinal purposes. In a few localities in England, the species has become a game animal.

## HUSBANDRY

The water deer is not yet known on "farms." However, it seems to be easily kept, has bred well in zoos, and has thrived in many British wildlife parks.

## ADVANTAGES

The assets of early maturity and high fecundity mean that the potential exists for rapid population expansion. Such an event occurred at Britain's Whipsnade Zoo. In 1929 and 1930, 32 deer were released into undeveloped pasture; by 1937, 120 fawns had been raised.

Because they are relatively small, and because in the wild state they aggregate only under exceptional conditions, water deer are unlikely to have any appreciable impact on vegetation in forests, farms, or gardens. Nonetheless, they can damage crops, and Chinese farmers, who consider them pests, often illegally kill them out of season.

## LIMITATIONS

These animals are swift and adept at escaping captivity. It is possible that because of territoriality only a single pair will live in a given area. Moreover, males are aggressive and must be kept apart.

They seem able to withstand chilly weather well, but a combination of wet and cold is harmful. Their heavy winter coat, essential to survival in the Far East, renders the animal susceptible to dehydration and heat exhaustion in comparatively mild climates, such as England's.

At birth, the tiny fawns are extremely vulnerable to a variety of predators, both birds and mammals. The species may need areas of dense cover or some shelter from wind.

## RESEARCH AND CONSERVATION NEEDS

The water deer has been successfully kept in semicaptivity for many years; however, for it to reach a level of domestication suitable for use as microlivestock, research is needed in the following areas:

- Reproduction;
- Performance under a range of environments;
- Grazing efficiency;
- Basic physiology;
- Captive breeding and domestication—measurements of growth rates, space requirements, and feed needs; and
- Modifying behavior to overcome territoriality—for instance, imprinting on humans, selection of docile specimens, hand rearing, and castration.

The water deer is not an endangered species; however, efforts should be made to preserve the populations in their native ranges and habitats.

Red Duiker

# 31

# Duikers

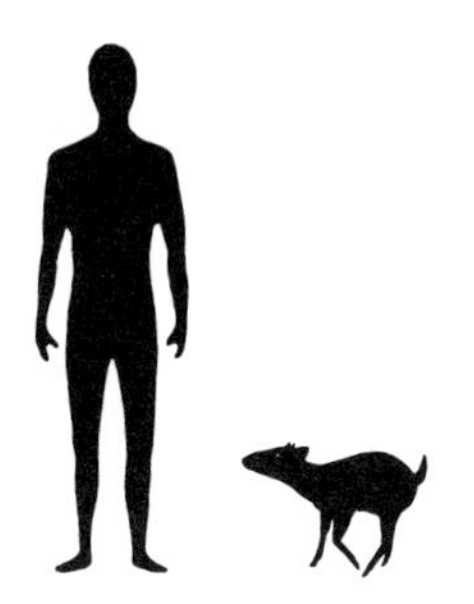

Duikers (*Cephalophus* and *Sylvicapra* species) are small African antelopes. Although they are ruminants like cattle, sheep, and goats, some are hardly bigger than hares or rabbits. One, the blue duiker, is less than 40 cm high and usually weighs a mere 4–6 kg. It and some of the slightly larger duikers might be suitable for household husbandry because their meat is an extremely popular food throughout much of Africa. In West Africa, for example, it is one of the most common meats sold in both rural and urban markets.

This idea, however, is highly speculative because, despite their popularity, little is known about these tiny animals. Their husbandry has been attempted only a few times, but the results were encouraging, and rearing duikers deserves further investigation. They are already being raised in captivity in the United States,[1] Zimbabwe,[2] Togo,[3] and Gabon. Researchers in Nigeria have bred blue duikers to the fourth generation and found that, if the animals were first handled by people while young, they remained docile.[4] Even blue duikers caught in the wild tame quickly if they are very young, but by the time they reach 3.5 months, they become barely tolerant of man's presence.

If duiker husbandry can be developed, it might provide not only a more regular source of meat, but also a lessening of the hunting pressures, thereby giving the wild populations a better chance of survival.

## AREA OF POTENTIAL USE

These antelopes are suitable for testing as microlivestock only in their native region, sub-Saharan Africa. Eventually, they might prove to have wider applicability.

[1] Information from R. Cowan.
[2] Chipangali Wildlife Trust, Bulawayo.
[3] Information from B. Chardonnet.
[4] Whittle and Whittle, 1977.

## APPEARANCE AND SIZE

Duiker species vary from about the size of a small dog to that of a small donkey. Most are similar in shape and are characterized by short front legs, arched back, and pointed hoofs. The tail is stubby, often with a terminal tuft. The coat varies from reddish brown to nearly black, although a few species are blue-gray and one is zebra striped.

Females are slightly larger than males, but the sexes look alike. In most species, both sexes bear small straight horns that project backward from the skull, frequently hidden in a long tuft of hair.

## DISTRIBUTION

Duikers inhabit virtually all regions of Africa below the Sahara—from Gambia in the west to Ethiopia in the east, and all countries as far south as South Africa.

## STATUS

Duikers are so shy that they are rarely encountered by people. But almost anywhere in Africa (other than North Africa), the observant traveler may glimpse them ducking into forests or thickets. Although there are still countless numbers, people are eating so many that in some localized areas the populations are fast heading toward extinction.

## HABITAT AND ENVIRONMENT

All but one species are found in rainforests or dense woodlands. The gray duiker, however, is found in savannas. If the vegetation is juicy, only a few of the species need a separate water source, so they can thrive in very dry sites.

## BIOLOGY

The main foods are fruits and seeds supplemented by leaves and shoots. Fruits, which they eat to a much greater degree than other antelopes, are an important part of their diet. Some rare species (for example, the red-flanked duiker, *C. rufilatus*) can graze. Occasionally (especially in captivity), duikers are also omnivorous, eating fish, crabs, insects, snails, frogs, small animals, or carrion; they also readily accept chopped meat.

Duikers reach sexual maturity at 9–15 months; gestation lasts about 7–8 months. In some species, females conceive a few days after calving on a 3- to 5-day postpartum estrus. Apparently one calf per birth is normal. A newborn blue duiker weighs between 0.4 and 0.7 kg.

Before one year of age, young duikers leave their parents to find their own mates and territories. Life expectancy is more than 10 years.

## BEHAVIOR

In spite of habitat differences, most duikers behave alike. In the wild they are nervous, shy, and retiring. When alarmed, they plunge into the protection of dense vegetation—hence the origin of the name duiker, which means "diver." Nonetheless, their behavior allows them to be easily netted. An experienced hunter can imitate duiker sounds and call the male out of the bush. Also, a startled animal freezes, thereby facilitating its capture.

Moving easily through dense vegetation, the head carried low, these tiny animals use regular runs. Forest duiker species are largely diurnal, although a few, such as the bay duiker, are nocturnal. Bush duikers are mainly nocturnal, feeding from early evening until morning. Such nocturnal species shelter during the day in holes (presumably dug by other animals) or inside fallen trees; the diurnal ones lie directly on the ground.

Blue duikers are the best-known species and are probably the most likely candidates for microlivestock (see page 332). They seem to be monogamous and apparently mate for life. Unlike most antelopes, their population densities can be high. The pairs reside in territories of 2–4 hectares, which both male and female stoutly defend against rivals. Other species appear to be polygamous and live in large territories (up to 80 hectares).

In captivity, the animals are generally calm. However, both males and females can be aggressive toward unfamiliar individuals of their own species. In an enclosure, one male can serve several females.

Large glands, located beneath each eye, exude a scent that is rubbed onto fences, trees, and other objects as territorial marking. In another form of marking the horns are rubbed against tree trunks.

## USES

As noted, duiker meat is much sought in many African countries, and the animals are regularly hunted. The meat is lean with little or no intramuscular fat (marbling).

Male blue duiker. (A.R. Devez, CNRS, Mission Biologique au Gabon)

Duikers also have promise as experimental animals. They are true ruminants, with four-part stomachs, and they produce cud. Some are only rabbit size, they need far less room or feed than sheep, and thus are potentially an efficient test animal for determining the nutritional value of forages. Blue duikers, for instance, have a digestion efficiency comparable to that of sheep, but, because of their small size, a test needs only four rabbit cages and 5–10 kg of feed. Sheep, by comparison, require much more spacious facilities and 150 kg of feed.

In Nigeria, blue duiker pelts are used in making karosses, a traditional dress. A single garment may contain up to 60 pelts.

## HUSBANDRY

Almost nothing is known about rearing duikers, but they seem to tame easily and perhaps may be kept in backyards like goats. Indeed, they reportedly make good house pets when hand raised. They are attractive, and from the day of capture young ones can be handled and petted.[5]

The Nigerian researchers who bred blue duikers to the fourth generation bottle-fed young specimens five times a day. Older animals were given feeds that included banana, plantain, and papaya; leaves of hibiscus, cassava, and banana; and dried corn. Variety seemed to be important, and the researchers could not predict the quantity of particular foods the animals would choose on any given day. In addition to varied vegetables, a small dish of salt or a salt lick was sometimes required.[6]

Duikers are unlikely to run away, except when startled. However, providing an enclosure is worthwhile. It enables them to establish a territory by marking poles, bushes, and fences. Although needing space in which to run, as little as 10 $m^2$ is reportedly sufficient for 2–4 animals.

Satisfactory shelters include an open-ended oil drum laid on its side, a lean-to made of palm frond, or a small hut made of local matting. Apart from providing shade and protection, shelters should be built so that excited animals can run through them. When cornered, duikers tend to either flee for shelter or jump upwards; a run-through shelter can prevent a frightened one from accidentally leaping over the fence.

Based on their own experiences, researchers at Pennsylvania State University in the United States report that blue duikers raised in captivity are easy to maintain, reproduce well, and are not fussy about

[5] Information from R. Cowan.

[6] Whittle and Whittle, 1977. The authors used the subspecies known as Maxwell's duiker.

## THE BLUE DUIKER

In one sense, the blue duiker is the most important animal in Africa. It is the only one found throughout the continent south of the Sahara. It occurs at a greater range of altitudes than most—as low as sea level in many places to almost 5,000 m elevation in Kenya. It occurs in habitats from dense rainforests to dryland savannas. And, in sub-Saharan Africa as a whole, the blue duiker is eaten more than any other animal (although in West Africa it is generally called Maxwell's duiker).

This very small antelope, caught by snare or net, can be found in the meat markets of villages, towns, and cities in all countries from Senegal to Madagascar. It is a source of food for tens of millions. Bushmen, Pygmies, Dinkas, and Mandingos, thousands of miles apart, all share the same fondness for duiker meat and for duiker-skin clothes.

Nevertheless, scientifically speaking, this is one of the world's least-known animals. And its numbers are diminishing rapidly. Areas that used to have plenty now have few or none. Overhunting and destruction of the rainforests are jointly contributing to their decline.

Despite the losses, people are snaring as many as they can, and there is no sense of concern—not even among most conservationists. However, in many locations there is already evidence that the animals won't be around much longer. Unless something is done—and soon—people will lose their major source of animal protein. If that happens, it is likely that they will move on to larger animals, such as gorillas, which would be an even worse disaster.

The best long-term solution is to organize duiker husbandry. Learning to rear duikers would benefit people throughout Africa. The blue duiker is the most suitable species; it is the most common and the most important. Also, it inhabits the edges of the forest and could therefore become a suitable species for ranching without denuding the forest.

Blue duikers are easy to maintain in captivity. They tame readily and like to shelter and sleep in boxes or cages. They are good converters of vegetation and produce top-quality lean meat. In addition, they are neither affected by tsetse flies nor are very susceptible to diseases.

The key now is to learn how to keep these very timid creatures under different conditions. We need to know their foods (especially foods that might be harvested from forests) and reproductive biology. We need to know the right numbers to house together. Most of all, we need projects aimed at rearing and breeding them in captivity under village conditions.

Vivian J. Wilson

environmental conditions. In fact, they say, blue duikers seem to enjoy living in cages.[7]

In order to raise duikers successfully, post-pubertal males must be separated. A female should be bred with the same male throughout her productive life span.

## ADVANTAGES

Many African countries already have a ready market for duiker meat. It is somewhat similar to goat meat, but most people agree that it is superior.

The animal can live on fibrous vegetation. Unlike conventional ruminant livestock, it is suitable for feeding an average family at one meal.

The ability to forage in undergrowth where other domestic livestock do not thrive makes duikers potential livestock for tropical forest and bushland regions. They can be raised for meat without cutting the trees or bushes to create pastures.

## LIMITATIONS

Duikers are easy prey for predators: eagles, pythons, wildcats, and people, among others. Thus, they probably require more sophisticated management than common livestock such as goats. However, the quality of their meat could more than compensate for the extra effort.

Some species are territorial, which means that they may do poorly in captivity, unless their social organization can be altered.

Under good conditions, the ideal slaughtering age reportedly occurs at 8–10 months, when the blue duiker can weigh 4 kg. Compared to rabbits and guinea pigs, production is relatively slow because of long gestation and lack of multiple births.

Duikers are resistant to trypanosomiasis.

One general problem is that duikers have short, sharp horns designed specifically for jabbing. This could be a potential danger, especially since the males of some species become aggressive when their females are receptive. However, the horns can easily be clipped and taped to limit the danger.

[7] Information from R. Cowan.

## RESEARCH AND CONSERVATION NEEDS

Because of the duikers' secluded lifestyle, much has still to be learned about their habits. Specific information on behavior and breeding is needed.

Animal scientists in Africa should gather small herds for comparative studies. This will provide insights into whether duiker temperaments facilitate or hinder their utilization. In addition, assessments of diet, growth rates, behavior in captivity, reproductive rate, adaptability, and future potential can be made. Management considerations include clipping horns, trimming hooves, and controlling lice and fleas.

Research of particular value would be chemical analysis of duiker milk and of other characteristic glandular secretions. The latter lend themselves especially to a study of animal communication.

Farming duikers might help rescue the wild populations by relieving hunting pressures. Programs in this area are recommended for locations where overhunting is occurring.

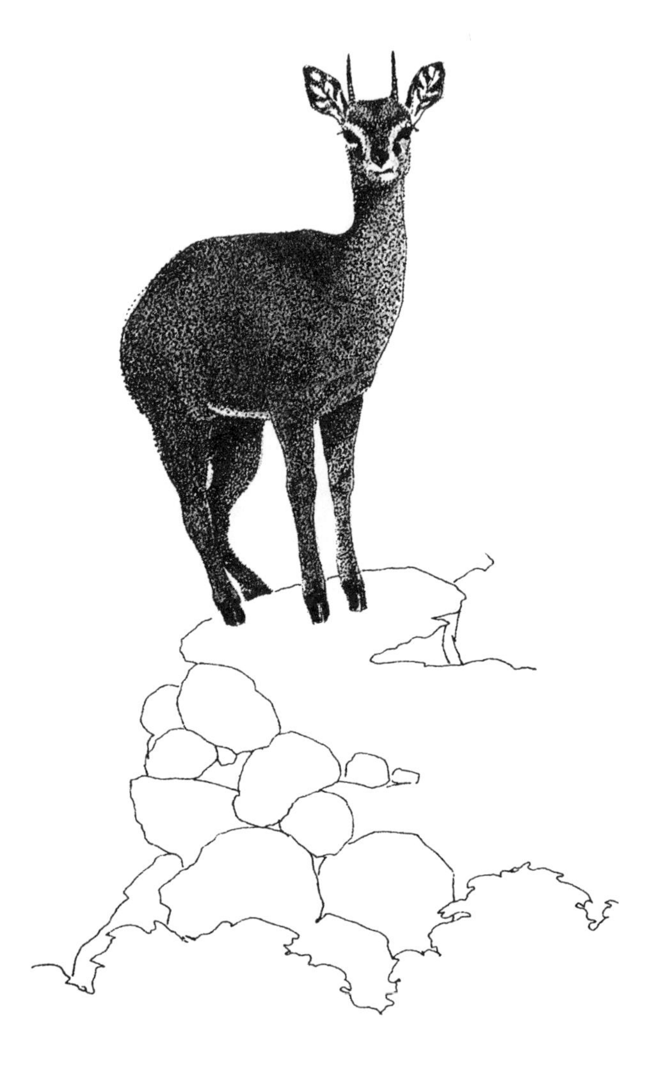

Klipspringer

# 32

# Other Small Antelope

The least known and most obscure of all antelope are the delicate African creatures called dikdik, suni, royal antelope, pygmy antelope, and klipspringer. The smallest is a West African form of the royal antelope that stands a mere 25 cm high and weighs less than 2 kg. The four-horned antelope of South Asia is a similarly tough, tractable animal that is also the size of a small dog.

None of these has previously been seriously considered for use as livestock, but they are possibly amenable to rearing in captivity and they provide some of the finest game meat in the world. Given New Zealand's experience with various deer species (see page 288) and Africa's experience with large antelopes, basic research to test out the possibility of organized dwarf-antelope production could prove to be rewarding.

## AREA OF POTENTIAL USE

The areas where these antelopes might be used are sub-Saharan Africa for the African species and South Asia for the four-horned antelope.

## APPEARANCE AND SIZE

Of these animals, some, such as the pygmy antelopes, have a crouched appearance with an arched back and short neck; others have a more upright posture with a long neck and a raised head. In all species, the males are smaller than the females and bear tiny spikelike horns.

## DISTRIBUTION

Collectively, these antelopes have native ranges covering huge areas of Africa and part of Asia.

**Dikdik:** Six species occur in two discontinuous distributions—one from Somalia and Ethiopia southward through Kenya and Tanzania, the other in Namibia and Angola.

**Sunis:** Eastern Africa, from Kenya to South Africa.

**Royal antelope:** West African forests.

**Pygmy antelope:** Central Africa from southeastern Nigeria to Zaire.

**Klipspringer:** Ethiopia to the Cape of Good Hope.

**Four-horned antelope:** India.

## STATUS

Many forms are protected by local laws, but none of the species is on the international endangered-species list.

## HABITAT AND ENVIRONMENT

The animals occupy habitats from dense, moist forests to dry, rocky outcrops and even to deserts. Their ranges have almost certainly been affected by humans—sometimes for the better, since many prefer the secondary growth that invades disturbed areas, notably after overgrazing or slash-and-burn agriculture.

Although dikdik and klipspringer usually frequent dry areas with scattered brush, the other dwarf antelopes normally stay in dense vegetation. All seem to live in definite areas and do not migrate. For instance:

- Dikdik live near streambanks.
- Sunis inhabit dry country with thick bush, but they can also be found in reed scrub along rivers and in forests up to 3,000 m elevation.
- Klipspringers live on stony mountain slopes, rocky outcrops, or the sides of steep gorges at altitudes from sea level to 4,000 m.
- Four-horned antelopes live in undulating or hill country and shelter in tall grass and open jungle, a terrain more common to deer than to antelope.

## BIOLOGY

Most of these antelope browse on shrubs. They are "concentrate selectors," taking easily digested vegetation such as buds, fruit, and succulent young leaves.

Also, most obtain much of their water requirement from dew and the vegetation they consume. Klipspringers, for example, are able to live for months without drinking. Sunis and four-horned antelope, on the other hand, drink regularly and seldom live far from water sources.

Little is known of the reproduction and general performance of these animals. Females become sexually mature at about 6 months in the smaller species and 10 months in the larger. Males become sexually mature at about 14 months. The young are born throughout the year, but births peak with the vegetation flush following the first rains. Where there are two rainy seasons a year, two birth peaks occur. Usually a single calf is born. Dikdik is the only one whose reproduction has been studied in detail. Its pregnancy lasts 172 days, one young is born at a time, and the birth weight is 600–800 g.

Their longevity is unknown but is probably in the range of 10–12 years.[1]

## BEHAVIOR

These tiny creatures have some of the habits of deer. They are shy and elusive and generally rely on concealment to escape detection. Their first response to a predator is to freeze, and then to flee like hares—dashing off in a series of erratic, zigzag leaps.

They live alone, in pairs, or in small family groups, but sometimes congregate in larger groups in thorn thickets. The species that have been studied most (dikdik and klipspringer) are strongly pair bonded. (A male, a female, and one or two young is typical, and a klipspringer rarely moves more than 5 m from its mate.) However, the royal and pygmy antelopes and the suni are more solitary in their behavior. Four-horned antelope are usually seen alone or in pairs.

These animals feed mostly in the early morning and late afternoon. Some species deposit dung and urine on particular sites. And they repeatedly daub secretions from glands in front of the eyes onto plant stems, where a sticky mass accumulates. Glands near the hooves mark the ground along frequently traveled pathways. Males also mark females with the scent, thus reinforcing the bond.

[1] Information from R. Dunbar.

The four-horned antelope has a whistling call, which helps keep the family group together. Males repeat it frequently in hot weather. Gestation is 8–8.5 months. If taken young, they reportedly tame easily.

## HUSBANDRY

Much more research needs to be done before attempts are made to convince anyone to domesticate these antelope. There are likely to be considerable difficulties. Guinea pigs, rabbits, and giant rats can successfully be kept in cages or small enclosures, but most antelope probably cannot. Larger enclosures will be needed.

The food habits and general behavior of these small animals must be studied closely. They are strictly monogamous, and it may be necessary to keep them in pairs. Reproduction, growth, and general performance must become understood under different environmental and nutritional conditions. Mixing species is another aspect to be examined—whether these antelope can be kept with other species in the same enclosure (typical of livestock farming in most poor nations) is not known.

To settle such questions, representative species of microantelope should be gathered for comparative assessment. Researchers should focus on the animals' social structure, on husbandry methods for maintaining them over generations, and on how best to breed them. If the findings are promising, then a campaign to domesticate these antelope could be mounted.

## USES

Throughout most of their ranges these animals are highly sought "bushmeats." The meat is lean and of extremely high quality. In Zaire's Ituri Forest, for example, pygmies net and kill large numbers of pygmy antelope, hanging the carcasses on sticks by the roadside for sale.

Because of their small size, these species might make good laboratory animals for ruminant studies. The dikdik, for instance, becomes a fully functioning ruminant at a body weight of about only 1.5 kg.[2] However, their digestive physiology is quite different from that of cattle, sheep, and goats, which makes them atypical ruminants.

Given organized production, it is likely that dwarf-antelope pelts could become commercially important. North Africa exports the hides

[2] P.P. Hoppe, 1984.

of medium-size antelopes to Europe for use in fine sueded leathers. Hides of the small species would almost certainly be in demand as well if a steady supply could be obtained as a by-product of meat production.

## ADVANTAGES

Microantelope provide some of the finest game meat. They are small and perhaps tractable. Most are already widely eaten and are being eliminated over broad areas of their range. Turning them into a sustainable, economic food source could provide motivation for their conservation.

These antelope can digest, and are adapted to, the indigenous forage over vast areas of Africa and South Asia. They are native to tropical habitats, where cattle and other livestock often grow poorly. They also appear to be resistant to trypanosomiasis.

## LIMITATIONS

Some African peoples (for example, the Kalahari Bushmen) have superstitions or social injunctions that prohibit the eating of some of these species.

Small antelope are probably not as efficient as larger ruminants in digesting fiber: the retention time in the rumen may be too short. On the other hand, quickly digestible cell walls of lush green plants can be used efficiently.[3]

The territorial behavior of most of these species may limit their rearing in large numbers under captive conditions.

## RESEARCH AND CONSERVATION NEEDS

As noted, preliminary research on dwarf antelope husbandry is required. Specifically, studies should be conducted to assess:

- Growth rates, feed efficiency, and reproductive rate;
- Carcass quality;
- Economics and the likely cost of production per animal unit; and
- Studies of digestion.[4]

[3] Information from R.E. Hungate.

[4] Hoppe, 1984. There is evidence that dikdiks may, by closing off the entrance to the rumen, swallow water-soluble nutrients directly into the later stomachs. This bypassing of the rumen fermentation may be an important nutritional advantage.

*The few statistics available today on the use of wildlife as food are probably much below actual consumption. Most food consumption surveys record food obtained by hunting or trapping under the indiscriminate heading of ''Bushmeat'' and neglect to include the many small animals that are normally collected by children. In Africa, an amazing variety of wildlife is eaten, including all wild ungulates, primates, all cats, and many species of birds and reptiles.*

Food and Agriculture Organization
*Ceres* magazine

*All the world's people* must *begin to overcome in themselves—and even more so in their children—senseless taboos about what is edible and what is not. Only then can we stop today's universal animal-protein wastage. How ironic it would be, in this scientific age, for mankind to starve largely because of a bunch of old wives' tales, irrational beliefs, silly associations, and the lack of a sufficient spirit of culinary and gustatory adventure.*

Calvin W. Schwabe
*Unmentionable Cuisine*

*Iguana is really good, a thousand times better than chicken.*

Omero Asinto, waiter
Pochote Bar and Restaurant
Barranca, Costa Rica

# Part VI

# Lizards

Large lizards have been important foods since prehistoric times and are still commonly hunted in parts of Asia, Africa, and Latin America.[1] Some (such as the monitor lizards seen in markets in Indochina) are carnivorous species that may be difficult to feed and raise economically. However, the iguanas of the Americas offer promise as microlivestock. They are herbivorous and feed primarily on leaves, flowers, and fruits, including many that are too high in the trees to be gathered by man or by other livestock.

Iguana meat is popular throughout much of Latin America, where consumers willingly pay more for it than for fish, poultry, pork, or beef. To fill the demand, several iguana species are hunted by rifle, slingshot, trap, and noose; they are even run down by trained dogs. Villagers (often small children) catch them for food for the family; professional hunters snare and sell them to vendors. Iguanas are hauled around in gunny sacks and wicker baskets by car, boat, horseback, and people on foot. In parts of El Salvador they have been known to arrive at the market by the truckload.

As a result of this indiscriminate hunting, iguanas are now steadily becoming more scarce, and the destruction of their habitat makes the situation even worse. At present, many parts of Latin America's tropical forests are being cleared. The green iguana in particular depends on trees, and as forests disappear its populations are destroyed.

However, iguanas are forest-edge species: they will grow well on ranches and farms as long as patches of trees are left standing. This

[1] A Spanish conquistador describing Yucatan in the mid-1500s reported that iguanas are "a most remarkable and wholesome food. There are so many of them that they are a great assistance to everyone during Lent. The Indians fish for them with snares which they fasten in trees and in their holes."

offers the hope that they can be raised as microlivestock. They reproduce so prolifically that, in principle, populations can build up exponentially. Mature females, for example, may produce 30 or more eggs a year for up to 10 years.

In the past few years, a notable research program in Panama and Costa Rica (see page 351) has laid the practical foundation for iguana farming. It has artificially incubated and raised thousands of green iguana hatchlings. With less than half a square meter of living space per animal, the hatchlings have grown as fast or faster than their wild counterparts. Unlike their kin in the wild—which in their first year of life suffer 95 percent mortality from birds, snakes, and other predators—the captive-raised iguanas show almost 100 percent survival. This research project demonstrates that ranching iguanas both for food or for repopulating depleted habitats is feasible. In addition, experimental projects for farming green iguanas have begun in Curaçao and El Salvador.

As livestock, lizards have advantages. Being cold-blooded they do not carry human diseases (except for those such as salmonella that result from gross mismanagement). They can be kept in captivity in fairly high densities without diseases breaking out. Although often aggressive in the wild, they coexist in dense populations with few problems as long as they are well fed.

There is an important conservation component as well. Farming these lizards may help check and may even reverse the downward trends of their populations by allowing large numbers to be released back to the wild at a size that inhibits predators.

There are three alternatives for utilizing iguanas:

- To manage wild stocks as game animals;
- To raise iguanas on farms, like chickens and pigs; or
- To raise young iguanas in captivity and then release them into the wild where they can grow to full size and later be harvested on a sustainable basis.

Although the green iguana has so far received the most attention, two other Latin American species are in farming trials. Black iguanas (garrobos) are being raised experimentally in El Salvador and Costa Rica. And the omnivorous tegu lizard, which produces a valuable leather, is beginning to be farmed in Argentina and is briefly described opposite. Both iguana species are described in the following chapters.

## TEGU

The tegus (*Tupinambis rufescens* and *T. teguixin*) are large lizards of South America. They are highly prized for their skins, which are made into leather for handbags and similar items. Tegus are heavily exploited; on average, Argentina exports over one million skins a year—about $15 million worth. The 50-year-old tegu industry is estimated to support as many as 30,000 people, including tannery employees and people in rural areas who hunt the lizards full- or part-time. Poor agricultural conditions make farming difficult in the Chaco, and sale of a single tegu skin is worth more than a day's wages for a farmhand. Some families also eat the tegu meat and use the fat for medicinal purposes. In many areas, the populations have already been driven to the verge of extinction. Husbandry, therefore, could be beneficial. Traders are hoping that captive operations run by families can eventually replace hunting.

Although usually associated with the arid regions of northern Argentina and Paraguay, the tegu's range actually covers much of South America—as far north as Colombia, including Trinidad and the Amazon basin. One species (*T. rufescens*) normally occurs in a dryland habitat, such as Argentina's Chaco region. It can occur in great numbers in pastures, probably because of the insects associated with cattle. But it also occurs in areas that are unsuitable for cattle. Some places where tegus are common are so dry that they can carry only a single cow in 10 hectares. There, farmers might find it profitable to raise tegus. This would maintain the native biological diversity while perhaps reducing the soil degradation that cattle cause.

The other species (*T. teguixin*) occurs in wet forests, such as those found in Argentina's Formosa Province.

Tegus are scientifically interesting. Little is known about their biology, and basic studies are needed. They are not much eaten, but in some areas the tail is considered a delicacy. Indeed, deteriorating economic conditions are already making them more important as a food resource. Convincing campesinos that they can increase their cash income and their meat production by rationally exploiting these large lizards should not be difficult.

Sustainable exploitation could also benefit tegu conservation. Large populations still exist in some areas in Argentina and Paraguay, but, overall, the species are declining. A well-designed management project could ensure the maintenance and reestablishment of large populations where numbers have drastically decreased.

# 33

# Green Iguana

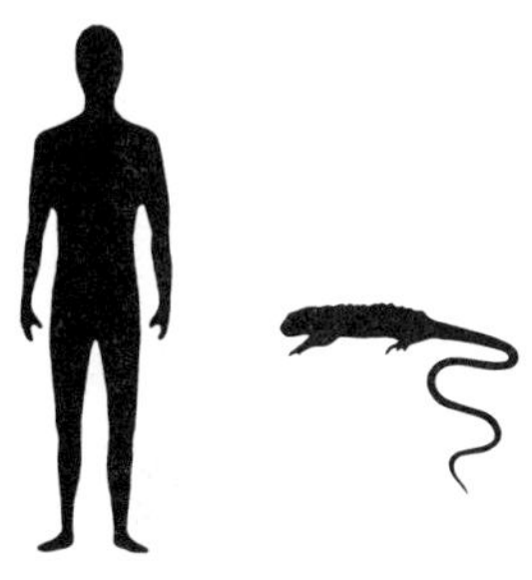

As noted, the green iguana (*Iguana iguana*) is consumed throughout Central America. It is already subject to heavy commercialization and its populations are plummeting. This is not likely to change: as human population increases, so does demand for iguana meat and eggs as well as for live iguanas, thousands of which are kept as house pets.

Because of its herbivorous diet, inoffensive nature, and high reproductive potential, this reptile is a prime species for intensive management. Alert, curious, and social, it is easily tamed from birth. Given minimal protection during the reproductive season, large populations can build up, and animals can be maintained in simple facilities.

However, although iguanas can be maintained as microlivestock, they may require three years to reach market size. Raising them to maturity entirely under captive conditions is thus probably uneconomic at present, when specimens can still be inexpensively (albeit often illegally) harvested from the wild. However, it has been found that if released into the wild, most will remain in nearby trees, especially if people erect simple feeding stations and keep them stocked with table scraps or weedy vegetation. This makes for very low-cost production during the years of waiting.

Given such findings, iguana farming has a promising future. With research, growth rates can undoubtedly be increased and maintenance costs reduced. Also, iguanas are becoming more valuable each year. Even now, farming them could be economic in some areas where wild iguanas are scarce, especially if the young are released into forested sites when they have grown to a size where mortality is low.

Although green iguanas will happily graze pastures, they need trees. Iguana farming, therefore, provides a way to keep tropical forests standing while still providing people with meat and income. By growing lizards, farmers don't have to fell the forests to create space for growing food crops or cattle. Iguana farming thus gives an economic incentive to preserve the beleaguered tracts of remaining trees. It may even render reforestation economically attractive. And it may promote the use of shelterbelts and contour strips of trees in cattle-ranching areas. Today's researchers estimate that 200–300 kg of meat can be produced per year per hectare from iguanas.

## APPEARANCE AND SIZE

The green iguana can grow to more than 2 m long, but more than half of that length is in its whiplike tail. Adults of breeding size generally weigh 2–4 kg. The scaly skin is green, yellowish, or golden brown with dark markings; color is a function of age and reproductive stage.

## DISTRIBUTION

This large lizard is indigenous to a vast region stretching from Mexico to northern Brazil and Peru, including a number of Caribbean islands. In Mexico, it occurs in tropical forests along both coasts.

## STATUS

Green iguanas were formerly abundant throughout Central America, but no longer. In the mangrove forests along Mexico's Pacific coast, for instance, only 5 percent of the former population remains. In Guatemala's Pacific lowlands, mere remnants are left. In El Salvador's jungles, the animals have declined to 1 percent of their former density. And in both Panama and Costa Rica, the species is now officially classified as endangered.

## HABITAT AND ENVIRONMENT

The green iguana usually lives near water in tropical lowland forests. It thrives only as long as some trees remain. The forests may be either humid or seasonally dry. The animal normally inhabits the treetops, feeding on tender shoots and fruits in the canopy; few other herbivores can convert such forest foliage into food for humans.

## BIOLOGY

Like cattle and other herbivorous mammals, iguanas have a specialized digestive system with an enlarged fermentation chamber in

Native habitat of the green iguana.

which bacteria break down plant cells. Vegetation is converted into meat about as effectively as in cattle, but not as quickly.

After reaching sexual maturity, at 2 or 3 years of age, females lay one clutch of 10–85 eggs each year. (For most specimens the average is probably about 35.)

## BEHAVIOR

The animals are generally slow moving and lethargic when cold, but during the heat of the day they become extremely alert and can run, swim, and climb with speed and agility.

They have a complex social structure and a defined annual reproductive cycle. Hatchlings adapt easily to captivity both when artifically incubated and when captured in the wild. However, captured adults, used to living free, are hard to keep in small enclosures.

## USES

Green iguanas are used mainly for food. The meat tastes somewhat like chicken and in Latin America is typically cooked in a spicy stew. The eggs are also consumed. Small and leathery shelled, they are considered special delicacies and are said to cure various ailments.

There is a sizable demand for live iguanas in the international pet trade.

Exploitation has been mainly for the flesh; the skins are wasted in most cases. However, given organized production, green iguana skin could perhaps become a farm by-product. It sells in the international reptile-leather market under the trade name "chameleon lizard." The

## IGUANA MAMA

The program that now offers hope for raising green iguanas as microlivestock is the brainchild of zoologist Dagmar Werner. She began in 1983 by collecting 700 iguana eggs and then learning what to do with them "on the job."

Through a combination of luck and intuition, coupled with determination, she discovered appropriate conditions for incubating the eggs on the first try—no less surprising because her incubators (dirt-filled wooden boxes warmed with light bulbs) were in an apartment in a high-rise building in Panama City. Nonetheless, most of the eggs hatched and the several hundred squirming young lizards were quickly trucked to a forest park near the Panama Canal. Here, for five years, assisted by a steady stream of enthusiastic Panamanian biology students, Werner experimented with cages, feeds, facilities, breeding, genetic selection, and the myriad aspects of management.

These dedicated researchers eventually devised what might be considered a production line, and they hatched and raised tens of thousands of green iguanas, reduced the animals' infant mortality rate, and created the basic underpinnings for economic production.

In 1988, administration for the project was shifted to neighboring Costa Rica, where Werner established another research farm as well as a fund-raising foundation. From there, she hopes to catalyze all Central America to take up iguana production.

Her immediate goal is, paradoxically, to preserve the lizards by putting an iguana in every pot. That will help reduce the indiscriminate hunting that is taking these animals toward extinction in the wild.

Her system is based on setting up feeding stations in the woods and releasing iguanas at seven months, an age at which they are virtually immune to predators. Iguanas need trees, and her ultimate goal is to help save the vanishing rainforests. With profitable iguana farming, she hopes to persuade farmers to save the trees as homes for iguanas, rather than clearing them for crops and cattle.

"Iguana management and an international marketing system will protect, rather than exterminate, the iguana," Werner says. "I don't think iguana farming will stop deforestation. But I do think it will contribute a great deal."

Opposite: Dagmar Werner with "Ignacio," her favorite of the thousands of iguanas she has raised.

hide is typically up to 20 cm wide. Thin and relatively fragile, it is glued to a fabric or cowhide backing to prevent it from tearing. The prime uses are for ladies' accessories, belts, wallets, and shoes. It is inappropriate for uses that involve flexure because the scales overlap and repeated flexing causes them to separate. The skin then becomes rough to the touch and loses its glossy finish.

## HUSBANDRY

The iguana farming project in Panama and Costa Rica (see sidebar) provides a model for what might be done elsewhere. In its first five years of operation, the project raised more than 10,000 green iguanas.

To create the farms, enclosures are constructed with sheet-metal walls sunk 30 cm in the ground.[1] Inside, the animals sleep in shelters made of bamboo and vegetation. Each shelter has an adjustable entrance slit through which young lizards can slither, but predators, which usually are larger, cannot. Most are set on stilts and food is served in the shade underneath. With this system, from 20 to 60 young iguanas are kept in an area of 10 $m^2$ (0.5–0.17 $m^2$ of land area per individual). In another "high-intensity" design, 30 hatchlings are kept in cages 1 $m^3$ in size (only 0.05 $m^2$ per individual).

The iguana farms also include an artificial nest consisting of a "tunnel" leading to a sand-filled egg-laying chamber. (Both tunnel and chamber are made of concrete blocks or other predator-safe material.) Female iguanas (at least in captivity) prefer this to digging their own tunnels, and it is an important advance in iguana breeding because it produces a hatching success of close to 100 percent. Recent versions require no human intervention: the eggs incubate and hatch by themselves and the hatchlings climb out of the nest through a hollow bamboo "pipe" and fall into plastic bags, which can be easily emptied twice a day.

The iguanas are fed mixtures of broken rice meal, meat meal, bone meal, fish meal, papayas, bananas, mangos, avocados, and a variety of leaves and flowers. Each day they receive fresh-cut leaves from plants such as beans, mustard, or hibiscus. Hatchlings are raised to an age of 6–10 months, when they are big enough to be released into forests, farmland with scattered trees, or into village backyards with almost no vulnerability to predators.

## ADVANTAGES

With each female producing an average of 35 eggs annually, these animals have inherently high reproductive potential. And if the young

[1] Information in this section provided by D. Werner.

are protected from predators during their first few months, populations can build up rapidly.

Iguanas adapt well to second-growth forest and to backyard conditions and can feed on the leaves of fruit trees or timber trees while the farmer can harvest the fruits or wood. Unless grossly overstocked, they are unlikely to affect the productivity of the trees.

## LIMITATIONS

These lizards may take three years to reach marketable size,[2] and the cost of raising them to usable size entirely in captivity is currently greater than their market value as food. Less-expensive methods of raising the animals must be found if large-scale commercial iguana farming is to become economically feasible. The main problem is the high cost of the enclosures, not the cost of food. The food costs no more than that for raising a chicken to marketable size, but iguanas do not grow as fast as chickens. For smallholders the only cost of raising them is labor, and that is often unimportant.

Free-ranging herbivorous lizards can damage home gardens.

## RESEARCH AND CONSERVATION NEEDS

For conservation purposes, the main need is to educate people throughout Latin America to the iguana's plight. For example, in Central America people catch pregnant females and cut out the eggs for food. (There is a widespread misconception that the females survive this brutality.) This is devastating to the iguana populations, yet by installing artificial nests, farmers could let the females live and still harvest the eggs. Moreover, people could eat half the eggs and incubate the other half to repopulate the trees around their farms.

It is also important to develop an understanding of iguana reproduction in the wild. At present, the longevity, growth rate, and age of sexual maturity are not well known. Such information would provide baseline data for creating conservation measures to reverse the depletion of this natural resource.

Further husbandry research is needed. Costs must be reduced. The effectiveness of artificial nests must be tested in village practice. Survival rates of captive-raised young after release must be studied. And harvesting and recruitment schemes should be developed to secure optimum exploitation of the repopulated forests.

[2] If they are introduced into reforestation areas, an economic return could be expected after 3–5 years, whereas the income from trees takes 15–20 years.

# 34

# Black Iguana

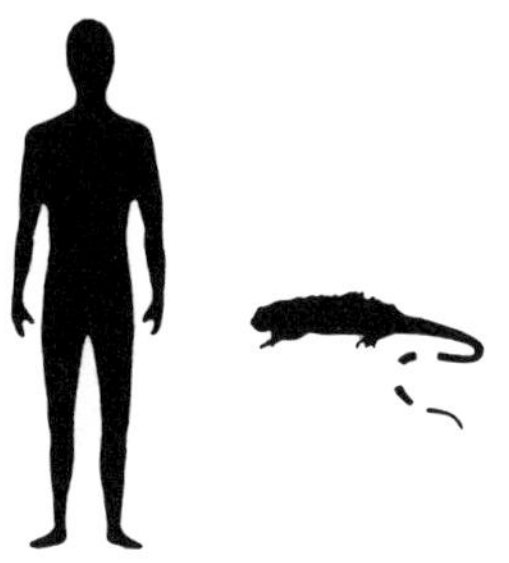

The work on green iguana husbandry may apply also to the black iguanas[1] (*Ctenosaura similis*, *C. acanthura*, *C. hemilopha*, and *C. pectinata*). These are similar lizards that live in dry habitats throughout Central America.

Exploitation of wild black iguanas for food began in ancient times and has continued in most places over hundreds of years without much harm to the natural populations. Until the 1970s, Latin Americans could obtain the meat and eggs with little effort, and the animals sold cheaply in city markets. In recent years, however, the lizards have become scarce and prices have risen sharply. For instance, in 1976 the central market of San Salvador was selling large specimens for the equivalent of 80 cents, but by 1979 prices were generally from three to six times as much. Today, black iguana meat costs more than fish, poultry, pork, or beef.

Eating small game animals is deeply rooted in the traditions of Central America and cannot readily be changed by legislation or education. Despite official edicts, campesinos on a subsistence diet will not willingly forego the little meat they get by capturing and eating lizards. Moreover, legislation is seldom effective because there are too few enforcement officials.

As with the green iguana, it seems likely that when the factors causing hatchling mortality are reduced, a large harvestable annual surplus can be produced. Black iguanas efficiently convert vegetation into high-grade protein suitable for human consumption. Young ones, however, are insectivorous and carnivorous rather than purely herbivorous, and during their first weeks of life may require a more expensive diet (perhaps meat scraps) than green iguanas require. Moreover, most insectivorous lizards require moving prey. They may or may not accept meat scraps as food.

[1] We use the name black iguana. In Mexico, it is called merely iguana; in English, it is commonly called spiny-tailed iguana. In Central America, it is called "garrobo" and "gallina de palo" (chicken of the trees).

## APPEARANCE AND SIZE

Black iguanas are smaller and more stocky than green iguanas, but weigh up to 3 kg. Their tails are spiny and short. The scales on the tail are enlarged, grow in spirals, and are sharp-pointed. Other than that (apart from being generally grayish black in color), the two animals are similar in appearance.

## DISTRIBUTION

The different black iguana species range from northern Mexico along both coasts of Central America to Panama and Colombia's Caribbean islands. Most tolerate moderate human presence well, often thriving around town garbage dumps and cemeteries.

## STATUS

As recently as 1981, black iguanas were shipped to markets by the truckload. Today, they are generally limited or even absent over much of their original range. Nevertheless, they are still *the* major game animals across extensive areas of Central America. Many of those taken are gravid females, which is disastrous for the populations. Excessive insecticide spraying is also thought to be reducing their populations in some areas.

## HABITAT AND ENVIRONMENT

Black iguanas thrive in dry, open woodland. They particularly like rocky hillsides, for they depend for shelter on crevices, rock piles, or soft soil in which they dig burrows.

## BIOLOGY

These omnivores feed on foliage, flowers, and fruits, but also on insects and small vertebrates. Adults spend some of their time climbing trees, but they are much less arboreal than green iguanas. Hatchlings are initially terrestrial but have a mostly arboreal stage in their early weeks of life.

Females lay one clutch of 20–90 eggs each year. The eggs are much smaller than those of the green iguana and therefore are not as popular a food.

Distribution of the black iguana.

## BEHAVIOR

These lizards live in burrows or in holes in trees. A typical burrow has several entrances and is 1–2 m long. Several females may combine efforts to form a complex communal burrow with several individual nest chambers. They are such diligent diggers that many are caught while absorbed in the task of adding a new room.

## USES

In some places, even where they are not common, black iguanas are still intensively hunted. Because their meat is valuable, the reward justifies the considerable effort involved in finding and killing them. Hence, where populations are so depleted that organized hunting is unprofitable, the animals are still subjected to relentless destruction by individuals.

## HUSBANDRY

Farming black iguanas is a novel idea. However, it is not a foolish one. In 1981 the Centro de Recursos Naturales (CENREN) in El

## OTHER IGUANAS

Other iguanas that deserve consideration for husbandry include the rock (rhinoceros) iguana (*Cyclura cornuta*), a herbivorous lizard of the Antillean region. This has been raised in considerable numbers at the National Zoo of Santo Domingo, Dominican Republic. Captive-bred specimens have reproduced at the age of 32 months, average clutch size has been about 15, and the average incubation time has been 82 days (at 31°C). The innate tameness of these island iguanas renders them better adapted for captive rearing than either the green or the black iguana, and their greater size is an advantage as well. On the other hand, the length of time required to reach maturity, and the relatively small egg clutch (about half that of the green iguana and one-third that of the black iguana) are definite disadvantages.

A related species, *Iguana delicatissima*, also deserves research and conservation attention. It occurs in the Leeward Islands and Martinique, and (as its specific name implies) is even better eating than the green iguana.

* Information from A. Ferreira.
** Lazal, 1973.

Salvador started a black iguana farming project. Since then, its researchers have collected data on growth rates, feeding patterns, and the maintenance and reproduction of captive adults. The program offers some promise of maintaining breeding stock in large outdoor enclosures, of producing large numbers of hatchlings, and of restocking depleted areas.

## ADVANTAGES

Latins believe that various ailments are cured or benefited by the flesh of these lizards, so they willingly pay much more for their meat than they would for equivalent amounts of other meats. In most places where the two occur together, the black iguana is preferred over the green iguana.

Compared with green iguana, the black iguana reproduces readily in captivity and has an even higher reproductive potential (averaging 43 eggs per clutch). It has the additional advantage that it thrives in deforested and altered habitats. It can survive near human settlements despite the attacks of dogs and cats, and it attains dense populations in suburban lots or open spaces. (It survives even downtown in cities such as Managua, but not in dense populations because dogs, cats, and humans take so many.) Hence, even in towns and cities, there are habitats capable of supporting it.

Black iguanas will feed on weedy vegetation or garbage, and adults seem easier to maintain in captivity than their green iguana counterparts.

## LIMITATIONS

The long delay (probably at least two years) before the animals reach marketable size might make it difficult for the grower to compete with common meats such as poultry or fish.

Certain parasitic worms can make the flesh inedible.

Although smaller than green iguanas, black iguanas are much more aggressive and will defend what they perceive to be their territory. They can inflict a painful bite. They also tend to escape from captivity more readily.

As noted, these omnivores depend on animal matter during part of their life cycle.

## RESEARCH AND CONSERVATION NEEDS

The biology of the black iguana deserves much more study.

Management techniques developed for the green iguana should be tested for their applicability for black iguana species. Harvesting and recruitment schemes must be developed to create sustainable populations in the wild. The specific needs for feeding the young also need to be addressed.

*I am convinced that in the Third World, it is only when poor people are assured of their livelihood that they will help us to safeguard their natural environments. So long as people remain hungry, it is very difficult to talk to them about conservation. As a result, I believe that development and conservation are inseparable. Only when conservation takes on a dimension of helping the poor, the downtrodden, the destitute, will it have an enduring impact.*

M.S. Swaminathan
International Union for
Conservation of Nature and Natural Resources

*I suspect that wildness is an inherited characteristic as was found for the turkey by Starker Leopold. So in attempting domestication one should expect to have difficulty until there are enough animals that you can select breeding stock for quietness and tractability.*

Ian McTaggart Cowan
Emeritus Professor of Zoology
University of British Columbia

*Cattle will never be extinct for the simple reason that man eats cattle. The best way to preserve wild species is by demonstrating that they, too, can be a valuable resource.*

Ernie Matteram
*New York Times Magazine*

*Wildlife utilization should be considered a legitimate form of land use, just as much as livestock husbandry. In fact, what has happened in the past is that Man has domesticated a limited number of animals, mainly in the temperate zones, while overlooking a considerable potential of other animals, which could be domesticated or used with equal validity.*

Antoon De Vos

# Part VII

# Others

As noted in the preface, this report by no means exhausts all the microlivestock possibilities. Lack of space and time precludes discussion of creatures such as edible insects, snails, worms, frogs, turtles, and bats, which in some regions are highly regarded foods. Similarly, we have not included fish, shrimp, and other aquatic life.

This is not to say that these are less worthy of consideration. The decision to leave them out was arbitrary, but with several recent breakthroughs in tropical beekeeping it seems prudent to include bees. Accordingly, the final chapter of this book describes the smallest livestock of all.

Bees are one of the most promising microlivestock. They forage on flowers that are otherwise little utilized and produce honey, wax, and other products of high value. They are important as plant pollinators and can greatly increase the production of some crops. Bees can be kept virtually anywhere with little disruption of other activities, and they are easily available.

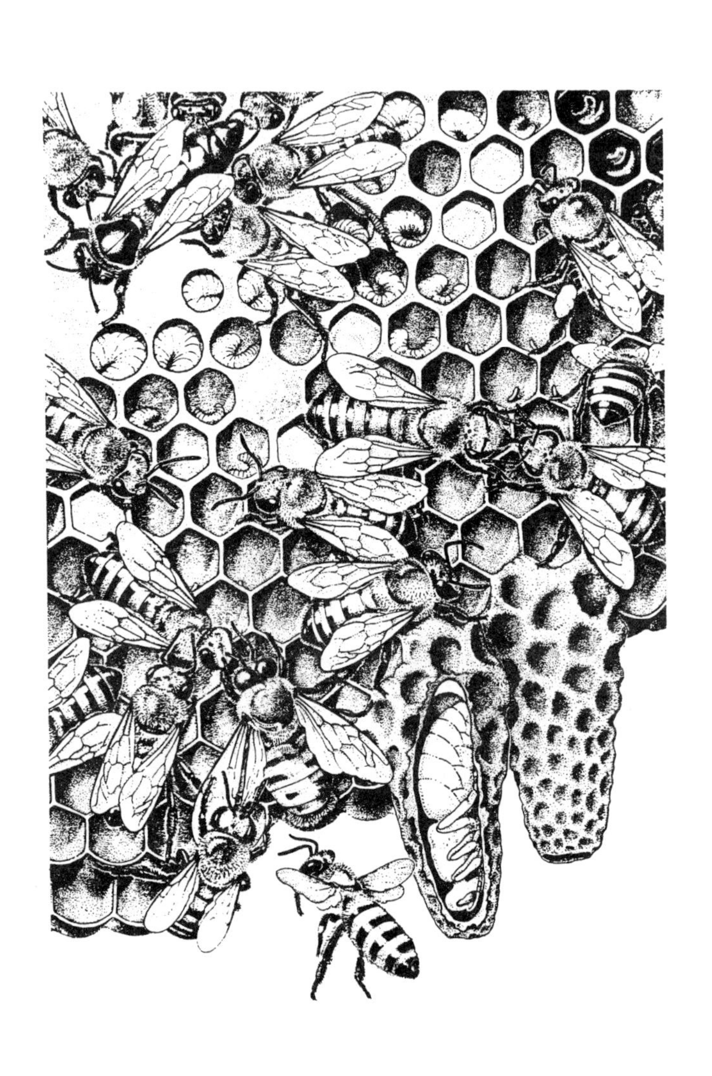

# 35

# Bees

Of all the livestock reviewed in this publication, bees are the smallest, the least demanding of space, probably the most familiar, and perhaps the most easily adapted to worldwide rural development efforts. For all that, however, they are an often forgotten component in agricultural programs. This is unfortunate because bees can be particularly valuable to tropical countries, providing pollination of crops, useful products, and a premium source of income.

Almost every village in every tropical country traditionally has had a beekeeper or two. Most use "seat-of-the-pants" methods and "rustic" hives, and this generally leads to low yields and inefficiencies. Today, numerous innovative methods and appropriate equipment are coming available. Many are still not widely known; however, their importance is slowly being recognized. Indeed, some developing countries are already turning bees into a valued natural resource.

For example, in less than 10 years Kenya has become self-sufficient in honey production and—despite increased local consumption—is now exporting. In seven years, Papua New Guinea has met its local honey requirements and now also exports its surplus. In only four years, Thailand increased yearly production from practically nothing to more than 1,000 tons. In Brazil, beekeeping is more widespread and production greater than before the outbreak of the African honey bee. These achievements can be largely attributed to the promotion of innovative equipment, modern beekeeping techniques, and extension support for small-scale beekeepers.

## APPEARANCE AND SIZE

Honey bees are generally easily recognized and need no description here. The major species and subspecies are all roughly similar in appearance and size.

## DISTRIBUTION

Among several hundred species of bees that store honey and pollen in harvestable amounts, only two "social" species, *Apis mellifera* and *Apis cerana*, produce multiple combs and can be kept in hives.[1] Given adequate forage and proper management, they can build up a honey surplus that can be harvested without harming their colonies.

*Apis mellifera*, the most widely distributed and exploited honey bee, comes from Europe and Africa. The subspecies from Europe (especially the Italian type, *A. mellifera ligustica*) is normally preferred because of its docility and high honey yields. It is now the predominant honey bee throughout the temperate zones of Europe, North America, Australasia, and China.

Of the many subspecies in Africa, *A. mellifera adansonii* and *A. mellifera scutellata* have the widest native ranges. The latter was accidently released in Brazil in 1957. It has become naturalized and has dominated the European bees formerly kept by beekeepers. By 1990, it had spread northwards to the southernmost areas of the United States and southwards deep into Argentina.

The docile Asian hive bee, *A. cerana*, is found in Asia from the Middle East to Japan and as far south as Indonesia. Although it produces much less honey per hive than *A. mellifera*, its overall production in many Southeast Asian countries may be greater.

## STATUS

Honey bees do not face extinction. However, genetic diversity is disappearing due to loss of habitat, insecticides, displacement by mass-produced, genetically uniform queens and exotic breeds, destructive harvesting, and the spread of diseases and pests such as protozoans, bacteria, insects, and mites.

## HABITAT AND ENVIRONMENT

Honey bees can exist in locations from deserts to rainforests and from near the Arctic to the tropics. They occur wherever there is nectar, pollen, tree resin (for nest building), shelter, and a little water. Heat, drought, and especially rain and humidity may curtail their activities, but a well-managed colony can survive periods of extreme adversity.

[1] Large amounts of honey and wax are also collected from wild species, particularly from the giant honey bee of Asia, *A. dorsata*. *A. florea*, the smallest honey bee, is "managed" somewhat in Oman, and hunted throughout the Middle East for small amounts of honey. Stingless bees, especially of the genera *Melipona* and *Trigona*, are also exploited. The scientific names follow those in C.D. Michener. 1974. *The Social Behavior of Bees*. Belknap. Cambridge, Mass.

## BIOLOGY

Honey bees live in rigidly hierarchical colonies. Normally there is a single queen. After mating, she begins laying hundreds of eggs a day. Those that are fertilized become sterile females, called workers; those that remain unfertilized develop into males (drones), whose only role is to fertilize future queens.

After two or three years the queen, worn-out, starts laying fewer and fewer eggs, and the colony may replace her. At that time a few female larvae are raised on royal jelly, a nutritious, little-understood secretion that causes them to develop into queens.

Workers perform different tasks as they mature. Young ones tend the queen, guard the hive, and raise the larvae. Older ones, comprising the vast majority of a colony's population, gather pollen and nectar and water. Pollen provides the protein and fats, and the nectar, converted into honey in the bee's body, provides carbohydrates to feed the colony.

## BEHAVIOR

Honey bee activities are dictated by weather, availability of food, genetics, and the overall strength of the colony. They are mediated by chemical interactions between the queen and the workers that control almost all behavior.

While the Asian and European hive bees are relatively docile, most of the African subspecies are unpredictable and may defend their colonies in great numbers and with great persistence. The threat of "killer bees" has been greatly exaggerated, however. Africans have provided them nests and hives—and harvested their honey—for thousands of years.

Occasionally, large numbers of the bees in a colony split off from their nest or hive. They usually cluster on a nearby tree or building, calmly waiting for "scouts" to find a suitable new home. These homeless swarms can be captured and will readily move into a hive—the simplest and cheapest way for beekeepers to acquire a colony.

## USES

Honey and beeswax are two of nature's best known and most valuable products. Honey can be employed in hundreds of foods. It is widely used in baking because, in addition to its flavor, it retains moisture better than sugar or syrup, and the product keeps longer.

Honey-based alcoholic beverages are popular in many parts of the world.

Beeswax also is used in many products, including candles, lubricants, polish, waterproofings, soaps, cosmetics, and electronics. Beeswax can be locally important for "lost-wax" metal-casting and sculptures as well as batik-dyed clothing.

Both bee pollen and royal jelly are used in cosmetics and can be eaten. Bee venom is used medicinally, particularly in Europe and the developing world. Bee larvae are eaten raw or fried in many parts of Asia and Africa and are considered a delicacy.

## HUSBANDRY

Keeping bees means managing a colony so that it produces surplus honey or wax. Specifically this requires:

- Providing a suitable hive;
- Obtaining bees by collecting a swarm, transferring a wild colony, or purchasing a colony complete with an active queen;
- Maintaining the colony free from natural enemies in an area that allows it to produce excess honey; and
- Harvesting the excess honey without weakening the colony or causing it to flee (abscond).

## ADVANTAGES

Few other livestock enterprises require less capital, less space, or less attention. Moreover, scarcely any other provides higher quality, more marketable products.

Beekeeping is a respected and traditional activity in most areas of the world. It promotes self-reliance and requires little, if any, land or money. It is an easy-entry cottage industry that can be started with minimal equipment or training. It can be done by any member of a family as it requires no special strength or size. It is especially appropriate for increasing women's income in the many areas where men are away working.

Beekeeping is suited to remote areas where many agricultural enterprises are at a competitive disadvantage because their products are bulky and far from the markets. Honey, wax, pollen, and other bee products can all be sold far from their point of origin, have high monetary value for their weight, and generally find a ready market. Local honeys often command premium prices (in part because many

## BETTER BEEHIVES

So many hives have been developed in recent years that a design now exists for almost any level of expertise. One of the most important for Third World beekeeping is the top-bar hive. This crate-like box, derived from an ancient Greek design, incorporates modern beekeeping principles but adds the innovation of sloping sides. Beneath its lid are removable boards (top bars) from which the bees hang the combs (because the sides are sloping the bees do not attach the edges of the comb to them). By lifting a top bar, the comb can be inspected and handled—and the honey harvested—with little disturbance to the colony.

This simple hive can make beekeeping accessible to even the poorest people. It is easy to build and use, and can be locally constructed from scrap lumber. It is well suited for raising most types of honey bees.

More elaborate and productive is the African long hive. In addition to having top bars, this square-sided box has removable frames within which the bees build their comb.

The most elaborate is the Langstroth hive—the type most common in temperate zones.* These yield the most honey. They can be made only where there are facilities for precise carpentry, where durable, nonwarping parts are available, and where there are good extension services to aid and assist beekeepers. Although traditional Langstroth hives are demanding to build and maintain, simplified designs have been created for use in developing countries.

* A Reverend Langstroth of Pennsylvania, USA, put together this hive in 1849. It has changed little in the past 140 years.

of them are considered medicinal). In addition, beekeeping encourages people to remain in rural areas rather than move to the city in search of an income.

Pollination increases the productivity of many crops, and therefore a few hives can boost local food production. By rotating hives among farmers' fields and orchards, a beekeeper performs a valuable service.

Beekeeping is also an important adjunct to reforestation and desert-reclamation projects. It can provide income during the long wait for the trees to reach marketable size. Many forests are potential reservoirs of honey and other products. In some cases, the bees also enhance the fruitfulness and standing value of forests. They also increase pollination in tree-seed orchards and tree nurseries. And in newly established forests, bees improve reseeding potential.

## STINGLESS BEES

By far the most common bees in the tropics, perhaps in the world, are bees that cannot sting. Their stingers are so atrophied as to be essentially nonexistent (but some can deliver a pretty fierce bite). They live in colonies and store their honey in wax pots, some as large as egg cups.

Stingless bees can be abundant. In a two-block area in downtown Panama City, 150 nests have been counted. However, many species depend on trees and their populations are plummeting as more and more tropical forests are felled.

Like honey bees, stingless bees have been "domesticated." For thousands of years in the tropical Americas, Indians have raised them in special hives made out of logs, gourds, clay pots, and other simple containers. Cortes reported in 1519 that Indians on the island of Cozumel, the now popular tourist spot off the east coast of Yucatan, practiced beekeeping. That was almost 300 years before the European honeybee was introduced. A popular Mayan drink was honey wine, and Mayan beekeepers carved stone earplugs to keep these bees out of their ears.

Honey from stingless bees has less sugar than normal honey. However, it is usually more tasty. It is used throughout the tropics: the Americas, Central Africa, and Southeast Asia, for instance. Normally, the nests are just robbed, which devastates the bees because the queen cannot fly, and when she is disturbed the colony dies. In some areas people open a little hole in the nest. By putting in a plug, they can then harvest honey a couple of times a year without destroying the colony.

Today, a few scientists are reconsidering stingless beekeeping. In the state of Maranhão in Brazil, an area of crushing poverty, biologist Warwick Kerr is harnessing stingless bees. He began after learning that peasants were spending a third of their meager incomes buying sugar. As part of his experiments, he keeps 60 hives stacked in his garage on the outskirts of the city of San Luis. (A notable feature of these bees is that there is no prohibition against keeping them in populated areas.)

Kerr has found that stingless bees can be made to produce well. After years of experimenting with different-sized boxes, he can now obtain more than 4.5 liters of honey per hive per year. He reports that the stingless bees are easy to maintain, and can be raised by poor people without land or equipment.

## LIMITATIONS

Compared to raising four-footed livestock, beekeeping is inexpensive and fairly trouble free, but it is not without problems. Many things can go wrong, and include, for example:

- Losses. Bees are susceptible to various predators, pests, and diseases. Although ways to avoid or control most afflictions are available, once a colony becomes infested it may have to be destroyed.
- Theft. Hives or combs—usually kept in secluded areas—that are full of honey are tempting targets and may be stolen.
- Swarming and absconding. Some or all of the bees may leave a hive and start a new nest, taking even the honey with them.[2]
- Bad management. Beekeeping requires certain skills and knowledge, and sometimes frequent attention (such as when the colony is stressed, diseased, or swarming).
- Lack of equipment. Beekeepers need hives, smokers (to quiet and repel the bees), hive tools, and—advisable but not absolutely necessary—gloves and a veil.
- Inadequate storage. Some areas lack the knowledge or the bottles in which to store liquid honey.[3]
- Pesticides and herbicides. Nectar- and pollen-collecting bees are vulnerable to insecticides, which farmers may apply (often inappropriately) at the time their crops are flowering. Herbicides can destroy important sources of bee forage.
- Neighborhood concerns. Bees can sting people and livestock. Although wild bees (and other insects) are the principal culprits, the beekeeper is often blamed.
- Stressful conditions. When rainfall, aridity, heat, or cold are excessive, bees often cannot produce surplus honey.

## RESEARCH AND CONSERVATION NEEDS

Surprisingly, much remains to be learned about the natural science of bees. Although some of the research requires sophisticated equipment and facilities, much can (or must) be performed locally by beekeepers because many factors—such as colony behavior, foraging

[2] Experienced beekeepers can usually anticipate and thwart this behavior.
[3] A new innovation is marketing honey in combs wrapped in plastic film. This avoids the expense of extractors and bottling, and—if harvested properly—reduces problems of adulteration and spoilage.

habits, and microclimatic adaptation—depend on the local conditions. This research may include the following:

- Adaptation. The provision of locally adapted bee varieties is an important basis for developing advanced beekeeping. This can be done by selecting an appropriate local queen. Further, local breeding of queens and workers lessens the probability of importing exotic diseases and pests. Recent developments promise to move the mass production of queens and workers from the realm of high technology to common practice.[4]
- Integration. Continued developments in beekeeping cannot succeed without research, promotion, education, training, and extension services. These are essential for integrating beekeeping with agriculture and reforestation efforts. Thus, descriptions of agronomic plants should always include pollination requirements as well as nectar and pollen potentials. Planting nectar- and pollen-producing firewood species would increase the number of bees, which would in turn help ensure forest survival.
- Bee plants. There is still much to be understood about the relative qualities that different plants bring to beekeeping. Recently, many valuable bee-forage plants have been identified.[5] These deserve special consideration in any reforestation or beautification projects. Broad introduction of these plants may also encourage beekeeping that could produce high-value "specialty" honeys or ensure more continuous production of honey. There may also be a place for "bee farms," where every plant is bee forage.
- New bee species. Nontraditional species of *Apis* as well as other members of the bee family (such as *Anthophora*, *Bombus*, *Megachilae*, *Nomia*, *Osmia*, *Xylocopa*, and especially *Trigona* and *Melipona*, which are stingless) should be studied to determine their role in pollination and—for some species—their further exploitation for honey.

[4] These breeding systems force a confined queen to lay her eggs in special plastic cups, from which the eggs can then be transferred to queen cells.
[5] See, for example, Crane, 1983 and Crane et al., 1984.

## Appendix A

# Selected Readings

A small selection follows of books and articles that are not too difficult to locate and that will help readers explore each topic further. Obscure documents are accompanied by an address from which readers can obtain a reprint or photocopy.

## GENERAL

Grzimek, B. 1975. *Animal Life Encyclopedia*. Vol. 1-13. Van Nostrand Reinhold Company, New York.

Robbins, C.T. 1983. *Wildlife Feeding and Nutrition*. Academic Press, Inc., New York.

## MICROBREEDS

Mason, I.L. 1988. *A World Dictionary of Livestock Breeds, Types and Varieties*. Commonwealth Agricultural Bureaux, Farnham Royal, Buckinghamshire, UK.

### Microcattle

Barnard, J.P. and J.P. Venter. 1983. *Indigenous and Exotic Beef Cattle in South West Africa—A Progress Report*. Available from Department of Agriculture and Nature Conservation, PB 13184, Windhoek 9000, Namibia. (Sanga)

Cheng Peilieu. 1984. *Livestock Breeds of China*. FAO Animal Production and Health Paper 46. Food and Agriculture Organization of the United Nations, Rome. 217 pp.

Cole, H.H. and W.N. Garrett, eds. 1974. *Animal Agriculture*. 2nd edition. W.H. Freeman and Company, San Francisco, California, USA.

Epstein, H. and I.L. Mason. 1984. Cattle. Pages 6-27 in *Evolution of Domesticated Animals*, I.L. Mason, ed. Longman, London.

Felius, M. 1985. *Genus* Bos: *Cattle Breeds of the World*. MSD-AGVET, Merck and Co., Rahway, New Jersey. 234 pp.

Gryseels, G. 1980. *Improving Livestock and Farm Production in the Ethiopian Highlands: Initial Results*. ILCA study. International Livestock Centre for Africa, Addis Ababa, Ethiopia.

ILCA. 1979. *Trypanotolerant Livestock in West and Central Africa*. Vol. 1, *General Study*. International Livestock Centre for Africa, Addis Ababa, Ethiopia.

Joshi, N.R. and R.W. Phillips. 1953. *Zebu Cattle of India and Pakistan*. FAO Agricultural Study No. 19. Food and Agriculture Organization of the United Nations, Rome. 297 pp.

Joshi, N.R., E.A. McLaughlin, and R.W. Phillips. 1957. *Types and Breeds of African Cattle*. FAO Agricultural Study No. 37. Food and Agriculture Organization of the United Nations, Rome. 297 pp.

Koger, M., T.J. Cunha, and A.C. Warnick, eds. 1973. *Crossbreeding Beef Cattle*. Series 2. University of Florida Press, Gainesville, Florida, USA.
Martinez Balboa, A. 1980. *La Ganadería en Baja California Sur*. Vol. 1. Editorial J.B. La Paz, BCS: Mexico. 229 pp. (Criollo)
Mason, I.L. and J.P. Maule. *The Indigenous Livestock of Eastern and Southern Africa*. Commonwealth Agricultural Bureaux, Farnham Royal, Buckinghamshire, UK.
Maule, J.P. 1989. *The Cattle of the Tropics*. Centre for Tropical Veterinary Medicine, Edinburgh University, Edinburgh, UK.
McDowell, L.R., ed. 1985. *Nutrition of Grazing Ruminants in Warm Climates*. Academic Press, Inc., New York.
McDowell, R.E. 1972. *Improvement of Livestock Production in Warm Climates*. W.H. Freeman and Company, San Francisco, California, USA.
McDowell, R.E. 1983. *Strategy for Improving Beef and Dairy Cattle in the Tropics*. Cornell International Agricultural mimeo. Cornell University, Ithaca, New York.
Miller, W.J. 1979. *Dairy Cattle Feeding and Nutrition*. Academic Press, Inc., New York.
National Research Council. 1980. *Animal Science in China*. CSCPRC Report No. 12. National Academy Press, Washington, D.C.
National Research Council. 1981. *Effect of Environment on Nutrient Requirements of Domestic Animals*. National Academy Press, Washington, D.C.
Perry, T.W. 1980. *Beef Cattle Feeding and Nutrition*. Academic Press, Inc., New York.
Rollinson, D.H.L. 1984. Bali cattle. Pages 28-34 in *Evolution of Domesticated Animals*, I.L. Mason, ed. Longman, London.
Rouse, J.E. 1970. *World Cattle*. Vol. 1, *Cattle of Europe, South America, Australia, and New Zealand*. University of Oklahoma Press, Norman, Oklahoma, USA.
Rouse, J.E. 1970. *World Cattle*. Vol. 2, *Cattle of Africa and Asia*. University of Oklahoma Press, Norman, Oklahoma, USA.
Rouse, J.E. 1973. *World Cattle*. Vol. 3, *Cattle of North America*. University of Oklahoma Press, Norman, Oklahoma, USA.
Rouse, J.E. 1977. *The Criollo-Spanish Cattle in the Americas*. University of Oklahoma Press, Norman, Oklahoma, USA. 303 pp.
Shirley, R.L. 1986. *Nitrogen and Energy Nutrition of Ruminants*. Academic Press, Inc., New York.
Thrower, W.R. 1954. *The Dexter Cow*. Revised third impression, 1980. The Spaulding Press, Bethel, Vermont, USA
Williamson, G. and W.J.A. Payne. 1965. *An Introduction to Animal Husbandry in the Tropics*. 2nd edition. Longman, London.

## Microgoat

Acharya, R.M. 1982. *Sheep and Goat Breeds of India*. FAO Animal Production and Health Paper No. 30. Food and Agriculture Organization of the United Nations, Rome.
Ademosun, A.A. 1985. Capra *spp.: The Emblem of Anarchy—The King-Pin of Man's Pastoral Life*. Ife Lectures No. 3. University of Ife, Ile-Ife, Nigeria.
Cheng Peilieu. 1984. *Livestock Breeds of China*. FAO Animal Production and Health Paper 46. Food and Agriculture Organization of the United Nations, Rome. 217 pp.
Cole, H.H. and W.N. Garrett. 1980. *Animal Agriculture*. W.H. Freeman and Company, San Francisco, California, USA.
Copland, J.W. 1985. *Goat Production and Research in the Tropics*. Proceedings of a

workshop held at the University of Queensland, Brisbane, Australia, February 6-8, 1984. Australian Centre for International Agricultural Research (ACIAR) Proceedings Series No. 7, 118 pp.

Devendra, C. and M. Burns. 1983. *Goat Production in the Tropics*. Commonwealth Agricultural Bureaux, Farnham Royal, Buckinghamshire, UK. 83 pp.

Devendra, C. and G.L. McLeroy. 1983. *Goat and Sheep Production in the Tropics*. 2nd edition. Longman, London. 271 pp.

Devendra, C. and J.E. Owen. 1983. Quantitative and qualitative aspects of meat production from goats. *World Animal Review* (FAO) 47:19-29.

Fielding, D. Goat-keeping in Mauritius. *Sonderdruck aus Zeitschrift für Tierzüchtung und Züchtungsbiologie*. Vol 97(1)(1980):21-27.

Gall, C., ed. 1981. *Goat Production*. Academic Press, London. 619 pp.

Haenlein, G.F.W. and D.L. Ace. 1983. *Goat Extension Handbook*. University of Delaware Cooperative Extension Service, Newark, Delaware, USA.

Husnain, H.V. 1985. *Sheep and Goats in Pakistan*. FAO Animal Production and Health Paper No. 56. Food and Agriculture Organization of the United Nations, Rome.

International Development Research Centre (IDRC). *Small Ruminant Production Systems in South and Southeast Asia*, C. Devendra, ed. Proceedings of a workshop held in Bogor, Indonesia, October 6-10, 1986, cosponsored by the Small Ruminant Collaborative Research Support Program. Ottawa, Ontario, Canada.

ILCA. 1983. *Peste des Petits Ruminants (PPR) in Sheep and Goats*, D.H. Hill, ed. Proceedings of the International Workshop held at the International Institute of Tropical Agriculture (IITA), Ibadan, Nigeria, September 24-26, 1980. International Livestock Centre for Africa, Addis Ababa, Ethiopia.

ILCA. 1985. *Sheep and Goats in Humid West Africa*. Proceedings of the International Workshop held in Ibadan, Nigeria, January 23-26, 1984. International Livestock Centre for Africa, Addis Ababa, Ethiopia.

ILCA. 1979. *Trypanotolerant Livestock in West and Central Africa*. Vol. 1, *General Study*. International Livestock Centre for Africa, Addis Ababa, Ethiopia.

King, J.W.B., ed. 1988. *Directory of Research on Sheep and Goats*. Commonwealth Agricultural Bureaux International, Wallingford, Oxford, UK. 271 pp.

Mason, I.L. 1981. Breeds. Pages 57-110 in *Goat Production*, C. Gall, ed. Academic Press, London.

Mason, I.L. 1984. Goat. Pages 85-99 in *Evolution of the Domesticated Animals*, I.L. Mason, ed. Longman, London.

Mason, I.L. and J.P. Maule. 1960. *The Indigenous Livestock of Eastern and Southern Africa*. Commonwealth Agricultural Bureaux, Farnham Royal, Buckinghamshire, UK.

National Research Council. 1980. *Animal Science in China*. CSCPRC Report No. 12. National Academy Press, Washington, D.C.

Phillips, R.W., R.G. Johnson, and R.T. Moyer. 1945. *The Livestock of China*. U.S. Department of State Publication No. 2249, Far Eastern Series 9. U.S. Department of State, Washington, D.C. 174 pp.

*Proceedings of the 3rd International Conference on Goat Production and Disease*. January 10-15, 1982, University of Arizona, Tucson, Arizona. Copies can be obtained from Kent Leach, Dairy Goat Journal, P.O. Box 1808, Scottsdale, Arizona 85252, USA.

Williamson, G. and W.J.A. Payne. 1965. *An Introduction to Animal Husbandry in the Tropics*. 2nd edition. Longman, London.

Wilson, R.T., T. Murayi, and A. Rocha. 1989. Indigenous African small ruminant strains with potentially high reproductive performance. *Small Ruminant Research* 2:107-117.

Winrock International. 1984. *Sheep and Goats in Developing Countries*. Winrock International Technical Paper. The World Bank, Washington, D.C.

## Microsheep

Acharya, R.M. 1982. *Sheep and Goat Breeds of India*. FAO Animal Production and Health Paper No. 30. Food and Agriculture Organization of the United Nations, Rome.

Carles, A.B. 1983. *Sheep Production in the Tropics*. Oxford University Press, London. 213 pp.

Cheng Peilieu. 1984. *Livestock Breeds of China*. FAO Animal Production and Health Paper 46. Food and Agriculture Organization of the United Nations, Rome. 217 pp.

Cole, H.H. and W.N. Garrett. 1980. *Animal Agriculture*. W.H. Freeman and Company, San Francisco, California, USA.

Devendra, C. and G.L. McLeroy. 1983. *Goat and Sheep Production in the Tropics*. Longman, London. 271 pp.

Fitzhugh, H.A. and G.E. Bradford, eds. 1983. *Hair Sheep of Western Africa and the Americas*. Westview Press, Boulder, Colorado, USA.

Husnain, H.V. 1985. *Sheep and Goats in Pakistan*. FAO Animal Production and Health Paper No. 56. Food and Agriculture Organization of the United Nations, Rome.

International Development Research Centre (IDRC). 1987. *Small Ruminant Production Systems in South and Southeast Asia*, C. Devendra, ed. Proceedings of a workshop held in Bogor, Indonesia, October 6-10, 1986, cosponsored by the Small Ruminant Collaborative Research Support Program. IDRC, Ottawa, Canada.

ILCA. 1983. *Peste des Petits Ruminants (PPR) in Sheep and Goats*, D.H. Hill, ed. Proceedings of the International Workshop held at the International Institute of Tropical Agriculture (IITA), Ibadan, Nigeria, September 24-26, 1980. International Livestock Centre for Africa, Addis Ababa, Ethiopia.

ILCA. 1985. *Sheep and Goats in Humid West Africa*. Proceedings of the International Workshop held in Ibadan, Nigeria, January 23-26, 1984. International Livestock Centre for Africa, Addis Ababa, Ethiopia.

ILCA. 1979. *Trypanotolerant Livestock in West and Central Africa*. Vol. 1, *General Study*. International Livestock Centre for Africa, Addis Ababa, Ethiopia.

King, J.W.B., ed. 1988. *Directory of Research on Sheep and Goats*. Commonwealth Agricultural Bureaux International, Wallingford, Oxford, UK. 271 pp.

Mason, I.L. 1980. *Prolific Tropical Sheep*. FAO Animal Production and Health Paper No. 17. Food and Agriculture Organization of the United Nations, Rome.

Mason, I.L. 1967. *Sheep Breeds of the Mediterranean*. Food and Agriculture Organization of the United Nations. Commonwealth Agricultural Bureaux, Farnham Royal, UK.

Mason, I.L. and J.P. Maule. 1960. *The Indigenous Livestock of Eastern and Southern Africa*. Commonwealth Agricultural Bureaux, Farnham Royal, Buckinghamshire, UK.

Phillips, R.W., R.G. Johnson, and R.T. Moyer. 1945. *The Livestock of China*. U.S. Department of State Publication No. 2249, Far Eastern Series 9. U.S. Department of State, Washington, D.C. 174 pp.

Ponting, K. 1980. *Sheep of the World*. Blandford Press, Ltd., Poole, Dorset, UK.

Ryder, M.L. 1984. Sheep. Pages 63-85 in *Evolution of Domesticated Animals*, I.L. Mason, ed. Longman, London.

Ryder, M.L. 1983. *Sheep and Man*. Monograph. Duckworth, London.

Terrill, C.E. 1970. Sheep breeds of the world. (505 breeds). USDA mimeo. U.S. Department of Agriculture, Washington, D.C. 163 pp.

Whitehurst, V.E., R.M. Crown, R.W. Phillips, and D.A. Spencer. 1947. Productivity of Columbia sheep in Florida and their use for crossing with native sheep. Florida Agricultural Experimental Station Bulletin 429. 34 pp.

Williamson, G. and W.J.A. Payne. 1965. *An Introduction to Animal Husbandry in the Tropics*. 2nd edition. Longman, London.

Wilson, R.T., T. Murayi, and A. Rocha. 1989. Indigenous African small ruminant strains with potentially high reproductive performance. *Small Ruminant Research* 2:107-117.
Winrock International. 1984. *Sheep and Goats in Developing Countries*. Winrock International Technical Paper. The World Bank, Washington, D.C.
Yalçin, B.C. 1979. *The Sheep Breeds of Afghanistan, Iran, and Turkey*. Food and Agriculture Organization of the United Nations, Rome.

## Micropig

Cheng Peilieu. 1984. *Livestock Breeds of China*. FAO Animal Production and Health Paper 46. Food and Agriculture Organization of the United Nations, Rome.
Clutton-Brock, J. 1981. *Domesticated Animals from Early Times*. British Museum (Natural History), London.
Cole, H.H. and W.N. Garrett. 1980. *Animal Agriculture*. W.H. Freeman and Company, San Francisco, California, USA.
Cunha, T.J. 1977. *Swine Feeding and Nutrition*. Academic Press, Inc., New York.
Epstein, H. and M. Bichard. 1984. Pigs. Pages 145-162 in *Evolution of Domesticated Animals*, I.L. Mason, ed. Longman, London.
Krider, J.L., J.H. Conrad, and W.E. Carroll. 1982. *Swine Production*. McGraw-Hill Company, New York.
Mason, I.L. and J.P. Maule. 1960. *The Indigenous Livestock of Eastern and Southern Africa*. Commonwealth Agricultural Bureaux, Farnham Royal, Buckinghamshire, UK.
National Research Council. 1983. *Little-Known Asian Animals with a Promising Economic Future*. National Academy Press, Washington, D.C.
National Research Council. 1980. *Animal Science in China*. CSCPRC Report No. 12. National Academy Press, Washington, D.C.
Panepinto, L.M. 1986. Miniature swine in biomedical research. *Lab Animal* 15(8)(November/December 1986):21-27.
Panepinto, L.M., R.W. Phillips, L.R. Wheeler, and D.H. Will. 1978. The Yucatan miniature pig as a laboratory animal. *Laboratory Animal Science* 28(3):308-313.
Phillips, R.W. and T.Y. Hou. 1944. Chinese swine and their performance compared with modern and crosses between Chinese and modern breeds. *Journal of Heredity* 35:365-79.
Pond, W.G. and J.H. Maner. 1984. *Swine Production and Nutrition*. AVI Publishing Company, Westport, Connecticut, USA.
Williamson, G. and W.J.A. Payne. 1965. *An Introduction to Animal Husbandry in the Tropics*. 2nd edition. Longman, London.

## POULTRY

The *World's Poultry Science Journal* is published 3 times a year by the World's Poultry Science Association, (c/o Institut für Kleintierzucht, Dornbergstrasse 25/27, Postfach 280, C-3100 Celle, Germany). It has also published a multilingual poultry dictionary.

*Poultry*, an international journal on poultry, is published by Misset International, a department of Uitgeversmaatschappij, C. Misset b.v., Doetinchem-The Netherlands. (Address: Misset International, P.O. Box 4, 7000 BA Doetinchem, Netherlands.)

*Poultry Science* is published monthly by the Poultry Science Association, Inc., 309 West Clark Street, Champaign, Illinois 61820, USA.

*British Poultry Science*, a quarterly, is available c/o Longman Group Ltd., Subscriptions (Journals) Department, Fourth Avenue, Harlow, Essex CM195AA, UK.

## Chicken

Attfield, H.H.D. 1990. *Raising Chickens and Ducks*. Volunteers in Technical Assistance, Arlington, Virginia, USA.

Crawford, R.D. 1984. Domestic fowl. Pages 298-311 in *Evolution of Domesticated Animals*, I.L. Mason, ed. Longman, London.

Ensminger, M.E. 1971. *Poultry Science*. Interstate Printers and Publishers, Inc., Danville, Illinois, USA.

Fielding, D. 1984. The silent scavengers. *African Farming* September/October 1984: 11-13.

Howman, K.C. 1980. *Pheasants: Their Breeding and Management*. K&R Books, UK.

Huchzermeyer, F.W. 1973. Free ranging hybrid chickens under African tribal conditions. *Rhodesian Agricultural Journal* 70:73-75.

Latif, M.A. 1985. *In-Depth Analysis on the Successful Experience of Poultry Production in Bangladesh*. Report submitted to FAO-APHC Regional Animal Production and Health Office. FAO/UN, Bangkok, Thailand.

Matthewman, R.W. 1977. *A Survey of Small Livestock Production at the Village Level in the Derived Savannah and Lowland Forest Zones of South West Nigeria*. M.Sc. thesis, University of Reading, UK.

Mérat, P. 1986. Potential usefulness of the Na (Naked Neck) gene in poultry production. *World's Poultry Science Journal* 42:124-142.

Nesheim, M.C., R.E. Austic, and L.E. Card. 1979. *Poultry Production*. 12th edition. Lea and Febiger, Philadelphia.

Phillips, R.W., R.G. Johnson, and R.J. Moyer. 1945. *The Livestock of China*. U.S. Department of State Publication No. 2249, Far Eastern Series 9. U.S. Department of State, Washington, D.C. 174 pp.

Sainsbury, D. 1980. *Poultry Health and Management*. Granada, London.

Scott, M.L., M.C. Nesheim, and R.J. Young. 1982. *Nutrition of the Chicken*. M.L. Scott and Associates, Ithaca, New York.

Smith, A.J. nd. Supplementation for scavenging animals. Unpublished paper. Copies available from A.J. Smith, Centre for Tropical Veterinary Medicine, Easter Bush, Roslin, Midlothian EH25 9RG, Scotland, UK.

Somes, R. 1984. International Registry of Poultry Genetic Stocks. Bulletin 469, University of Connecticut Agricultural Experiment Station, Storrs, Connecticut, USA.

## Araucanian

Wilhelm, G.O.E. 1965-1966. La Gallina Araucana (*Gallus inauris* castelloi 1914). Universidad de Concepción, Concepción, Chile, Boletía de Concepción 40:5-26.

## Duck

A film, *Duck Farming—an Indonesian Tradition*, is available from the CSIRO Film and Video Unit, 314 Albert Street, East Melbourne 3002, Victoria, Australia. Soundtrack is in English or Indonesian and it is available in 16 mm or VHS/Beta Video.

Attfield, H.H.D. 1990. *Raising Chickens and Ducks*. Volunteers in Technical Assistance, Arlington, Virginia, USA.

Clayton, G.A. 1984. Common duck. Pages 334-339 in *Evolution of Domesticated Animals*, I.L. Mason, ed. Longman, London.
Farrell, D. and P. Stapleton, eds. 1986. *Duck Production, Science and World Practice*. Proceedings of a duck production workshop, November 18-22, 1985, Bogor, Indonesia. Central Research Institute for Animal Science. Available from Department of Biochemistry, Microbiology and Nutrition, University of New England, Armidale, NSW 2351, Australia. 430 pp.
Holderread, D. 1978. *The Home Duck Flock*. The Hen House, P.O. Box 492, Corvallis, Oregon 97330, USA.
Kingston, D.K., D. Kosasih, and I. Ardi. 1979. *The Rearing of Alabio Ducklings and Management of the Laying Duck Flocks in the Swamps of South Kalimantan*. Report No. 9. Centre for Animal Research and Development, Ciawi, Bogor, Indonesia.
Petheram, R.J. and A. Thahar. 1983. Duck egg production systems in West Java. *Agricultural Systems* 2:75-86.
Phillips, R.W., R.G. Johnson, and R.J. Moyer. 1945. The Livestock of China. U.S. Department of State Publication No. 2249, Far Eastern Series 9. U.S. Department of State, Washington, D.C. 174 pp.
Sainsbury, D. 1980. *Poultry Health and Management*. Granada, London.
Scott, M.L. and W.F. Dean. 1991. *Nutrition and Management of Ducks*. M.L. Scott of Ithaca, Publisher (P.O. Box 4464, Ithaca, New York 14852, USA)
Warren, A.G. 1972. Ducks and geese in the tropics. *World Animal Review* 3:35-36.
Yi Yung and Yu-Ping Zhon. 1980. The Pekin duck in China. *World Animal Review* 34:11-14.

## Geese

Crawford, R.D. 1984. Goose. Pages 345-349 in *Evolution of Domesticated Animals*, I.L. Mason, ed. Longman, London.
FAO. 1983. *The Goose and its Possible Use for Controlling Weeds*. Small Animals for Small Farms. GAN-1. FAO Regional Office for Latin America and the Caribbean, Casilla 10095, Santiago, Chile.
Holderread, D. 1981. *The Book of Geese—a Complete Guide to Raising the Home Flock*. Hen House Publications, P.O. Box 492, Corvallis, Oregon 97339, USA.
Phillips, R.W., R.G. Johnson, and R.J. Moyer. 1945. *The Livestock of China*. U.S. Department of State Publication No. 2249, Far Eastern Series 9. U.S. Department of State, Washington, D.C. 174 pp.
Soames, B. 1980. *Keeping Domestic Geese*. Blandford, London. 159 pp.
USDA Extension Service. 1983. *Raising Geese*. Farmer's Bulletin No. 2251. U.S. Department of Agriculture, Washington, D.C.
Warren, A.G. 1972. Ducks and geese in the tropics. *World Animal Review* 3:35-36.

## Guinea Fowl

Many research papers have been published in French, Italian, Spanish, and Russian, but few in English.

Ayeni, J.S.O. 1983. Studies of Grey Breasted Helmet Guineafowl (*Numida meleagris galeata* Pallas) in Nigeria. *World's Poultry Science Journal* 39(2):143-151.
Belshaw, R.H.H. 1985. *Guinea Fowl of the World*. Nimrod Book Services, P.O. Box No. 1, Liss, Hampshire, UK.

Blum, J.C., J. Guillaume, and B. Leclercq. 1975. Studies of the energy and protein requirements of the growing guinea-fowl. *British Poultry Science* 16:157-168.
Cauchard, J.C. 1971. *La Pintade*. Henri Peladan, Uzès, France.
Chemillier, J. 1984. *L'Alimentation de la Pintade*. F. Hoffmann-La Roche and Cie, 52, Boulevard du Parc, 92521 Neuilly-sur-Seine, Cedex, France.
Commonwealth Agricultural Bureaux. 1968. *Guinea Fowl Breeding and Management*. Annotated Bibliography No. 85. The Commonwealth Bureau of Animal Breeding and Genetics, King's Buildings, Edinburgh 9, Scotland, UK.
Leclercq, B. 1982. Alimentation des Futurs Reproducteurs Pintades. Pages 101-111 in *Fertilité et Alimentation des Volailles*. I.N.R.A. Edit., Versailles, France.
Leclercq, R. and B. Sauveur. 1982. L'alimentation des reproducteurs pintades. Pages 113-135 in *Fertilité et Alimentation des Volailles*, I.N.R.A. edit., Versailles, France.
Lukefahr, S. Raising guinea fowl. Paper available from author, c/o International Small Livestock Research Center, Alabama Agricultural and Mechanical University, P.O. Box 264, Normal, Alabama 35762, USA.
Ministry of Agriculture, Fisheries and Food. 1985. *Small-scale guinea-fowl production*. Pamphlet 934. Ministry of Agriculture, Fisheries and Food (Publications), Lion House, Alnwick, Northumberland NE66 2PF, UK.
Mongin, P. and M. Plouzeau. 1984. Guinea-fowl. Pages 322-324 in *Evolution of Domesticated Animals*, I.L. Mason, ed. Longman, London.
USDA. 1970. *Raising Guinea Fowl*. Leaflet No. 519. U.S. Department of Agriculture, Washington, D.C.

## Muscovy

Clayton, G.A. 1984. Muscovy duck. Pages 340-344 in *Evolution of Domesticated Animals*, I.L. Mason, ed. Longman, London.
De Corville, H. 1972. *La Production du Canard de Barbarie*. Copies available from Institut nacional de la recherce agronomique (INRA), Nouzilly, France.
Farrell, D. and P. Stapleton, eds. 1986. *Duck Production, Science and World Practice*. Proceedings of a duck production workshop, November 18-22, 1985, Bogor, Indonesia. Central Research Institute for Animal Science. Copies available from D. Farrell, University of New England, Armidale, NSW 2351, Australia.
Jeffreys, M.D.W. 1956. The Muscovy duck. *The Nigerian Field* 21(3):108-11.
Kingston, D.J., D. Kosasih, and I. Ardi. 1978. *The Use of the Muscovy Duck for Hatching of Alabio Duck Eggs in the Swamplands of Kalimantan*. Centre Report No. 7. Centre for Animal Research and Development, Bogor, Indonesia.
Richet, M. 1985. La production du canard de Barbarie. *Techniques Agricoles* 3770.

## Pigeon

Abs, M., ed. 1983. *Physiology and Behavior of the Pigeon*. Academic Press, Inc., New York.
Hawes, R.O. 1984. Pigeons. Pages 351-356 in *Evolution of Domesticated Animals*, I.L. Mason, ed. Longman, London.
Hollander, W.F. and W.J. Miller. 1981. Hereditary variants of behavior and vision in the pigeon. *Iowa State Journal of Research* 55(4):323-331.
Levi, W.M. 1965. *Encyclopedia of Pigeon Breeds*. T.F.H. Publications, Jersey City, New Jersey, USA.

Levi, W.M. 1981. *The Pigeon*. 2nd edition revised. Levi Publishing, Sumter, South Carolina, USA.

Murton, R.K., R.P.J. Thearle, and J. Thompson. 1972. Ecological studies of the feral pigeon (*Columba livia*). I. Population, breeding biology and methods of control. II. Flock behaviour and social organisation. *Ecology* 9:835-889.

University of Wisconsin. 1977. *Pigeons*. Bulletin 4-H 135. University of Wisconsin, Madison, Wisconsin, USA.

USDA. 1963. *Squab Raising*. Farmers Bulletin No. 684. U.S. Department of Agriculture, Washington, D.C.

## Quail

Bourquin, O. 1980. *Biology of the Quail (*Coturnix coturnix *Linnaeus 1758)*. Ph.D. thesis, University of Natal, Pietermaritzburg, Republic of South Africa.

Katoh, H. and N. Wakasugi. 1980. Studies on the blood groups in the Japanese quail: detection of three antigens and their inheritance. *Developmental and Comparative Immunology* 4:99-110.

Kawahara, T. 1973. Comparative study of quantitative traits between wild and domestic Japanese quail (*Coturnix coturnix japonica*). *Experimental Animals* 22(supplement): 138-50.

Marsh, A. *Quail Manual*. Marsh Farms, 7171 Patterson Drive, Garden Grove, California 92641, USA.

National Academy of Sciences. 1969. *Coturnix (*Coturnix coturnix japonica*): Standards and Guidelines for the Breeding, Care, and Management of Laboratory Animals*. National Academy of Sciences, Washington, D.C.

*Quail Production Technology*. Leaflet No. 13/1982. Central Avian Research Institute, Izatnagar, India.

Somes, R.G., Jr. 1981. *International Registry of Poultry Genetic Stocks*. Storrs Agricultural Experiment Station, University of Connecticut, Storrs, Connecticut, USA.

Varghese, S.K. 1981. Coturnix International—at home and in the Caribbean. *Poultry Science* 60:1747.

Varghese, S.K. 1982. Coturnix production proves to be a profitable business. *Poultry Science* 61:1560.

Varghese, S.K. 1983. Processing and yield of coturnix (Japanese quail). *Poultry Science* 62:1517.

Wakasugi, N. and K. Kondo. 1973. Breeding methods for maintenance of mutant genes and establishment of strains in the Japanese quail. *Experimental Animals* 22(Supplement):151-9.

Woodard, A.E., A. Abplanalp, W.O. Wilson, and P. Vohra. 1973. *Japanese Quail Husbandry in the Laboratory*. Department of Avian Sciences, University of California, Davis, California 95616, USA.

## Turkey

Berg, R. and D. Halvorson. 1985. *Turkey Management Guide*. Minnesota Turkey Growers Association, 678 Transfer Road, St. Paul, Minnesota 55114, USA.

Crawford, R.D. 1984. Turkey. Pages 325-334 in *Evolution of Domesticated Animals*, I.L. Mason, ed. Longman, London.

Ensminger, M.E. 1971. *Poultry Science*. Interstate Printers and Publishers, Inc., Danville, Illinois, USA.
Sainsbury, D. 1980. *Poultry Health and Management*. Granada, London.

## RABBITS

The Rabbit Research Center has been established at Oregon State University, Corvallis, Oregon 97331, USA. It publishes the *Journal of Applied Rabbit Research*, whose purpose is to convey current research information to those with an interest in commercial rabbit production. The journal describes research conducted at the center and reviews rabbit research results reported in the world's scientific literature.

Founded in 1976 in Paris, the World Rabbit Science Association (Tyning House, Shurdington, Cheltenhem, Gloucestershire, GL51 5XF, UK) publishes a newsletter of coming events and chronicles the progress of rabbit farming in most countries of the world. Its aims are to facilitate the advancement of the various branches of the rabbit industry, disseminate knowledge, and study problems of production and marketing. Its members represent individuals and associations in 20 countries.

The American Rabbit Breeder's Association (1925 South Main Street, Bloomington, Illinois 61701, USA) has published a book, *Guide to Producing Better Rabbits*, which contains an abundance of information concerning management and diseases of the domestic rabbit. The organization also publishes a bimonthly magazine, *Domestic Rabbit*, that includes a great deal of updated information on rabbit raising.

Asociación Española de Cunicultura (Spanish Rabbit Science Association), Now, 23 08785 Ballbona d'Anoia, Barcelona, Spain, has a set of slides for training, covering breeds, handling, pathology, and commercialization.

*The Report of the Workshop on Rabbit Husbandry*. Copies of this report are available free of charge from the International Foundation for Science, Sibyllegatan 47, S-114 42 Stockholm, Sweden.

*A Working Rabbit Literature Resources File*. This material is directed at scientists working on some aspect of domestic rabbit research. It lists methods of computer searching for scientific reports in the world's scientific literature, various types of computer literature searches, including commercially available ones, and government services. It also includes a list of prepared abstracts. The address is D.D. Caveny and H. L. Enos, Colorado State University, Fort Collins, Colorado 80523, USA.

### Domestic Rabbit

Ministry of Livestock Development. 1981. *Rabbit Production*. Kenya Ministry of Livestock Development. Copies can be obtained from the Agricultural Information Centre, P.O. Box 14733, Nairobi, Kenya.
Attfield, H.H.D. 1977. *Rabbit Raising*. Volunteers in Technical Assistance, 1815 Lynn Street, Arlington, Virginia 22209, USA. 90 pp.
Bennett, R. 1975. *Raising Rabbits the Modern Way*. Garden Way Publishing, Charlotte, Vermont 05445, USA.
Cheeke, P.R. 1987. *Rabbit Feeding and Nutrition*. Academic Press, Inc., New York.
Cheeke, P.R., N.M. Patton, S.D. Lukefahr, and J.I. McNitt. 1986. *Rabbit Production*. Sixth edition. The Interstate, Danville, Illinois, USA.

FAO. 1986. *Cooking Rabbits*. Food and Agriculture Organization of the United Nations, Regional Office for Latin America and the Caribbean, Santiago, Chile.
FAO. 1986. *Self-Teaching Manual on Backyard Rabbit Rearing*. Food and Agriculture Organization of the United Nations, Regional Office for Latin America and the Caribbean, Santiago, Chile.
FAO. 1986. *Backyard Rabbit Rearing: Some Basic Husbandry Practices*. Small Animals for Small Farms. GAN-16. Food and Agriculture Organization of the United Nations, Regional Office for Latin America and the Caribbean, Santiago, Chile.
Schlolaut, W., ed. 1985. *A Compendium of Rabbit Production*. Gesellschaft für Technische Zusammenarbeit (GTZ) Publication No. 169. TZ-Verlagsgesellschaft mbH, Postfach 36, D6101 Rossdorf, Germany.
Sicivaten, J. and D. Stahl. 1982. *A Complete Handbook on Backyard and Commercial Rabbit Production*. CARE/Philippines.
Vogt, D.W. 1982. *Raising Rabbits in Hawaii*. Circular 499, Cooperative Extension Service, College of Tropical Agriculture and Human Resources, University of Hawaii at Manoa.

# RODENTS

## Agouti

A report summarizing preliminary results in a captive-breeding project in Mexico is available from A. D. Cuarón (see Research Contacts for address).

Deutsch, L.A., and G. Santos. 1984. Contribuiçao para o conhecimento do gênero *Dasyprocta*—reproduçao de cutias—entre cruzamente das espécies *Dasyprocta aguti*, *Dasyprocta azarae* e *Dasyprocta fuliginosa*. Congresso Brasileiro de Zoologia, 11, Belém, 1984. *Resumos*. . .Belém, SBZ/UFPa/MPEC, p. 383-4/(Resumos 390). Copies available from the authors, see Research Contacts.
Clark, M.M. and B.G. Galef, Jr. 1977. Patterns of agonistic interaction and space utilization by agoutis, *Dasyprocta punctata*. *Behavioral Biology* 20:135-140.
Leopold, A.S. 1959. Pages 388-393 in *Wildlife of Mexico*. University of California Press, Berkeley and Los Angeles.
Meritt, D.A., Jr. 1982. Bibliography of *Dasyprocta* (Rodentia). *Brenesia* 19/20:595-612.
Meritt, D.A., Jr. 1983. Preliminary observations on reproduction in the Central American agouti, *Dasyprocta punctata*. *Zoo Biology* 2:127-131.
Morris, D. 1962. The behaviour of the green acouchi (*Myoprocta pratti*) with special reference to scatter hoarding. *Proceedings Zoological Society of London* 139:701-732.
Smith N. 1974. Agouti and babassu. *Oryx* 12(5):581-582.
Smythe, N. 1978. *The Natural History of the Central American Agouti* (Dasyprocta punctata). Smithsonian Contributions to Zoology No. 257. Smithsonian Institution, Washington, D.C., USA. 52 pp.
Tribe, C.J., R. Leher, M.C.O. Doglio, M.I.L. Rebello, D.S. Mello, and E.M.M. Guimar es. 1985. Uma tentativa de esclarecimento do uso de espaço e estrutura social na cutia *Dasyprocta a. aguti* (Rodentia). Pages 296-7 in *Congresso Brasileiro de Zoologia*, 12, Campinas, 1985. *Resumos*. . .Campinas, SBZ/UNICAMP, 1985. (Resumo 614.)

## Capybara

A capybara bibliography, *Bibliografia sobre Chiguires* (Hydrochoerus hydrochaeris), is available from E. Gonzalez-Jimenez and José Roberto de Alencar Moreira (see Research Contacts for addresses).

Aliaga Rodriguez, L. 1979. *Produccion de Cuyes*. Universidad Nacional del Centro del Peru, Puno.

Baldizan, A., R.M. Dixon, and R. Parra. 1983. Digestion in the capybara (*Hydrochoerus hydrochaeris*). *South African Journal of Animal Science* 13(1):27-28.

Gonzalez-Jimenez, E. 1977a. The capybara, an indigenous source of meat in tropical America. *World Animal Review* 21:24-30.

Gonzalez-Jimenez, E. 1977b. Digestive physiology and feeding of capybaras. Pages 167-177 in *Handbook Series in Nutrition and Food*. Section G, *Diets, Culture Media and Food Supplements*, M. Rechcigl, ed. CRC Press, Cleveland, Ohio, USA.

Gonzalez-Jimenez, E. and R. Parra. 1975. The capybara, a meat-producing animal for the flooded areas of the tropics. Pages 81-86 in *Proceedings of the Third World Conference on Animal Production*, R.L. Reid, ed. Sydney University Press.

Gonzalez-Jimenez, E. 1984. Capybara. Pages 258-259 in *Evolution of Domesticated Animals*, I.L. Mason, ed. Longman, London.

Macdonald, D.W., K. Krantz, and R.T. Aplin. 1984. Behavioural, anatomical and chemical aspects of scent marking amongst Capybaras (*Hydrochoerus hydrochaeris*) (Rodentia:Caviomorpha). *Journal of Zoology* (London) 202:341-360.

Mones, A. and J. Ojasti. 1986. *Hydrochoerus hydrochaeris*. *Mammalian Species* 264: 1-7.

Ojasti, J. 1968. Notes on the mating behavior of the capybara. *Journal of Mammalogy* 49:345-374.

Ojasti, J. 1973. *Estudio Biologico del Chiguire o Capybara*. Fondo Nacional de Investigaciones Agropecuaria, Caracas, Republica de Venezuela. 27 pp.

Zara, J.L. 1973. Breeding and husbandry of the capybara at Evansville Zoo. *International Zoo Yearbook* 13:137-139.

## Coypu

Christen, M.F. 1978. Evaluacion nutritiva de cuatro dietas mono especificas en la alimentacion del coipo (*Myocastor coypus* Molina 1782). Tesis. Faculty of Veterinary Science, Universidad Austral de Chile, Valdivia, Chile.

Deems, E.F., Jr., and D. Pursley. 1978. North American Fur bearers. Maryland Wildlife Administration, Department of Natural Resources, Cheltenham, Maryland, USA. pp. 155.

CORFO. n.d. Pieles finas. Especie coipos. Tomo I, II and III. Santiago, Chile.

Murua, R., O. Newman, and J. Propelmann. 1981. Food habits of *Myocastor coypus* (Molina) in Chile. Pages 544-558 in *Proceedings of the Worldwide Furbearer Conference*, J.A. Chapman and D. Pursley, eds. Vol. 1. Worldwide Furbearer Conference, Inc., Frostburg, Maryland, USA.

Willner, G.R. 1982. Nutria; *Myocastor coypu*. Pages 1059-1076 in *Wild Mammals of North America*, J.A. Chapman and G.A. Feldhamer, eds. The John Hopkins University Press, Baltimore, Maryland, USA.

## Giant Rat

Ajayi, S.S. 1975. *Domestication of the African Giant Rat.* Department of Forest Resources, University of Ibadan, Ibadan, Nigeria.

Faturoti, E.O., O.O. Tewe, and S.S. Ajayi. 1982. Crude fibre tolerance by the African giant rat (*Cricetomys gambianus* Waterhouse) (Potential sources of protein, Nigeria.) *African Journal of Ecology* 20(4):289-292.

Faturoti, E.O., O.O. Tewe, and S.S. Ajayi. 1983. Effect of varying dietary oil levels on the performance of the African giant rat (*Cricetomys gambianus* Waterhouse). *Nutrition Reports International* 27(3):525-530.

Den Hartog, A.P. and A. de Vos. 1973. The use of rodents as food in tropical Africa. *Nutrition Newsletter* (FAO) 11 (2):1-14.

Halcrow, J.G. 1958. The giant rat of East Africa. *Nature* (London) 181:649-650.

Matthewman, R.W. 1977. *A Survey of Small Livestock Production at the Village Level in the Derived Savanna and Lowland Forest Zones of South West Nigeria.* Development Studies 24. Department of Agriculture and Horticulture, University of Reading, UK.

Tewe, O.O., S.S. Ajayi, and E.O. Faturoti. 1984. Giant rat and cane rat. Pages 291-293 in *Evolution of Domesticated Animals*, I.L. Mason, ed. Longman, London.

## Grasscutter

Asibey, E.O.A. 1979. Some problems encountered in the field study of the grasscutter (*Thryonomys swinderianus*) population in Ghana. Pages 214-7 in *Wildlife Management in Savannah Woodland*, S.S. Ajayi and L. B. Halstead, eds. Cambridge University Press, Cambridge, UK.

Baptist, R. and G.A. Mensah. 1986. The Cane-Rat—Farm animal of the future? *World Animal Review* 60:2-6.

Matthewman, R.W. 1977. *A Survey of Small Livestock Production at the Village Level in the Derived Savanna and Lowland Forest Zones of South West Nigeria.* Development Study 24. Department of Agriculture and Horticulture, University of Reading, UK.

Pich, S. and K.J. Peters. 1985. *Possibilities of using the cane cutter for meat production in Africa.* Unpublished manuscript. Copies available from K.J. Peters, Institute of Animal Breeding, Gottingen, Germany.

Tewe, O.O., S.S. Ajayi, and E.O. Faturoti. 1984. Giant rat and cane rat. Pages 291-293 in *Evolution of Domesticated Animals*, I.L. Mason, ed. Longman, London.

## Guinea Pig

"Let's Raise Guinea Pigs," a filmstrip on guinea pig farming, is available from World Neighbors International Headquarters, 5116 North Portland Avenue, Oklahoma City, Oklahoma 73112, USA.

Aliaga Rodriguez, L. 1979. *Producción de Cuyes.* Universidad Nacional del Centro del Peru, Huancayo, Peru.

Aliaga Rodriguez, L. 1983. *Improvement of Guinea Pig Breeding as a Means of Increasing the Productivity and Production of Meat for Consumption by the Rural Population in Peru.* National University of Central Peru, Huancayo, Peru.
Bolton, R. 1979. Guinea pigs, protein, and ritual. *Ethnology* 18(3):229-252.
Huss, D.L. 1982. Small animals for small farms in Latin America. *World Animal Review (FAO)* 43:24-29.
FAO. 1982. *The Guinea Pig and a Hypothetical Development Centre.* Small Animals for Small Farms. FAO, Regional Office for Latin America and the Caribbean, Santiago, Chile.
Loetz, E. and C. Novoa. 1983. Meat from the guinea pig. *Span* 26(2):84-86.
Muller-Haye, B. 1984. Guinea-pig or cuy. Pages 252-257 in *Evolution of Domesticated Animals*, I.L. Mason, ed. Longman, London.
Stansfield, S.K., C.L. Scribner, R.M. Kaminski, T. Cairns, J.B. McCormick, and K.M. Johnson. 1982. Antibody to ebola virus in guinea pigs: Tandala, Zaire. *The Journal of Infectious Diseases* 146(4):483-486.

## Hutia

Anderson, S., C.A. Woods, G.S. Morgan, and W.L.R. Oliver. 1983. *Geocapromys brownii. Mammalian Species* 201:1-5.
Canet, R.S. and V. Berovides Alvares. 1984. Ecomorfología y rendimiento de la jutia conga (*Capromys pilorides* Say). *Poeyana* 279:1-19.
Canet, R.S. and V. Berovides Alvares. 1984. Reproducción y ecología de la jutia conga (*Capromys pilorides* Say). *Poeyana* 220:1-20.
Howe, R., and G.C. Clough. 1971. The Bahamian hutia *Geocapromys ingrahami* in captivity. *International Zoo Yearbook* 11:89-93.
Johnson, M.L., R.H. Taylor, and N.W. Winnick. 1975. The breeding and exhibition of capromyd rodents at Tacoma Zoo. *International Zoo Yearbook* 15:53-56.
Oliver, W.L.R. 1977. The hutias (Capromyidae) of the West Indies. *International Zoo Yearbook* 17:14-20.
Oliver, W.L.R. 1985. *The Jamaican hutia or Indian Coney* (Geocapromys brownii). *A model programme for captive breeding and reintroduction?* Symposium of the Association of British Wild Animal Keepers, 10:35-52. (Copies available from author, see Research Contacts.)
Rowlands, I.W., and B.J. Weir, eds. 1974. *The Biology of Hystricomorph Rodents.* Symposium of the Zoological Society of London, 34:1-482.

## Mara

Dubost, G. and H. Genest. 1974. Le comportement social d'une colonide maras, *Dolichotis patagonum* Z. dans le Parc de Branféré. *Zeitschrift für Tierpsychologie* 35:225-302.
Taber, A.B. and D.W. Macdonald. 1984. Scent dispensing papillae and associated behaviour of the mara, *Dolichotis patagonum* (Rodentia: Caviomorpha). *Journal of Zoology* (London) 203:298-302.

## Paca

A report summarizing progress in the captive-breeding project in Panama is available from Nicholas Smythe, Smithsonian Tropical Research Institute, Box 2071, Balboa, Panama. A report summarizing preliminary results in the captive-breeding project in Mexico is available from Alfredo D. Cuaron, (see Research Contacts).

Collett, S.F. 1981. Population characteristics of *Agouti paca* (Rodentia) in Colombia. Michigan State University, Publications of the Museum, *Biological Series* 5(7): 485-602.

Matamoros, Y. 1982. Investigaciones preliminares sobre la reproducción, comportamiento, alimentación y manego del tepezcuinte (*Cuniculus paca*, Brisson) en cautiverio. Pages 961-992 in *Zoologia Neotropical*, Actas del VIII Congreso Latinoamericano de Zoologia, P.J. Salinas, ed. (Copies available from Y. Matamores, see Research Contacts.)

## Vizcacha

Jackson, J.E. nd. *Growth rates in vizcacha (*Lagostomus maximus*) in San Luis, Argentina*. Instituto Nacional de Tecnología Agropecuaria, San Luis, Argentina. (Copies available from author, see Research Contacts.)

Llanos, A.C., and J.A. Crespo. 1952. Ecología de la vizcacha (*Lagostomus maximus maximus* Blainv.) en el nordeste de la provincia de Entre Ríos. *Revista de Investigaciones Agrícolas* (Buenos Aires) 6:289-378.

## Other Rodents

Butynski, T.M. 1973. Life history and economic value of the springhare (*Pedetes capensis* Forster) in Botswana. *Botswana Notes Record* 5:209-213.

Butynski, T.M. 1979. Reproductive ecology of the springhaas *Pedetes capensis* in Botswana. *Journal of Zoology* 189:221-232.

Butynski, T.M. 1984. Nocturnal ecology of the springhare, *Pedetes capensis*, in Botswana. *African Journal of Ecology* 22:7-22.

Butynski, T.M. and R. Mattingly. 1979. Burrow structure and fossorial ecology of the springhare *Pedetes capensis* in Botswana. *African Journal of Ecology* 17:205-215.

Kofron, C.P. 1987. Seasonal reproduction of the springhare, *Pedetes capensis*, in southeastern Zimbabwe. *African Journal of Ecology* 25.

Rosenthal, M.A., and D.A. Meritt. 1973. Hand-rearing springhaas at Lincoln Park Zoo. *International Zoo Yearbook* 18:206-208.

Smithers, R.H.N. 1983. *The Mammals of the Southern African Subregion*. University of Pretoria, Pretoria.

Velte, F.F. 1978. Hand-rearing springhaas *Pedetes capensis* at Rochester Zoo. *International Zoo Yearbook* 18:206-208.

## DEER AND ANTELOPE

*The Deer Farmer*, a quarterly magazine, is published by the New Zealand Deer Farmers' Association, P.O. Box 2678, Wellington, New Zealand.

*Deer Farming* is the quarterly magazine of the British Deer Farmers Association (22 Levat Road, Inverness IV2 3NS, Scotland, UK).

Chaplin, R.E. 1977. *Deer*. Blandford Press, Poole, Dorset, UK.

Hoffman, R.R. 1985. Digestion and feeding in deer: their morphophysiological adaptation. In *Proceedings International Conference on Deer Biology and Production*, eds. P.F. Fennessy and K.R. Drew. Royal Society of New Zealand Bulletin 22, Wellington, New Zealand.

Fennessy, P.F. and K.R. Drew, eds. 1985. *Biology of Deer Production*. Proceedings of an international conference held February 13-18, 1983, Dunedin, New Zealand. Royal Society of New Zealand Bulletin 22, Wellington, New Zealand. Available from Royal Society of New Zealand, Private Bag, Wellington, New Zealand.

Wemmer C., ed. 1987. *The Biology and Management of the Cervidae*. Smithsonian Institution Press, Washington, D.C., USA.

Yerex, D. 1982. *The Farming of Deer*. Agricultural Publishing, Box 176, Carterton, New Zealand.

### Mouse Deer

Kay, R.N.B. 1987. The comparative anatomy and physiology of digestion in tragulids and cervids and its relation to food intake. In *The Biology and Management of the Cervidae*, C. Wemmer, ed. Smithsonian Institution Press, Washington, D.C.

Dubost, G. 1975. Le comportement du chevrotain africain, *Hyemoschus aquaticus* Ogilby (Artiodactyla, Ruminantia). *Zeitschrift für Tierpsychologie* 37:403-501.

Dubost, G. 1978. Un aperçu sur l'écologie du chevrotain africain, *Hyemoschus aquaticus* Ogilby, Artiodactyle Tragulide. *Mammalia* 42:1-62.

Ralls, K., C. Barasch, and K. Minkowski. 1975. Behaviour of captive mouse deer, *Tragulus napu*. *Zeitschrift Tierpsychology* 37:356-378.

Robin, K.P. 1979. *Zum Verhalten des Kleinkantschils (*Tragulus javanicus*)*. Ph.D. dissertation, University of Zurich. (Copies available from K.P. Robin, see Research Contacts.)

Vidyadaran, M.K., M. Hilmi, and R.A. Sirimane. 1982. The gross morphology of the stomach of the Malaysian lesser mousedeer (*Tragulus javanicus*). *Pertanika* 5(1): 34-38.

Vidyadran, M.K., S. Vellayan, and R. Karuppiah. 1983. Muscle weight distribution of the Malaysian lesser mousedeer (*Tragulus javanicus*). *Pertanika* 6(2):63-69.

Wharton, D.C. 1987. Captive Management of Tragulids at New York Zoological Park. In *Biology and Management of the Cervidae*, C. Wemmer, ed. Smithsonian Institution Press, Washington, D.C., USA.

### Muntjac

Anderson, J. 1981. Studies on digestion in *Muntiacus reevesi*. M.Phil. thesis. Cambridge University.

Barrette, C. 1977. The social behavior of captive muntjac *Muntiacus reevesi* (Ogilby 1839). *Zeitschrift für Tierpsychologie* 43:188-213.
Chaplin, R.E. 1977. *Deer*. Blandford Press, Poole, Dorset, England
Chapman, D.I., N.G. Chapman, J.G. Mathews, and D.H. Wurster-Hill. 1983. Chromosome studies of feral muntjac (*Muntiacus* sp.) in England. *Journal of Zoology, London* 201:557-559.
Dansie, O. 1970. *Muntjac*. British Deer Society, Welwyn Garden City, England. 22 pp.
Dubost, G. 1970. L'organisation spatiale et sociale de *Muntiacus reevesi* Ogilby 1839 en semi-liberte. *Mammalia* 34:331-355.
Dubost, G. 1971. Observations éthologiques sur le muntjac (*Muntiacus muntjak* Zimmermann 1780 et *Muntiacus reevesi* Ogilby 1839) en captivité et semi-liberté. *Zeitschrift für Tierpsychologie* 28:387-427.
Jackson, J.E., D.I. Chapman, and O. Dansie. 1977. A note on the food of muntjac deer (*Muntiacus reevesi*). *Journal of Zoology* (London) 183:546-548.
Lu Ho-gee and Sheng He-lin. 1984. Status of the black muntjac *Muntiacus crinifrons*, in eastern China. *Mammal Review* 14(1):29-36.
Oli, Madan K. 1986. Studies on stereotyped behavior of barking deer (*Muntiacus muntjak*). M.S. dissertation, Institute of Science and Technology, Tribhauvan University, Kathmandu, Nepal. 84 pp.
Shi Liming, Ye Yingying, and Duan Xinsheng. 1980. Comparative cytogenetic studies on the red muntjac, Chinese muntjac, and their F1 hybrids. *Cytogenetics and Cell Genetics* 26:22-7.
Yonzon, Pralad B. 1978. Ecological studies on *Muntaicus muntjack*. *Journal of Natural History Museum* (Nepal) 2(2):91-100.

## Musk Deer

Anonymous. 1974. Feeding musk deer in captivity and collecting musk from the live animal. *Dongwuxue Zhazi, China* 2:11-14.
Anonymous. 1975. Preliminary experience in raising the survival rate of musk deer. *Dongwuxue Zhazi, China* 1:17-19.
Bannikov, A.G., S.K. Ustinov, and P.N. Lobanov. 1980. The musk deer *Moschus moschiferus* in the USSR. International Union for Conservation of Nature and Natural Resources (IUCN), Gland, Switzerland.
Bi, S.Z., Y.H. Yan, Z.X. Qing, P.T. Sheng, Y.M. Wu, C.F. Chen, G.K. Yang, T.B. Yin, and Y.J. Lu. 1980. Dissection and analysis of the musk gland of *M. moschiferus* and a preliminary investigation into its histology. *The Protection and Use of Wild Animals* (China) 1:14-19.
Green, M.J.B. 1978. Himalayan musk deer (*Moschus moschiferus moschiferus*). Pages 56-64 in *Threatened Deer*. International Union for Conservation of Nature and Natural Resources (IUCN), Morges, Switzerland.
Green, M.J.B. 1985. The musk trade, with particular reference to its impact on the Himalayan population of *Moschus chrysogaster*. In *Conservation in Developing Countries*. Bombay Natural History Society, Bombay.
Green, M.J.B. 1985. *Aspects of the Ecology of the Himalayan Musk Deer*. Ph.D. dissertation, University of Cambridge, U.K. 280 pp.
Groves, C.P. 1978. The taxonomy of *Moschus*, with particular reference to the Indian Region. *Journal of the Bombay Natural History Society* 72:662-676.
Holloway, C.W. 1973. Threatened deer of the world: conservation status. *Biological Conservation* 5(4).

IUCN. 1974. *Red Data Book: Mammalia*. International Union for Conservation of Nature and Natural Resources (IUCN), Morges.
Jamwal, P.S. 1972. Collection of deer musk in Nepal. *Journal of Bombay Natural History Society* 69(3):647-649.
Seth, S.D., A. Muktiopadtiyay, K. Raghunatham, and R.B. Arora. 1975. *Pharmodymamics of Musk*. Central Council for Research on Indian Medicine and Homoeopathy, New Delhi, India.
Zhang B. 1983. Musk-deer. Their capture, domestication and care according to Chinese experience and methods. *Unasylva* 35(139):16-24.
Zhang, B.L., F.M. Dang, and B.S. Li. 1979. *The Farming of Musk Deer*. Agricultural Publishing Company, Peking.

## Water Deer

Chaplin, R.E. 1977. *Deer*. Blandford Press, Poole, Dorset, England.
Cooke, A. and L. Farrell. 1983. *Chinese water deer (*Hydropotes inermis*)*. British Deer Society Publication No. 2. Elvy and Gibbs, Canterbury, UK.
Feer, F. 1982. Quelques observations éthologiques sur l'Hydropote de chine, *Hydropotes inermis* (Swinhoe, 1870) en captivité. *Zeitschrift für Saeugetierkunde* 47:175-185.
Sheng Helin and L. Housl. 1984. A preliminary study on the river deer (*Hydropotes inermis*) population of Shoushau Island (China) and adjacent islets. *Acta Theriologica Sinica* 4:161-166.

## Duikers

Dittrich, L. 1972. Beobachtungen bei der Haltung von *Cephalophus*-Arten sowie zur Fortpflanzung und Jugendentwicklung von *C. dorsalis* und *C. rufilatus* in gefangenschaft. *Zoologische Garten, Lpz (N.F.)* 42:1-16.
Dubost, G. 1980. L'écologie et la vie sociale du céphalophe bleu (*Cephalophus monticola* Thunberg), petit ruminant forestier africain. *Zeitschrift für Tierpsychologie* 54: 205-266.
Dunbar, R.I.M. and E.P. Dunbar. 1979. Observations on the social organisation of common duiker in Ethiopia. *African Journal of Ecology* 17:249-252.
Dunbar, R.I.M. and E.P. Dunbar. 1980. *Animal Behaviour* 28:219.
Ketelhodt, H.F. Von. 1977. The lambing interval of the blue duiker, *Cephalophus monticola* Gray in captivity, with observations on its breeding and care. *South African Journal of Wildlife Research* 7:41-43.
Ralls, K. 1975. Agonistic behavior in Maxwell's duiker, *Cephalophus maxwelli*. *Mammalia* 39:241-249.
Schweers, S. 1984. The reproductive biology of the banded duiker *Cephalophus zebra* in comparison with other species of *Cephalophus*. *Zeitschrift für Saeugetierkunde* 49(1):21-36.
Whittle, C. and P. Whittle. 1977. Domestication and breeding of Maxwell's duiker. *The Nigerian Field* 42(4):13-21.

## Antelope

Spinage, C.A. 1986. *The Natural History of Antelopes*. Christopher Helm, UK.

## Klipspringer

Dunbar, R. and P. Dunbar. 1974. *Zeitschrift für Tierpsychologie* 35:481-493.

Dunbar, R.I.M. 1979. Energetics, thermoregulation and the behavioural ecology of klipspringer. *African Journal of Ecology* 17:217-230.

Dunbar, R.I.M. and E.P. Dunbar. 1980. The pairbond in klipspringer. *Animal Behavior* 28:219-229.

## Dikdik

Hendricks, H. 1975. *Zeitschrift für Tierpsychologie* 38:55-69.

Feer, F. 1979. Observations écologiques sur le neotrague de Bates (*Neltragus batesi* de Winton, 1903, Artiodactyle, ruminant, Bovide) du nord-est du gabon. *Terre et vie* 33:159-239.

Hoppe, P.P. 1975. *Acta Physiol. Scand.* 95, 9A.

Hoppe, P.P. 1976. Tritiated water turnover in Kirk's dikdik, *Madoqua* (Rhynchotragus) *kirki*, Günther, 1880. *Zeischrift für Saügetierkunde, Mitt.* 24:318-319.

Hoppe, P.P. 1977. How to survive heat and aridity: ecophysiology of the dikdik antelope. *Veterinary Medical Review* 1:77-86.

Hoppe, P.P. 1977. Comparison of voluntary food and water intake and digestion in Kirk's dik-dik and suni. *East African Wildlife Journal* 15:41-48.

Hoppe, P.P. 1977. Rumen fermentation and body weight in African ruminants. Pages 141-150 in *Proceedings 13th International Congress of Game Biologists*, T.J. Peterle, ed. Wildlife Society, Washington, D.C., USA.

Hoppe, P. 1984. Strategies of digestion in African herbivores. Pages 222-243 in *Herbivore Nutrition in the Tropics and Subtropics*. F.M.C. Gilchrist and R.I. Machie, eds. The Science Press, South Africa.

Hungate, R.E., G.D. Phillips, A. McGregor, D.P. Hungate, and H.K. Buechner. 1959. Microbial fermentation in certain mammals. *Science* 130:1192-1194.

Morat, P. and M. Nording. 1978. Maximum food intake and passage of markers in the alimentary tract of the lesser mouse-deer. *Malaysian Applied Biology* 7:11-17.

Yalden, D.W., M.J. Largen, and D. Kock. 1984. Catalogue of the mammals of Ethiopia. 5. Artiodactyla. *Monitore Zoologico Italiano Supplemento* 19:67-221.

# LIZARDS

## Green Iguana

Burghardt, G.M. and A.S. Rand, eds. 1982. *Iguanas of the World: Behavior, Ecology, and Conservation*. Noyes, Park Ridge, New Jersey, USA.

Fitch, H.S. and R.W. Henderson. 1977. *Age and Sex Differences, Reproduction and Conservation of* Iguana iguana. Milwaukee Public Museum Press. Milwaukee, Wisconsin, USA.

Hirth, H.F. 1963. Some aspects of the natural history of *Iguana iguana* on a tropical strand. *Ecology* 44:613-615.

Klein, E.H. 1982. Reproduction of the green iguana (*Iguana iguna* L.) in the tropical dry forest of southern Honduras. *Brenesia* 19/20:301-310.

Lazell, J.D., Jr. 1973. The lizard genus *Iguana* in the Lesser Antilles. *Bulletin of the Museum of Comparative Zoology, Harvard University* 145(1):1-28.

Tamsitt, J.R. and D. Valdivieso. 1963. The herpetofauna of the Caribbean Islands San Andres and Providencia. *Revista de Biología Tropical* 11(2):131-139. (Copies available from G. Lardé, see Research Contacts.)

Van Devender, R.W. 1982. Growth and ecology of spiny-tailed and green iguanas in Costa Rica, with comments on the evolution of herbivory and large body size. Pages 162-183 in *Iguanas of the World. Behavior, Ecology, and Conservation*. Noyes, Park Ridge, New Jersey, USA.

Werner, D. and T. Miller. 1984. Artificial nests for female green iguanas. *Herpetological Review* 15(2):57-58.

Werner, D. 1986. Iguana management in Central America. *BOSTID Developments* 6(1). 4 pp. (Available from BOSTID Publications, HA 476, National Research Council, 2101 Constitution Avenue, N.W., Washington, D.C. 20418.)

## Black Iguana

Burghardt, G.M. and A.S. Rand, eds. 1982. *Iguanas of the World: Behavior, Ecology, and Conservation*. Noyes, Park Ridge, New Jersey, USA.

Fitch, H.S. and R.W. Henderson. 1978. Ecology and exploitation of *Ctenosaura similis*. *University of Kansas Science Bulletin* 51:483-500. (Copies available from the authors, see Research Contacts.)

Fitch, H.S., R.W. Henderson, and D.M. Hillis. 1982. Exploitation of Iguanas in Central America. In *Iguanas of the World: Behavior, Ecology, and Conservation*. Noyes, Park Ridge, New Jersey, USA.

Iverson, J.B. 1979. Behavior and ecology of the rock iguana, *Cyclura carinata*. *Bulletin of the Florida State Museum Biological Science* 24(3):175-358.

Martínez, M.G. 1986. *Hábitos alimenticios de iguanas y garrobos*. La Prensa Gráfica, San Salvador, May 5:55. (Copies available from G. Lardé, see Research Contacts.)

Sanchez, S.A. 1985. El garrobo. *El Diario de Hoy, San Salvador* June 18:36. (Copies available from G. Lardé, see Research Contacts.)

## Tegu

Donadio, O.E. and J.M. Gallardo. 1984. Biología y Conservación de las Especies del Género Tupinambis (Squamata, Sauria Teiidae) en la Republica Argentina. Revista del Museo Argentino de Ciencias Naturales "Bernardino Rivadavia" Zoologia, 13(11):117-127. Agosto.

Revision Bibliografica de algunos aspectos de la Biología conservación y explotación de los lagartos del género topinambis (Daudin, 1803) by Laura Venturino, Dr. Thesis, Universidad de la Republic (Uruguay) 1983).

Report to World Wildlife Fund, USA, Biología y aprovechamiento de una población de *Tupinambis* references, by Claudio Blanco and Leohor Pessina (1985).

## BEES

In addition to the publications listed, many countries run beekeeping programs, and have literature and extension services available based on local experiences.

The International Bee Research Association (IBRA) publishes a *Newsletter for Beekeepers in Tropical and Subtropical Countries* twice each year. Its purpose is to provide a forum for exchange of information about beekeeping. The *Newsletter* is funded by the United Kingdom Overseas Development Agency (ODA) and is distributed free of charge to those in developing countries involved with beekeeping. Write to the International Bee Research Association, 18 North Road, Cardiff CF1 3DY, United Kingdom, for subscription information. The IBRA can also provide information on forthcoming conferences, grant programs, and the names of other beekeepers in your country.

The International Agency for Apiculture Development (IAAD) publishes a quarterly newsletter called *Cornucopia* that contains articles from beekeepers around the world, especially from the developing world. It provides information on producing, selling, and manufacturing hive products on a local level and discusses agroforestry and the pest and pesticide problems of honeybees. Write to *Cornucopia*, c/o M. Coleman, 6N 909 Roosevelt Avenue, St. Charles, Illinois 60174, USA, for a free sample. Subscriptions are $12 a year.

Adjare, S. 1984. *The Golden Insect: a Handbook on Beekeeping for Beginners.* Technology Consultancy Centre, University of Science and Technology, Kumasi, Ghana, in association with Intermediate Technology Publications, Ltd., 9 King Street, Covent Garden, London, WC2 E8HN, UK. 103 pp.

Anderson, R.H., B. Buys, and M.F. Johannsmeier. 1983. *Beekeeping in South Africa.* Department of Agriculture Bulletin No. 394. Pretoria, South Africa. 207 pp.

Bradbear, N. and D. De Jong. 1985. The management of africanized honeybees. IBRA Leaflet No. 2. International Bee Research Association (IBRA). Available in English or Spanish from Hill House, Gerrards Cross, Buckinghamshire SL9 ONR, UK. 4 pp.

Clauss, B. 1982. A Beekeeping Handbook. Agricultural Information Service, Gaborne, Botswana. 65 pp.

Crane, E., P. Walker, and R. Day. 1984. Honey plants of the world. International Bee Research Association, London.

FAO. 1984. Proceedings of the expert consultation on beekeeping with *Apis mellifera* in tropical and sub-tropical Asia. 9-14 April, 1984. FAO Regional Office for Asia and the Tropical Region, Paliwan Mansion, Phra Atit Road, Bangkok, Thailand.

Gentry, C. 1984. Small Scale Beekeeping. Manual M-17. Peace Corps, Washington, D.C.

IBRC. 1976. *Apiculture in Tropical Climates.* E. Crane, ed. Full report of the First Conference. International Bee Research Association, London.

IBRC. 1982. *Proceedings of the Second International Conference on Apiculture in Tropical Climates.* New Delhi, 1980. International Bee Research Association, London.

IBRC. 1985. *Proceedings of the Third International Conference on Apiculture in Tropical Climates.* Nairobi, Kenya, November 5-9, 1984. International Bee Research Association, London.

IBRC. 1988. *Proceedings of the Fourth International Conference on Apiculture in Tropical Climates.* Cairo, Egypt, November 1988. International Bee Research Association, London.

Kigatiira, K.I. 1974. Hive Designs for Beekeeping in Kenya. *Proceedings Entomological Society of Ontario* 105:118-128.

McGregor, S.E. 1976. Insect Pollination of Cultivated Crop Plants. U.S. Department of Agriculture Handbook No. 496. Government Printing Office, Washington, D.C. 411 pp.

Razafindrakoto, C. 1972. *Beekeeping in Madagascar.* Université Paul-Sabaier de Toulouse, France. 123 pp.

Sammataro, D. 1985. *Beekeeping in the Philippines.* 2nd edition. International Bee Research Association, London.

Sammataro, D. and A. Avitabile. 1986. *Beekeepers Handbook.* 2nd edition. Collier Books, MacMillan Co., New York.

Saubolle, B.R, and A. Bachmann. 1979. *Beekeeping: an Introduction to Modern Beekeeping in Nepal.* Sahayogi Prahashan, Tripureshwar, Kathmandu, Nepal.

Silberrad, Roger E.M. n.d. *Beekeeping in Zambia.* Apimondia. Available through the International Bee Research Association, London.

Singh, S. 1964. *Beekeeping in India.* Silver Jubilee Publication of the Entomological Society of India. Entomological Society of India, New Delhi.

Smith, Francis G. 1960. *Beekeeping in the Tropics.* Western Printing Service, Bristol, UK.

Townsend, G.F. 1978. Preparation of Honey for Market. Ministry of Agriculture and Food Publication 544. Ministry of Agriculture and Food, Ontario, Canada.

Townsend, G.F. 1976. Transitional hives for use with the tropical African bee *Apis mellifera adansonii.* Pages 181-189 in *Apiculture in Tropical Climates.* E. Crane, ed. Full report of the First Conference. International Bee Research Association, London.

# Appendix B

# Research Contacts

This section includes the names and addresses of individuals and institutions who are familiar with the animal listed. Each person has agreed to answer requests for information and advice. This personal follow-through is the main way this book can stimulate the development of microlivestock.

## MICROBREEDS

### Microcattle

B.K. Ahunu, Animal Science Department, University of Ghana, P.O. Box 25, Legon, Ghana

D.O. Alonge, Meat Hygiene and Preservation Laboratory, Faculty of Veterinary Medicine, University of Ibadan, Ibadan, Nigeria

M. Andres Poli, EEA San Luis INTA, CC 17, 5730 Mercedes (SL), Argentina

F.B. Bareeba, Department of Animal Science, Makerere University, P.O. Box 7062, Kampala, Uganda

J.A. Bennett, Department of Animal, Dairy, and Veterinary Sciences, Utah State University, Logan, Utah 84322-4815, USA

J.M. Berruecos V., School of Veterinary Medicine and Zoology, Universidad Nacional Autónoma de México, Ciudad Universitaria, Alvaro Obregón, 04510, Mexico D.F., Mexico

Cai Li, Animal Science and Veterinary Research Institute, Southwest College of Minority Nationalities, Chengdu, People's Republic of China

Centro de Investigaciones Pecuarias (CENIP), A.P. 227-9, Santo Domingo, Dominican Republic

G.L. Chan, B4-5 Nanyang Mansion, Shenzhen, Guangdong, People's Republic of China

Y. Cheneau, Institut d'Elevage et de Médecine Vétérinaire des Pays Tropicaux, Laboratoire de Farcha, BP 433, N'Djamena, Chad

Cheng Peilieu, Institute of Animal Science, Chinese Academy of Sciences, Beijing, People's Republic of China

W.R. Cockrill, 29 Downs Park West, Bristol BS6 7QH, UK

J.H. Conrad, Department of Animal Science, University of Florida, Gainesville, Florida 32611, USA

T.J. Cunha, P.O. Box R, La Verne, California 91750, USA

D.E. Deppner, Trees for the Future, 11306 Estona Drive, Silver Spring, Maryland 20902, USA (breeds of the Himalayas, East Indies)

C. Devendra, Agriculture Food and Nutrition Sciences Division, International Development Research Centre (IDRC), Tanglin P.O. Box 101, 9124 Singapore

N.I. Dim, Department of Animal Science, Faculty of Agriculture, Institute for Agricultural Research, Ahmadu Bello University, P.M.B. 1044, Zaria, Nigeria

Director, Agriculture and Forestry, PB 13184, Windhoek 9000, Namibia

V.M.S. Duarte, Instituto de Investigação Veterinária, C.P. 7, Huambo, Angola

M. Felius, Kerkhoflaan 47C, 3034 TA Rotterdam, The Netherlands

D. Fleharty, American Dexter Cattle Association, R.R. #1, Box 378, Concordia, Missouri 64020, USA

G. Garcia Lagombra, Animal Husbandry Department, Universidad Nacional "Pedro Henríquez Ureña," Decanato Ciencias Agropecuarias, Km 5½ Aut. Duarte, Santo Domingo, Dominican Republic
F. Gebreab, College of Veterinary Science, Addis Ababa University, Addis Ababa, Ethiopia
G.F.W. Haenlein, Department of Animal Science and Agricultural Biochemistry, University of Delaware, Newark, Delaware 19717-1303, USA
Z.B. Jensen, Animal and Dairy Science Research Institute, PB X02, Irene, Transvaal 1675, South Africa
Jin Xizhen, Animal Science and Veterinary Research Institute, Southwest College of Minority Nationalities, Chengdu, People's Republic of China
G. Kitsopanidis, Faculty of Agriculture, Aristotelian University of Thessaloniki, University Campus, Thessaloniki, Greece
G.E. Kodituwakku, Department of Veterinary Clinical Studies, University of Peradeniya, Peradeniya, Sri Lanka
N.M. Konnerup, 74 North Sunrise Boulevard, Camano Island, Washington 98292, USA
L.Y.L. Li, Department of Animal Husbandry and Veterinary Science, South China Agricultural College, Ghangzhou, People's Republic of China
J.K. Loosli, 406 S.W. 40th Street, Gainesville, Florida 32607, USA
A.B. Lwoga, Sokoine University of Agriculture, P.O. Box 3000, Morogoro, Tanzania
A. Martinez Balboa, Legaspy 1025, Apartado Postal 43, 23000, La Paz, Baja California Sur, Mexico
A. Masam, Livestock Training Institute, Tengeru, P.O. Box 3101, Arusha, Tanzania
J.P. Maule, Hannahfield Quarry House, 578 Lanark Road West, Balerno, Midlothian EH14 7BN, UK
M. Mazid, Faculty of Agriculture, University of Aleppo, Aleppo, Syria
L.R. McDowell, Department of Animal Sciences, University of Florida, Gainesville, Florida 32611, USA
J.A. Morgan, College of Natural Resources and Environmental Studies, University of Juba, P.O. Box 82, Juba, Sudan
D.B. Mpiri, Livestock Production Research Institute, Private Bag, Mpwapwa, Tanzania
Ni Shicheng, Animal Husbandry and Veterinary Science, Zhejiang Agricultural University, Hangzhou, People's Republic of China
B. Nikolov, Institute of Cattle and Sheep Breeding, 6000 Stara Zagora, Bulgaria
T. Olson, University of Florida, Gainesville, Florida 32611, USA
T.W. Perry, Department of Animal Sciences, Purdue University, West Lafayette, Indiana 47907, USA
R.W. Phillips, The Representative, Apartment 810, 1101 South Arlington Ridge Road, Arlington, Virginia 22202, USA
F. Pirchner, Technical University of Munich, D-8050 Munich-Weihenstephan, Germany
A. Ramamohana Rao, Postgraduate Studies, Andhra Pradesh Agricultural University, Rajendranagar, Hyderabad, Andhra Pradesh 500 030, India
Soedomo Reksohadiprodjo, Department of Animal Nutrition, Faculty of Animal Husbandry, Universitas Gadjah Mada, Sekip Blok N-42, Yogyakarta, Indonesia
A.Y. Robles, Dairy Training and Research Institute, University of the Philippines at Los Baños, College, Laguna 3732, Philippines
Sie Chenxia, Department of Animal Science, Nanjing Agricultural University, Nanjing, People's Republic of China
A.D. Tillman, 523 West Harned Place, Stillwater, Oklahoma 74074, USA
J.C.M. Trail, Livestock Productivity and Trypanotolerance Group, International Livestock Centre for Africa (ILCA), P.O. Box 30709, Nairobi, Kenya
A.C. Warnick, Department of Animal Science, University of Florida, Gainesville, Florida 32611, USA

## Microgoat

E.A. Adebowale, Institute of Agricultural Research and Training, University of Ife, Moor Plantation, P.M.B. 5029, Ibadan, Nigeria
A.A. Adeloye, Faculty of Agriculture, Adeyemi College of Education, Onjo, Nigeria
A.A. Ademosun, Department of Animal Science, University of Ife, Ile-Ife, Nigeria
I.F. Adu, N.A.P.R.I., Ahmadu Bello University, Zaria, Nigeria
I.A. Akinbode, Faculty of Agriculture, University of Ife, Ile-Ife, Nigeria
A.O. Akinsoyinu, Department of Animal Science, University of Ibadan, Ibadan, Nigeria
J.A. Akonji, Institut de Recherches Zootechniques, Office National de la Recherche Scientifique et Technique (ONAREST), B.P. 65, Yaounde, Cameroon
D.O. Alonge, Faculty of Veterinary Medicine, University of Ibadan, Ibadan, Nigeria
Baik Dong Hoon, Department of Animal Science, College of Agriculture, Jeonbug National University, JEONJU, 520, Republic of Korea
J.S.F. Barker, Department of Animal Science, University of New England, Armidale, New South Wales 2351, Australia
K. Boateng, Cooperative Extension Service, University of the Virgin Islands, Box L, Kingshill, St. Croix, Virgin Islands 00850, USA
G.E. Bradford, University of California, Davis, California 95616, USA
M.S. Browne, Nutrition Consulting Services, P.O. Box 788, St. Thomas, Virgin Islands 00801, USA
T.D. Bunch, Department of Animal, Dairy, and Veterinary Sciences, Utah State University, Logan, Utah 84322-4815, USA
T.C. Cartwright, Department of Animal Science, Texas A&M University, College Station, Texas 77843, USA
Centro de Investigaciones Pecuarias (CENIP), A.P. 227-9, Santo Domingo, Dominican Republic
C. Devendra, Agriculture Food and Nutrition Sciences Division, International Development Research Centre (IDRC), Tanglin P.O. Box 101, 9124 Singapore, Singapore
N.I. Dim, Department of Animal Science, Faculty of Agriculture, Institute for Agricultural Research, Ahmadu Bello University, P.M.B. 1044, Zaria, Nigeria
D. Fielding, Centre for Tropical Veterinary Medicine, Royal (Dick) School of Veterinary Studies, University of Edinburgh, Easter Bush, Roslin, Midlothian EH25 9RG, Scotland, UK
H.A. Fitzhugh, Winrock International, Route 3, Petit Jean Mountain, Morrilton, Arkansas 72110, USA
W.C. Foote, Utah State University International Sheep and Goat Institute, UNC 48, Utah State University, Logan, Utah 84322, USA
G.R. Gerona, Department of Animal Science and Veterinary Medicine, Visayas State College of Agriculture, Baybay, Leyte, Philippines
G.F.W. Haenlein, Department of Animal Science and Agricultural Biochemistry, University of Delaware, Newark, Delaware 19717-1303, USA
A. Essa Haid, Essa International, Inc., 2550 M Street, Suite 405, Washington, D.C. 20037, USA (small East African)
W. Harding, Agrarian Society, Palazzo de la Salle, Valletta, Malta
M.A. Hasnath, Department of Animal Husbandry, Bangladesh Agricultural University, P.O. Agricultural University, Mymensingh, Bangladesh
A.S. Hoversland, Division of Research Grants, National Institutes of Health, 5333 Westbard Avenue, Bethesda, Maryland 20205, USA
L. Hulse, National Pygmy Goat Association, RFD #1, Fern Avenue, Amesbury, Massachusetts 01913, USA
T.P. Husband, Department of Natural Resource Science, University of Rhode Island, Kingston, Rhode Island 02881, USA (bezoar, agrimi)

D.L. Huss, P.O. Box 426, Menard, Texas 76859, USA
B. Hutchins, 2535 Halsted Road, Apartment 1, Rockford, Illinois 61103, USA
N. Ibàñez Herrera, Paraguay 160—Urbanización, Apartado 315, El Recreo, Trujillo, Peru
S. Karam Shah, Livestock Experimental Station Rakh Khairewala, District Muzaffergarh, Pakistan
C. Katongole, Faculty of Veterinary Medicine, Makerere University, PO Box 7062, Kampala, Uganda
G. Kiwuwa, Department of Animal Science, Makerere University, PO Box 7062, Kampala, Uganda
L.Y.L. Li, Department of Animal Husbandry and Veterinary Science, South China Agricultural College, Ghangzhou, People's Republic of China
D.R. Lincicome, Animal Parasitology Institute, U.S. Department of Agriculture, Building 1180, BARC-East, Beltsville, Maryland 20705, USA
Liu Chengxiang, Animal Husbandry and Veterinary Science Research Institute, Guangdong Agricultural Academy, Guangdong, People's Republic of China
J.K. Loosli, 406 S.W. 40th Street, Gainesville, Florida 32607, USA
M. Mahyuddin Dahan, Department of Animal Sciences, Universiti Pertanian Malaysia, Serdang, Selangor, Malaysia
G.C.H. Mandari, Livestock Research Centre, West Kilimanjaro, P.O. Box 103, Hai, Tanzania
J. Mann, Institute of Tropical Veterinary Medicine, University of Giessen, Germany
A. Martinez Balboa, Legaspy 1025, Apartado Postal 43, 23000, La Paz, Baja California Sur, Mexico
I. Mecha, Department of Animal Science, University of Nigeria, Nsukka, Anambra State, Nigeria
J. Metcalf, School of Medicine, University of Oregon, Eugene, Oregon 97403, USA
K. Mitchell, Baptist Bible Institute, P.O. Box 301, Basseterre, St. Kitts, West Indies
M. Khusahry Mohd Yusuff, Malaysian Agricultural and Research Institute (MARDI), G.P.O. Box 12301, Kuala Lumpur 01-02, Malaysia
A. Monsi, Departments of Animal Science/Food Science and Technology, Rivers State University of Science and Technology, P.M.B. 5080, Port Harcourt, Nigeria
G. Montsma, Klomperweg 79, 6741 PD, Lunteren, Netherlands
J. Moore, Department of Animal Science, University of Florida, Gainesville, Florida 32611, USA
C.M. Munyabuntu, Department of Animal Science, P.O. Box 7062, Makerere University, Kampala, Uganda
Th. Murayi, Institut des Sciences Agronomiques du Rwanda, P.O. Box 138, Butare, Rwanda
W.N.M. Mwenya, School of Agricultural Sciences, University of Zambia, Box 32379, Lusaka, Zambia
A. Nivyobizi, Faculté des Sciences Agronomiques, Université du Burundi, BP 2940, Bujumbura, Burundi
G.A. Norman, Tropical Development Research Institute, 56/62 Gray's Inn Road, London WC1X 8LU, UK
K. Okello-Lapenga, Department of Veterinary Pathobiology, Makerere University, PO Box 7062, Kampala, Uganda
S. Ramahefarison, Département de Recherches Zootechniques et Vétérinaires (CENRADERU), BP 904, Amatobe, Antananarivo, Madagascar
A. Ramamohana Rao, Postgraduate Studies, Andhra Pradesh Agricultural University, Rajendranagar, Hyderabad, Andhra Pradesh 500 030, India
R.K. Rastogi, Department of Livestock Science, University of the West Indies, St. Augustine, Trinidad, West Indies

G. Rasul, District Forest Officer & Wildlife Warden, P.O. Box 501, Gilgit (northern area), Pakistan

D. Rattner, Institute of Animal Research, Kibbutz Lahav, Israel (Sinai goat, ya-ez hybrids)

Soedomo Reksohadiprodjo, Department of Animal Nutrition, Faculty of Animal Husbandry, Universitas Gadjah Mada, Sekip Blok N-42, Yogyakarta, Indonesia

A. Repollo, Jr., Palawan National Agricultural College, Aborlan, Palawan, Philippines

A.Y. Robles, Dairy Training and Research Institute, University of the Philippines, Los Baños, Laguna 3732, Philippines

A. Rocha, Instituto de Reproduçao e Melhoramento Animal, C.P. 25, Matola, Mozambique

C. Rodriguez, Centro de Investigaciones Pecuarias (CENIP), Autopista Duarte Km 24, Apartado 227-9, Santo Domingo, Dominican Republic

M. Nasim Siddiqi, Zoology Department, University of Peshawar, N.W.F.P., Pakistan

M. Shelton, Texas A & M University, College Station, Texas 77843, USA (criollo)

A. Shkolnik, Department of Zoology, Tel Aviv University, Tel Aviv, Israel

O.C. Simpson, Cameroon Road, Prairie View A&M University, Prairie View, Texas 77445, USA

Sie Chenxia, Department of Animal Science, Nanjing Agricultural University, Nanjing, People's Republic of China

R.G. Wahome, Department of Animal Production, University of Nairobi, P.O. Box 29053, Nairobi, Kenya

R. Watts, Animal Science Department, Natural Resources Development College, Lusaka, Zambia

R.T. Wilson, Bartridge House, Umberleigh, North Devon EX37 9AS, UK

H. Wood, Highland Game Farm, Route 4, Box 450, Alexandria, Indiana 46001, USA

T.P. Husband, Department of Natural Resources Science, University of Rhode Island, Kingston, Rhode Island 02881, USA

## Microsheep

E.A. Adebowale, Institute of Agricultural Research and Training, University of Ife, Moor Plantation, P.M.B. 5029, Ibadan, Nigeria

I.F. Adu, N.A.P.R.I., Ahmadu Bello University, Zaria, Nigeria

I.A. Akinbode, Faculty of Agriculture, University of Ife, Ile-Ife, Nigeria

A.O. Akinsoyinu, Department of Animal Science, University of Ibadan, Ibadan, Nigeria

M. Amin, Jammu and Kashmis State Sheep and Shearing Products Development Board, Nowshera, Srinigar, Kashmir 190 011, India

Ahmad Mustaffa B. HJ. Babjee, Veterinary Services Department, Ibu Pejabat Perkhidmatan Haiwan, Kementerian Pertanian, Malaysia, Jalan Mahameru, Kuala Lumpur 10-02, Malaysia (grazing in rubber plantations)

J.S.F. Barker, Department of Animal Science, University of New England, Armidale, New South Wales, 2351, Australia (Indonesian breeds)

K. Boateng, Cooperative Extension Service, College of the Virgin Islands, Box L, Kingshill, St. Croix, Virgin Islands 00850, USA

G.E. Bradford, University of California, Davis, California 95616, USA

R. Breckle, U.S. Forest Service, Alsea, Oregon 97324, USA (grazing sheep in forests)

T. Bunch, Department of Animal, Dairy, and Veterinary Sciences, Utah State University, Logan, Utah 84322-4815, USA (Iranian thin-tailed, Navajo, wild-sheep hybrids)

M. Carpio P., Departamento de Producci6n Animal, Universidad Nacional Agraria, Apartado 456, La Molina, Lima, Peru

Centro de Investigaciones Pecuarias (CENIP), A.P. 227-9, Santo Domingo, Dominican Republic
S. Cottrell, Pippin Farm, 3081 Goshen Road, Bellingham, Washington 98226, USA
C. Courtney, Veterinary Faculty, University of Florida, Gainesville, Florida 32611, USA
D. Deller, Virgin Islands Department of Agriculture, St. Thomas, Virgin Islands 00820, USA
A. Dettmers, 5620 Longwood Avenue, Port Charlotte, Florida 33953, USA (West African dwarf)
C. Devendra, Agricultural Food and Nutrition Sciences Division International Development Research Centre (IDRC), Tanglin P.O. Box 101, 9124 Singapore, Singapore
N.I. Dim, Department of Animal Science, Faculty of Agriculture, Institute for Agricultural Research, Ahmadu Bello University, P.M.B. 1044, Zaria, Nigeria
K. Drew, Invermay Agricultural Research Centre, Private Bag, Mosgiel, New Zealand
F. El Amin, Department of Animal Husandry, Faculty of Veterinary Science, University of Khartoum, P.O. Box 32, Khartoum North, Sudan
Gholam-Hossein Erabi, Faculty of Agriculture, University of Tabriz, Tabriz 51664, Iran
H.A. Fitzhugh, Winrock International Institute for Agricultural Development, Route 3, Petit Jean Mountain, Morrilton, Arkansas 72110, USA
W. Foote, Animal, Dairy, and Veterinary Science Department, UNC 48, Utah State University, Logan, Utah 84322-4815, USA (Virgin Islands white hair, wild-sheep hybrids)
E.-H. Gueye, Centre de Recherches Zootechniques de Kolda, Institut Sénégalais de Recherches Agricoles, (ISRA), B.P. 53, Kolda, Senegal
G. Garcia Lagombra, Animal Husbandry Department, Universidad Nacional "Pedro Henríquez Ureña," Decanato Ciencias Agropecuarias, Km 5½ Aut. Duarte, Santo Domingo, Dominican Republic
G.F.W. Haenlein, Department of Animal Science and Agricultural Biochemistry, University of Delaware, Newark, Delaware 19717-1303, USA
E. Henson, Bemborough, Guiting Power, Cheltenham, Gloucester, UK (rare breeds)
D.L. Huss, P.O. Box 426, Menard, Texas 76859, USA
N. Ibàñez Herrera, Paraguay 160—Urbanización, Apartado 315, El Recreo, Trujillo, Peru
P. Jewell, Department of Physiology, University of Cambridge, Cambridge CB2 1TN, UK
Jones Sheep Farm, RR 2 Box 185-55, Peabody, Kansas 66866, USA
P.E. Loggins, Animal Science Department, 108-D Suite, University of Florida, Gainesville, Florida 32611, USA
J.K. Loosli, 406 S.W. 40th Street, Gainesville, Florida 32607, USA
G.C.H. Mandari, Livestock Research Centre, West Kilimanjaro, P.O. Box 103, Hai, Tanzania
J.P. Maule, Hannahfield Quarry House, 578 Lanark Road West, Balerno, Midlothian EH14 7BN, UK
L.K. McNeal, Animal, Dairy, and Veterinary Science Department, UNC 48, Utah State University, Logan, Utah 84322-4815, USA
I. Mecha, Department of Animal Science, University of Nigeria, Nsukka, Anambra State, Nigeria
K. Mitchell, Baptist Bible Institute, P.O. Box 301, Basseterre, St. Kitts, West Indies
A. Monsi, Department of Animal Science, Food Science and Technology, Rivers State University of Science and Technology, P.M.B. 5080, Port Harcourt, Nigeria
C.M. Munyabuntu, Department of Animal Science, Makerere University, P.O. Box 7062, Kampala, Uganda

Th. Murayi, Institut des Sciences Agronomiques du Rwanda, P.O. Box 138, Butare, Rwanda

W.N.M. Mwenya, School of Agriculture Sciences, University of Zambia, Box 32379, Lusaka, Zambia

M. Ngendahayo, Institut des Sciences Agronomiques du Rwanda, B.P. 138, Butare, Rwanda

R. Njwe, Département des Sciences Animales, Ecole Nationale Supérieure Agronomique, Centre Universitaire de Dschang, B.P. 138, Yaoundé, United Republic of Cameroon

W.J.A. Payne, 91 Bedwardine Road, London SE19 3AY, UK

R.W. Phillips, The Representative, Apartment 810, 1101 South Arlington Ridge Road, Arlington, Virginia 22202, USA

G.V. Raghavan, Animal Science Department, College of Veterinary Science, Andhra Pradesh Agricultural University, Rajendranagar, Hyderabad 500 030, Andhra Pradesh, India

A. Ramamohana Rao, Postgraduate Studies, Andhra Pradesh Agricultural University, Rajendranagar, Hyderabad, Andhra Pradesh 500 030, India

R.K. Rastogi, Department of Livestock Science, University of the West Indies, St. Augustine, Trinidad, West Indies

Soedomo Reksohadiprodjo, Department of Animal Nutrition, Faculty of Animal Husbandry, Universitas Gadjah Mada, Sekip Blok N-42, Yogyakarta, Indonesia

A.Y. Robles, Dairy Training and Research Institute, University of the Philippines at Los Baños, College, Laguna 3732, Philippines

A. Rocha, Instituto de Reprodução e Melhoramento Animal, C.P. 25, Matola, Mozambique

M.L. Ryder, c/o H.F.R.O., Bush Estate, Penicuik, Edinburgh, UK (Barbados blackbelly, wool)

Sie Chenxia, Department of Animal Science, Nanjing Agricultural University, Nanjing, People's Republic of China

C. David Simpson, Range and Wildlife Department, Texas Tech University, P.O. Box 4349, Lubbock, Texas 79409, USA

C.E. Terrill, U.S. Department of Agriculture, Building 005, BARC-West, Beltsville, Maryland 20705, USA

H.N. Turner, Genetics Research Laboratories, CSIRO, P.O. Box 184, North Ryde, New South Wales 2113, Australia (Indonesian breeds)

R. Vaccaro, Universidad Nacional Agraria, La Molina, P.O. Box 456, Lima, Peru

R.G. Wahome, Department of Animal Production, P.O. Box 29053, Nairobi, Kenya

S. Wildeus, College of the Virgin Islands, Agricultural Experiment Station, R.R. 2, Box 10051 Kingshill, St. Croix, Virgin Islands 00850, USA

R.T. Wilson, Bartridge House, Umberleigh, North Devon EX37 9AS, UK

Wu Jitang, Animal and Veterinary Science, Jiangsu Province Academy of Agricultural Sciences, Nanjing, People's Republic of China

A. Yenikoye, Faculté d'Agronomie, Université de Niamey, B.P. 10662, Niamey, Niger

## Micropig

I.A. Akinbode, Faculty of Agriculture, University of Ife, Ile-Ife, Nigeria

D.O. Alonge, Meat Hygiene and Preservation Laboratory, Faculty of Veterinary Medicine, University of Ibadan, Ibadan, Nigeria

J.M. Andrews, Mesoamerican Ecology Institute, Tulane University, New Orleans, Louisiana 70118, USA (Yucatan pig)
G. Ashton-Peach, Box 17, Ohaeawai, Bay of Islands, New Zealand (kunekune)
M. Berruecos V. Muz, Facultad de Medicina Veterinaria y Zootecnia, Ciudad Universitaria, Mexico D.F., Mexico
R. Blakely, Sedgwick County Zoo, Wichita, Kansas 67203, USA (Ossabaw)
K. Boateng, Cooperative Extension Service, University of the Virgin Islands, Box L., Kingshill, St. Croix, Virgin Islands 00850, USA
I.L. Brisbin, Savannah River Ecology Laboratory, P.O. Drawer E, Aiken, South Carolina 29801, USA (Ossabaw)
L. Bustad, Veterinary School, Washington State University, Pullman, Washington, 99163, USA
C. Catibog-Sinha, Forest Products Research and Development Institute, College, Laguna 4031, Philippines
Centro de Investigaciones Pecuarias (CENIP), A.P. 227-9, Santo Domingo, Dominican Republic
Charles River Laboratories, Inc., 251 Ballardvale Street, Wilmington, Massachusetts 01887-0630, USA
J.H. Conrad, Department of Animal Science, University of Florida, Gainesville, Florida 32611, USA
T.J. Cunha, PO Box R, LaVerne, California 91750, USA
F. Dederer, 712 Brown Street, Philadelphia, Pennsylvania 19123, USA
A. Dettmers, 5620 Longwood Avenue, Port Charlotte, Florida 33953, USA (miniature pigs)
C. Devendra, Agriculture Food and Nutrition Sciences Division, International Development Research Centre (IDRC), Tanglin P.O. Box 101, 9124 Singapore
N.I. Dim, Department of Animal Science, Faculty of Agriculture, Ahmadu Bello University, P.M.B. 1044, Zaria, Nigeria
G. Edgar, Greenhill, 1 R.D., Balfour, Southland, New Zealand (kunekune)
H. Hendricks, Department of Animal, Dairy, and Veterinary Sciences, Utah State University, Logan, Utah 84322-4815, USA
B.J. Hepburn, Swine Program, Charles River Laboratories, 251 Ballardvale Street, Wilmington, Massachusetts 01887-0630, USA (laboratory pigs)
P. Holbrook, Aria Road, via Te Kuiti, New Zealand (kunekune)
R.R. Hook, Jr., Sinclair Medical Research Farm, University of Missouri, Route 3, Columbia, Missouri 65203, USA (laboratory pigs)
J.O. Ilori, Faculty of Agriculture, University of Ife, Ile-Ife, Nigeria
G.E. Kodituwakku, Department of Veterinary Clinical Studies, University of Peradeniya, Peradeniya, Sri Lanka
J.K. Loosli, 406 S.W. 40th Street, Gainesville, Florida 32607, USA
J.H. Maner, Winrock International, Route 3, Petit Jean Mountain, Morrilton, Arkansas 72110, USA
J.J. Mayer, Savannah River Ecology Laboratory, P.O. Drawer E, Aiken, South Carolina 29801, USA (Ossabaw)
K. Mitchell, Baptist Bible Institute, P.O. Box 301, Basseterre, St. Kitts, West Indies
A. Monsi, Departments of Animal Science/Food Science and Technology, Rivers State University of Science and Technology, P.M.B. 5080, Port Harcourt, Nigeria
J. Mtimuni, Livestock Production Department, Bunda College of Agriculture, University of Malawi, P.O. Box 219, Lilongwe, Malawi
D.G. Mugisa, c/o Insure Consult, P.O. Box 30002, Nakivubo, Kampala, Uganda
K. Mull, 2834 Hamner Avenue #208, Norco, California 91760, USA
W.L.R. Oliver, Pigs and Peccaries Specialist Group, Jersey Wildlife Preservation Trust, Les Augrès Manor, Jersey, Channel Islands, UK (pigmy hog)

L. Panepinto, Redstone Canyon, P.O. Box 105, Masonville, Colorado 80541, USA (miniature pigs)

W.J.A. Payne, 91 Bedwardine Road, London SE19 3AY, UK

C.C. Perdomo, Centro Nacional de Pesquisa de Suinos e Aves, BR 153, km 110, CP D-3, 89700 Concordia, SC, Brazil

R.W. Phillips, The Representative, Apartment 810, 1101 South Arlington Ridge Road, Arlington, Virginia 22202, USA (Chinese breeds)

W.G. Pond, Meat Animal Research Center, U.S. Department of Agriculture, Box 166, Clay Center, Nebraska 68933, USA

H. Popenoe, International Programs in Agriculture, 3028 McCarty Hall, University of Florida, Gainesville, Florida 32611, USA (New Guinea pigs)

B.S. Reddi, College of Veterinary Science, Andhra Pradesh Agricultural University, Tirupati 517 502, India

S. Avendañó Reyes, Instituto Nacional de las Investigaciones sobre los Recursos Biotécnicos, (INIREB), Apartado Postal 97, Mérida, Yucatan, Mexico (Yucatan)

C. Robinson, Robinson's Racing Pigs, 5809 20th Avenue South, Tampa, Florida 33619-5457, USA (miniature pigs)

C. Arellano Sota, Regional Office for Latin America and the Caribbean, Food and Agriculture Organization of the United Nations, Casillo 10095, Santiago, Chile (Cuino)

A. Sriramamurty, College of Veterinary Science, Andhra Pradesh Agricultural University, Tirupati 517 502, India

M.M. Swindle, Comparative Medicine/Lab Animal Resources, Medical University of South Carolina, 171 Ashley Avenue, Charleston, South Carolina 29425, USA (laboratory pigs)

N. Tilakaratne, Veterinary Research Institute, Gannoruwa, Peradeniya, Sri Lanka

R.G. Wahome, Department of Animal Production, P.O. Box 29053, Nairobi, Kenya

Wu Jitang, Animal and Veterinary Science, Jiangsu Provincial Academy of Agricultural Sciences, Nanjing, People's Republic of China

Yong Shen Lee, Department of Animal Husbandry and Veterinary Medicine, South China Agricultural University, Guangzhou, People's Republic of China

## POULTRY

Information on poultry can generally be obtained from local ministries of agriculture or veterinary and livestock services. However, another source is the World's Poultry Science Association (c/o Institut für Kleintierzucht, Dornbergstrasse 25/27, Postfach 280, C-3100 Celle, Germany). It has over 5,000 members from 40 national groups. Its objectives are (1) to advance poultry science and the poultry industry, (2) to disseminate and facilitate the exchange of knowledge pertaining to all branches of the poultry industry, (3) to encourage the promotion of world poultry congresses and regional conferences, and (4) to cooperate with other international organizations. It maintains working groups to explore and assess research work on tropical poultry problems. The *World's Poultry Science Journal* is published 3 times a year. It has also published a multilingual poultry dictionary.

### Chicken

A.O. Aduku, Department of Animal Science, Faculty of Agriculture, Ahmadu Bello University, P.M.B. 1044, Zaria, Nigeria

D.O. Alonge, Meat Hygiene and Preservation Laboratory, Faculty of Veterinary Medicine, University of Ibadan, Ibadan, Nigeria

A. Ben-David, Poultry Production Specialist, 5, Azar Street, Holon 58291, Israel

I.L. Brisbin, Jr., Savannah River Ecology Laboratory, P.O. Drawer E, Aiken, South Carolina 29801, USA

J.K. Camoens, Senior Livestock Specialist, Asian Development Bank, P.O. Box 789, Manila 2800, Philippines

J.R. Couch, Poultry Science Department, Texas A&M University, College Station, Texas 77843, USA

R.D. Crawford, Department of Animal and Poultry Science, University of Saskatchewan, Saskatoon, Saskatchewan S7N 0W0, Canada

M. Cuca G., Institución de Enseñanza e Investigación en Ciencias Agrícolas, Colegio de Postgraduados, Chapingo, Mexico

N.I. Dim, Animal Science Department, Faculty of Agriculture, Ahmadu Bello University, P.M.B. 1044, Zaria, Nigeria

D. Fielding, Centre for Tropical Veterinary Medicine, Royal (Dick) School of Veterinary Studies, University of Edinburgh, Easter Bush, Roslin, Midlothian EH25 9RG, Scotland, UK

R.O. Hawes, Department of Animal and Veterinary Sciences, University of Maine, Orono, Maine 04469, USA

D. Holderread, P.O. Box 492, Corvallis, Oregon 97339, USA

C.H. Huang, Food and Fertilizer Technology Center for the Asian and Pacific Region, Fifth Floor, 14 Wenchow Street, Taipei, Taiwan

E.H. Ketelaars, Department of Tropical Husbandry, P.O. Box 338, 6700 A.H. Wageningen, The Netherlands

M.A. Latif, Department of Poultry Science, Bangladesh Agricultural University, Mymensingh, Bangladesh

H. Martoyo, Faculty of Animal Science, Institut Pertanian Bogor, Jalan Raya Pajajaran, Bogor, West Java, Indonesia

K. Mitchell, Baptist Bible Institute, P.O. Box 301, Basseterre, St. Kitts, West Indies

E.T. Moran, Jr., Poultry Science Department, Auburn University, Auburn, Alabama 36849-4291, USA

S. Ochetim, School of Agriculture, University of the South Pacific, Alafua Campus, Apia, Western Samoa

S. Parks, Route 1, Box 47-C, Brookshire, Texas 77423, USA (Araucanian chicken)

R.W. Phillips, The Representative, Apt. 810, 1101 South Arlington Ridge Road, Arlington, Virginia 22202, USA

H.L. Polley, Programme Développement Plateau Bateke, CBZO, B.P. 4728, Kinshasa 2, Zaire

A. Ramakrishman, Center for Advanced Studies in Poultry Science, Kerala Agricultural University, Mannuthy 680 651, Kerala, India

C.V. Reddy, Faculty of Veterinary Science, Andhra Pradesh Agricultural University, Rajendranagar, Hyderabad 500 030, India

E. Ross, Department of Animal Science, University of Hawaii, Honolulu, Hawaii 96822, USA

M.L. Scott, 16 Spruce Lane, Ithaca, New York 14850, USA

T.J. Sexton, Agricultural Research Service, U.S. Department of Agriculture, Beltsville Agricultural Research Center, Building 262, Beltsville, Maryland 20705, USA

R.G. Somes, Nutritional Sciences, Box U-17, University of Connecticut, Storrs, Connecticut 06268, USA

M. Sunde, Poultry Science Department, University of Wisconsin, Madison, Wisconsin 53706, USA

R.T. Wilson, Bartridge House, Umberleigh, North Devon EX37 9AS, UK

## Duck

R.E. Abdelsamie, Department of Biochemistry, Microbiology, and Nutrition, University of New England, Armidale, New South Wales 2351, Australia

S. Ahmed, Department of Poultry Science, Bangladesh Agricultural University, Mymensingh, Bangladesh

A. Bande, PATI, Popondetta Agricultural College, Box 131, Popondetta, Oro Province, Papua New Guinea

R.S. Bird, Cherry Valley Farms Ltd., Rothwell, Lincoln LN7 6BR, UK

V.D. Bulbule, Central Duck Breeding Farm, Government of India, Hessaraghatta, Bangalore 560 088, India

J.K. Camoens, Asian Development Bank, P.O. Box 789, Manila 2800, Philippines

K.M. Cheng, Avian Genetics Laboratory, Department of Animal Science, University of British Columbia, Vancouver, British Columbia V6T 2A2, Canada

W.F. Dean, Cornell University Duck Research Laboratory, P.O. Box 217, Eastport, New York 11941, USA

N.I. Dim, Animal Science Department, Faculty of Agriculture, Ahmadu Bello University, P.M.B. 1044, Zaria, Nigeria

P. Edwards, Agricultural and Food Engineering Division, Asian Institute of Technology, P.O. Box 2754, Bangkok 10501, Thailand

L. Fishelson, Department of Zoology, Tel Aviv University, Ramat-ave, Tel Aviv 69978, Israel

W.F. Floresca, Department of Animal Science and Veterinary Medicine, Visayas State College of Agriculture, Baybay, Leyte 7127, Philippines

H. Godoy, Ingeniero Agronomío, Simon Bolivar 2397, Santiago, Chile

P. Hardjosworo, Fakultas Peternakan, Institut Pertanian Bogor, Bogor, Indonesia

R. Henry, Cherry Valley Farms Limited, Divisional Offices, North Kelsey Moor, Lincoln LN7 6HH, UK

D. Holderread, P.O. Box 492, Corvallis, Oregon 97339, USA

W.F. Hollander, Route 4, Box 168, Ames, Iowa 50010, USA

E. Hoffmann, Canning, Nova Scotia B0P 1H0, Canada

C.H. Huang, Food and Fertilizer Technology Center for the Asian and Pacific Region, Fifth Floor, 14 Wenchow Street, Taipei, Taiwan

G. Ibañez H., Bolivar 451, Trujillo, Peru

Z. Jungkaro, Research Institute for Inland Fisheries, Jalan Sempur No. 1, Indonesia

G.A.R. Kamar, Animal Production Department, Faculty of Agriculture, Cairo University, Giza, Egypt

E.H. Ketelaars, Department of Tropical Husbandry, P.O. Box 338, 6700 A.H. Wageningen, The Netherlands

I.P. Kompiang, Bali Penelitian Ternak, P.O. Box 123, Bogor 16001, West Java, Indonesia

M.A. Latif, Department of Poultry Science, Bangladesh Agricultural University, Mymensingh, Bangladesh

H. Martoyo, Faculty of Animal Science, Institut Pertanian Bogor, Jalan Raya Pajajaran, Bogor, Indonesia

P. Mongin, Station de Recherches Avicoles, I.N.R.A., Nouzilly 37380 Monnaie, France

E.T. Moran, Jr., Poultry Science Department, Auburn University, Auburn, Alabama 36849-4291, USA

S. Ochetim, School of Agriculture, University of the South Pacific, Alafua Campus, Apia, Western Samoa

R.W. Phillips, The Representative, Apartment 810, 1101 South Arlington Ridge Road, Arlington, Virginia 22202, USA

H. Pingel, Karl-Marx-Universität, Sektion TV, Lehrstuhl Geflügel- und Pelztierzucht, 7010 Leipzig, Stephanstrasse 12, Germany

S. Poernomo, Research Institute for Veterinary Science, Jalan R.E. Martadinata 32, P.O. Box 52, Bogor, West Java, Indonesia
A. Ramakrishman, Center for Advanced Studies in Poultry Science, Kerala Agricultural University, Mannuthy 680 651, Kerala, India
B. Retailleau, Grimaud Frères S.A., La Corbière, 49450 Roussay, France
T.S. Sandhu, Cornell University Duck Research Laboratory, P.O. Box 217, Eastport, New York 11941, USA
M.L. Scott, 16 Spruce Lane, Ithaca, New York 14850, USA
B.K. Shingari, Department of Animal Science, Punjab Agricultural University, Ludhiana 141 004, India
C. Tai, Taiwan Livestock Research Institute, 112 Farm Road Shin-hua, Tainan, Taiwan 71210
B. Tangendjaja, Bali Penelitian Ternak, P.O. Box 123, Bogor 16001, Indonesia
Tian Fuh Shen, Department of Animal Husbandry, National Taiwan University, Taipei 10764, Taiwan
S.S. Tsai, Laboratory of Veterinary Pathology, National Pintung Institute of Agriculture, Pintung, Taiwan
A. Romero Villanueva, Centro Tropical de Capacitación Agropecuaria y Forestal, Apartado Postal #45, Macuspana 86700 Tabasco, Mexico
R.T. Wilson, Bartridge House, Umberleigh, North Devon EX37 9AS, UK
Wildfowl Trust, Slimbridge, UK
Yeong Shue Woh, Malaysian Agricultural Research and Development Institute (MARDI), G.P.O. No. 12301, 50774 Kuala Lumpur, Malaysia
Cha Tak Yimp, Poultry and Swine Unit, Department of Veterinary Services, Jalan Mahameru, 50630 Kuala Lumpur, Malaysia

## Geese

Many universities have a department of animal or poultry science that should know about the culture and management of these birds.

A. Bande, PATI, Popondetta Agricultural College, Box 131, Popondetta, Oro Province, Papua New Guinea
J.K. Camoens, Asian Development Bank, P.O. Box 789, Manila 2800, Philippines
F.D. Chesterley, 4502 Loomis Trail, Blaine, Washington 98230, USA
R.D. Crawford, Department of Animal and Poultry Science, University of Saskatchewan, Saskatoon, Saskatchewan S7N 0W0, Canada
N.I. Dim, Animal Science Department, Faculty of Agriculture, Ahmadu Bello University, P.M.B. 1044, Zaria, Nigeria
D. Holderread, P.O. Box 492, Corvallis, Oregon 97339, USA
S. Jackson, P.O. Box 3, Greenbank, Washington 98253, USA (weeder geese)
Liu Fuan, Department of Animal Husbandry and Veterinary Medicine, South China Agricultural University, Guangzhou, People's Republic of China
P. Mongin, Station de Recherches Avicoles, I.N.R.A., Nouzilly 37380, Monnaie, France
E. Ross, Department of Animal Science, University of Hawaii, Honolulu, Hawaii 96822, USA
B.K. Shingari, Department of Animal Science, Punjab Agricultural University, Ludhiana 141 004, India
H.R. Wilson, Poultry Science Department, University of Florida, Gainesville, Florida 32611, USA

## Guinea Fowl

A.O. Aduku, Department of Animal Science, Faculty of Agriculture, Ahmadu Bello University, P.M.B. 1044, Zaria, Nigeria

American Guinea Club, c/o Angie Papp, 620 Payne Road, New Albany, Indiana 47150, USA

J.S.O. Ayeni, Kainji Lake Research Institute, P.M.B. 666, New Bussa, Kwara State, Nigeria

A. Ben-David, 5, Azar Street, Holon 58291, Israel

G.S. Brah, Department of Animal Science, Punjab Agricultural University, Ludhiana 141 004, India

X. Colonna C., 14 Cotham Road, Bristol BS6 6DR, UK

N.I. Dim, Department of Animal Science, Faculty of Agriculture, Ahmadu Bello University, P.M.B. 1044, Zaria, Nigeria

R. Dumont, Ecole Nationale Supérieure des Sciences Agronomiques Appliquées, 26, boulevard Docteur Petitjean 21100 Dijon, France

R. Etches, Department of Animal and Poultry Science, University of Guelph, Guelph, Ontario N1G 2W1, Canada

D. Holderread, P.O. Box 492, Corvallis, Oregon 97339, USA

A.F. Hutton, Garaina Farms, Scrubby Creek Road, Gympie MS:115, Queensland 4570, Australia

D.N. Johnson, Natal Parks, Game and Fish Preservation Board, P.O. Box 661, Pietermaritzburg 3200, South Africa

S.D. Lukefahr, International Small Livestock Research Center, Department of Food Science and Animal Industries, Alabama A&M University, P.O. Box 264, Normal, Alabama 35762, USA

W. Michie, Poultry Husbandry Division, North of Scotland College of Agriculture, Parkhead, Craibstone, Bucksburn, Aberdeen AB2 9SX, Scotland, UK

P. Mongin, Station de Recherches Avicoles, I.N.R.A., Nouzilly, 37380 Monnaie, France

S. Ochetim, School of Agriculture, University of the South Pacific, Alafua Campus, Apia, Western Samoa

A.N. Okaeme, Wildlife/Range Division, Kainji Lake Research Institute, P.M.B. 666, New Bussa, Kwara State, Nigeria

J.S. Sandhu, Poultry Breeder, Department of Animal Science, Punjab Agricultural University, Ludhiana 141 004, India

B.K. Shingari, Department of Animal Science, Punjab Agricultural University, Ludhiana 141 004, India

R.G. Somes, Nutritional Sciences, Box U-17, University of Connecticut, Storrs, Connecticut 06268, USA

R.T. Wilson, Bartridge House, Umberleigh, North Devon EX37 9AS, UK

## Muscovy

R.E. Abdelsamie, Department of Biochemistry, Microbiology, and Nutrition, University of New England, Armidale, New South Wales 2351, Australia

G.P. Bilong, Department of Primary Industry, Monogastric Research Centre, P.O. Box 73, Lae, Papua New Guinea

J.K. Camoens, Asian Development Bank, P.O. Box 789, Manila 2800, Philippines

H. de Carville, Station de Recherches Avicoles, I.N.R.A., Centre de Tours, Nouzilly 37380, Monnaie, France

F.D. Chesterley, 4502 Loomis Trail, Blaine, Washington 98230, USA
X. Colonna C., 14 Cotham Road, Bristol BS6 6DR, UK
N.I. Dim, Animal Science Department, Faculty of Agriculture, Ahmadu Bello University, P.M.B. 1044, Zaria, Nigeria
R. Dumont, Ecole Nationale Supérieure des Sciences Agronomiques Appliquées, 26, boulevard Docteur Petitjean 21100 Dijon, France
G.R. Gerona, Department of Animal Science and Veterinary Medicine, Visayas State College of Agriculture, Baybay, Leyte, Philippines
Barry Glofcheskie, Department of Environmental Biology, University of Guelph, Guelph, Ontario N1G 2W1, Canada (fly control)
D. Holderread, P.O. Box 492, Corvallis, Oregon 97339, USA
W.F. Hollander, Route 4, Box 168, Ames, Iowa 50010, USA
E. Hoffmann, Canning, Nova Scotia B0P 1H0, Canada
C.H. Huang, Food and Fertilizer Technology Center for the Asian and Pacific Region, Fifth Floor, 14 Wenchow Street, Taipei, Taiwan (mule duck)
G.A.R. Kamar, Animal Production Department, Faculty of Agriculture, Cairo University, Giza, Egypt
W. Michie, Poultry Husbandry Division, North of Scotland College of Agriculture, Parkhead, Craibstone, Bucksburn, Aberdeen AB2 9SX, Scotland, UK
Donald E. Mock, Department of Extension Entomology, 239 Waters Hall, Kansas State University, Manhattan, Kansas 66506-4004, USA (fly control)
P. Mongin, Station de Recherches Avicoles, I.N.R.A., Nouzilly 37380 Monnaie, France
S. Ochetim, School of Agriculture, University of the South Pacific, Alafua Campus, Apia, Western Samoa
H. Pingel, Karl Marx Universität, Sektion TV, Lehrstuhl Geflügel- und Pelztierzucht, 7010 Leipzig, Stephanstrasse 12, Germany
B. Retailleau, Grimaud Frères S.A., La Corbière, 49450 Roussay, France
I. Romboli, Le Quericole, Universita Delgi Studi di Pisa 56010 S. Piero A Grado, Italy
T.S. Sandhu, Cornell University Duck Research Laboratory, P.O. Box 217, Eastport, New York 11941, USA
G.A. Surgeoner, Department of Environmental Biology, University of Guelph, Guelph, Ontario N1G 2W1, Canada (fly control)
C. Tai, Taiwan Livestock Research Institute, 112 Farm Road Shin-hua, Tainan, Taiwan 71210
R.T. Wilson, Bartridge House, Umberleigh, North Devon EX37 9AS, UK

## Pigeon

A.O. Aduku, Department of Animal Science, Faculty of Agriculture, Ahmadu Bello University, P.M.B. 1044, Zaria, Nigeria
D.O. Alonge, Meat Hygiene and Preservation Laboratory, Faculty of Veterinary Medicine, University of Ibadan, Ibadan, Nigeria
J.K. Camoens, Asian Development Bank, P.O. Box 789, Manila 2800, Philippines
K.M. Cheng, Avian Genetics Laboratory, Department of Animal Science, University of British Columbia, Vancouver, British Columbia V6T 2A2, Canada
R. Dillinger, 3201 Huffman Boulevard, Rockford, Illinois 61103, USA
N.I. Dim, Animal Science Department, Faculty of Agriculture, Ahmadu Bello University, P.M.B. 1044, Zaria, Nigeria
R.O. Hawes, Department of Animal and Veterinary Sciences, University of Maine, Orono, Maine 04469, U.S.A.
D. Holderread, P.O. Box 492, Corvallis, Oregon 97339, USA (carrier pigeons)

W.F. Hollander, Route 4, Box 168, Ames, Iowa 50010, USA
M.A. Latif, Department of Poultry Science, Bangladesh Agricultural University, Mymensingh, Bangladesh
Y. Le Hénaff, Laboratoire de l'Hôpital Avranches, Avranches, Cotentin, France (carrier pigeons)
S.D. Lukefahr, International Small Livestock Research Center, Department of Food Science and Animal Industries, Alabama A&M University, P.O. Box 264, Normal, Alabama 35762, USA
G.A.R. Kamar, Animal Production Department, Faculty of Agriculture, Cairo University, Giza, Egypt
W.J. Miller, Department of Genetics, Iowa State University, Ames, Iowa 50011, USA
G. Owen, 6520 Wiscasset Road, Bethesda, Maryland 20816, USA
G. Peña Pérez, Apartado Postal 23, La Nucía, Alicante, Spain
E. Ross, Department of Animal Science, University of Hawaii, Honolulu, Hawaii 96822, USA
M.D. Roush, Green Valley Squab Farm, 1348 Green Valley Road, Folsom, California 95630, USA
T.S. Sandhu, Cornell University Duck Research Laboratory, P.O. Box 217, Eastport, New York 11941, USA
J.S. Sim, Department of Animal Science, University of Alberta, Edmonton, Alberta T6G 2P5, Canada
R.T. Wilson, Bartridge House, Umberleigh, North Devon EX37 9AS, UK

## Quail

Many people raise Japanese quail. Probably the best way to locate a local source is to consult local pet stores, the county farm advisor's office, university faculty of agriculture, game-bird fanciers, and the like. Finding people with local research experience is usually most useful to beginners.

Coturnix International, Inc., is a nonprofit organization with primary interests in promoting small domesticated food animal production for both youth training and improved nutrition. The use of coturnix (Japanese quail) for these purposes is already well established in a number of Michigan schools and in the Dominican Republic, with special emphasis on science projects, especially where nutritional inadequacies exist. The address is Coturnix International Inc., International Headquarters, 1111 Michigan Avenue, Box 2500, East Lansing, Michigan 48823, USA.

Avian Sciences Department, University of California, Davis, California 95616, USA
J.K. Camoens, Asian Development Bank, P.O. Box 789, Manila 2800, Philippines
K.M. Cheng, Avian Genetics Laboratory, Department of Animal Science, University of British Columbia, Vancouver, British Columbia V6T 2A2, Canada
D.E. Deppner, Trees for the Future, 11306 Estona Drive, Silver Spring, Maryland 20902, USA
D. Holderread, P.O. Box 492, Corvallis, Oregon 97339, USA
S.D. Lukefahr, International Small Livestock Research Center, Department of Food Science and Animal Industries, Alabama A&M University, P.O. Box 264, Normal, Alabama 35762, USA
H.L. Marks, Genetics Unit, Room 107 Livestock-Poultry Building, University of Georgia, Athens, Georgia 30602, USA
A. Marsh, Marsh Farms, 7171 Patterson Drive, Garden Grove, California 92641, USA
H. Martoyo, Faculty of Animal Science, Institut Pertanian Bogor, Jalan Raya Pajajaran, Bogor, West Java, Indonesia

Z. Abidin bin Mohd Noor, Department of Veterinary Services, Jalan Sultan Salahuddin, 50630 Kuala Lumpur, Malaysia (chicken-quail hybrids)
P.K. Pani, Indian Veterinary Research Institute, Izatnagar 243 122, Bareilly, Uttar Pradesh, India
H. Pingel, Karl Marx Universität, Sektion TV, Lehrstuhl Geflügel- und Pelztierzucht, 7010 Leipzig, Stephanstrasse 12, Germany
D. Polin, Department of Animal Science, Michigan State University, East Lansing, Michigan 48824, USA
C.V. Reddy, Faculty of Veterinary Science, Andhra Pradesh Agricultural University, Rajendranagar, Hyderabad 500 030, India
Rodale Research Center, RD #1, Box 323, Kutztown, Pennsylvania 19530, USA
E. Ross, Department of Animal Science, University of Hawaii, Honolulu, Hawaii 96822, USA
B.K. Shingari, Department of Animal Science, Punjab Agricultural University, Ludhiana 141 006, India
R.G. Somes, Nutritional Sciences, Box U-17, University of Connecticut, Storrs, Connecticut 06268, USA
R. Tempest, Heifer Project International, P.O. Box 808, Little Rock, Arkansas 72203, USA
S.K. Varghese, Department of Animal Science, Michigan State University, East Lansing, Michigan 48824-1225, USA
N. Wakasugi, Laboratory of Animal Genetics, Nagoya University, Furo-cho, Chikusa-ku, Nagoya, Japan
H.R. Wilson, Poultry Science Department, Archer Road, University of Florida, Gainesville, Florida 32611, USA

## Turkey

M. Berruecos V. Muz, Facultad de Medicina Veterinaria y Zootecnia, Ciudad Universitaria, Mexico D.F., Mexico
R.D. Crawford, Department of Animal and Poultry Science, University of Saskatchewan, Saskatoon, Canada
M. Cuca G., Institución de Enseñanza e Investigación en Ciencias Agrícolas, Colegio de Postgraduados, Chapingo, Mexico
K.E. Nestor, Ohio Agricultural Research and Development Center, Ohio State University, Wooster, Ohio
H. Pingel, Karl-Marx-Universität, Sektion TV, Lehrstuhl Geflügel- und Pelztierzucht, 7010 Leipzig, Stephanstrasse 12, Germany
M.L. Scott, 16 Spruce Lane, Ithaca, New York 14850, USA
T.J. Sexton, Agricultural Research Service, U.S. Department of Agriculture, Beltsville Agricultural Research Center, Building 262, Beltsville, Maryland 20705, USA
R.G. Somes, Nutritional Sciences, Box U-17, University of Connecticut, Storrs, Connecticut 06268, USA

## Potential New Poultry

J.E. Duckett, P.O. Box 12378, 50776 Kuala Lumpur, Malaysia (barn owls)
W.R. Marion, Department of Wildlife and Range Science, University of Florida, Gainesville, Florida 32611, USA (chachalaca)
C. Smal, Wildlife Service, Sidmonton Place, Bray County, Wicklow, Ireland (barn owls)

# RABBITS

## Domestic Rabbit

A.O. Aduku, Department of Animal Science, Faculty of Agriculture, Ahmadu Bello University, P.M.B. 1044, Zaria, Nigeria
J. Baranga, Department of Zoology, Makerere University, P.O. Box 7062, Kampala, Uganda
M. Baselga, Departamento de Ciencia Animal, Escuela Técnica Superior de Ingenieros Agrónomos, Camino de Vera 14, 46022 Valencia, Spain
C. de Blas, Departamento Alimentación Animal, Escuela Técnica Superior de Ingenieros Agrónomos, Ciudad Universitaria, 28040 Madrid, Spain
J.A. Castelló, Real Escuela Oficial y Superior de Avicultura (REOSA), Calle Plana Paraíso, s/n Arenys de Mar, Barcelona, Spain
C.B. Chawan, International Small Livestock Research Center, Department of Food Science and Animal Industries, Alabama A&M University, P.O. Box 264, Normal, Alabama 35762, USA
P.R. Cheeke, Rabbit Research Center, Department of Animal Science, Oregon State University, Corvallis, Oregon 97331, USA
S. Fekete, University of Veterinary Science, Department of Animal Nutrition, Budapest, Pf.2, H-1400 Hungary
J. Fernandez C., Departamento de Ciencia Animal, Escuela Técnica Superior de Ingenieros Agrónomos, Universidad Politécnica de Valencia, Camino de Vera, 14, 46022 Valencia, Spain
C. Cervera Fras, Departamento de Ciencia Animal, Universidad Politécnica de Valencia, Apartado 22012, Valencia, Spain
D.J. Harris, Pel-Freez Rabbit Meat, Inc., P.O. Box 68, Rogers, Arkansas 72756, USA
J.A.B. Hundleby, Ciskei Agricultural Corporation Ltd., P.O. Box 59, Bisho, Ciskei, South Africa
D.L. Huss, P.O. Box 426, Menard, Texas 76859, USA
R. de Jong, Department of Tropical Animal Production, Marykeweg 40, 6709 P.G. Wageningen, The Netherlands
L.A. Kamwanja, Bunda College of Agriculture, University of Malawi, P.O. Box 219, Lilongwe, Malawi
M. Leyún, Instituto Técnico Gestión Porcino (ITGP), Sección Conejos, Ctra. del Sadar, s/n 31006 Pamplona (Navarra), Spain
H. Löliger, Lisztstrasse 20, 3100 Celle, Germany
S.D. Lukefahr, International Small Livestock Research Center, Department of Food Science and Animal Industries, Alabama A&M University, P.O. Box 264, Normal, Alabama 35762, USA
T.P.E. Makhambera, Bunda College of Agriculture, University of Malawi, P.O. Box 219, Lilongwe, Malawi
J.I. McNitt, Center for Small Farm Research, Southern University, P.O. Box 11170, Baton Rouge, Louisana 70813-1170, USA
M. Mgheni, Sokoine University of Agriculture, Animal Science and Production Department, P.O. Box 3004, Morogoro, Tanzania
K. Mitchell, Baptist Bible Institute, P.O. Box 301, Basseterre, St. Kitts, West Indies
G.A. Norman, Tropical Products Institute, Culham, Abingdon, Oxfordshire OX14 3DA, UK
L.N. Odonker, P.O. Box 11, Tsito-Awudome, Volta Region, Ghana
O.C. Onwudike, Department of Animal Science, University of Ife, Ile-Ife, Nigeria

G.G. Partridge, Rowett Research Institute, Bucksburn, Aberdeen AB2 9SB, Scotland, UK
N.M. Patton, Rabbit Research Center, Oregon State University, Corvallis, Oregon 97331, USA
M.J. Paya, Departamento de Microbiología, Facultad de Veterinaria, Avenida Puerta de Hierro, Ciudad Universitaria, 28040 Madrid, Spain
T. Pérez, Departamento de Reproducción Animal, Instituto Nacional de Investigaciones Agrarias (INIA), Avenida Puerta de Hierro, s/n, 28040 Madrid, Spain
H.L. Polley, Programme Développement Plateau Bateke, CBZO, B.P. 4728, Kinshasa 2, Zaire
R. Ramchurn, School of Agriculture, University of Mauritius, Réduit, Mauritius
T.E. Reed, Box 98, Markle, Indiana 46770, USA
E. Respaldiza, Instituto Nacional de Investigaciones Agrarias (INIA), Departamento de Higiene y Sanidad Animal, Calle Embajadores, 68, 28012 Madrid, Spain
B. Retailleau, Grimaud Frères S.A., La Corbière, 49450 Roussay, France
A. Rodríguez, Departamento Microbiología e Inmunología, Facultad de Veterinaria, Calle Miguel Servet, numero 117, 50013 Zaragoza, Spain
R. Rodriguez de L., "Conejos" Centro de Investigación Cientifica del Estado de México A.C., Camino a Huexotla No. 5, San Miguel Coatlinchan, 56170 Texcoco, Estado de México, Mexico
W. Schlolaut, Hessen Institute for Animal Production, Neu-Ulrichstein, 6313 Homberg, West Germany
W.D. Semuguruka, Department of Veterinary Pathology, Faculty of Veterinary Medicine, Sokoine University of Agriculture, P.O. Box 3018, Chuo Kikuu Morogoro, Tanzania
I. Sierra, Departamento Producción Animal, Facultad de Veterinaria, Calle Miguel Servet, numero 117, 50013 Zaragoza, Spain
R. Valls, Institut de Recerce i Tecnologia Agroalimentària (IRTA), Torre Marimon, 08140 Caldes de Montbuí, Barcelona, Spain
P. Zaragoza, Departamento de Genética y Mejora, Facultad de Veterinaria, Calle Miguel Servet, numero 117, 50013 Zaragoza, Spain

## RODENTS

### Agouti

C.J.R. Alho, Departamento de Biologia Animal, Universidade de Brasilia, 70910 Brazilia, DF, Brazil
R.H. Baker, 302 North Strickland Street, Eagle Lake, Texas 77434-1841, USA
H. Bonilla Encizo, ICA Tibaitata, Apartado Aereo 151123, Bogotá, Colombia
A.D. Cuarón O., Instituto de Historia Natural, Apartado Postal No. 6, Tuxtla Gutierrez, Chiapas 29000, Mexico
L.A. Deutsch, Parque Zoologico de São Paulo, São Paulo, Brazil
G. Dubost, Laboratorie d'Ecologie, 4 avenue Petit Château, 91800 Brunoy, France
J.K. Frenkel, Department of Pathology, University of Kansas Medical Center, Kansas City, Kansas 66013, USA
A. Escajadillo, Gorgas Memorial Laboratory, P.O. Box 935, APO Miami, Florida 34002-0012, USA OR Avenida Justo Arosemena 34-30, Apartado 6991, Panama 5, Panama
W. Hallwachs, Department of Biology, University of Pennsylvania, Philadelphia, Pennsylvania 19104-6018, USA

G. Hislop, Wildlife Section, Forest Division, Ministry of Food Production, Port-of-Spain, Trinidad and Tobago
S. Ingrand, I.N.R.A., Campus agronomique, B.P. 709, 97387 Kourou Cedex, French Guiana
D.H. Janzen, Department of Biology, University of Pennsylvania, Philadelphia, Pennsylvania 19104, USA
D. Kleiman, National Zoological Park, Washington, D.C. 20008, USA
A. Lavorenti, Animal Sciences Department, ESALQ Universidade de São Paulo, C.P. 9, Piracicaba, São Paulo 13.400, Brazil
D.A. Meritt, Jr., 2200 North Cannon Drive, Chicago, Illinois 60614, USA
J.R. de Alencar Moreira, EMBRAPA-CPATU, Agricultural Research Center of the Humid Tropics, Caixa Postal 48, Belém, Para 66000, Brazil
C.Ma. Rojas G., CATIE, 7170 Turrialba, Costa Rica
G. Santos, Fundação Parque Zoologico de São Paulo, São Paulo, Brazil
N. Smith, Department of Geography, University of Florida, Gainesville, Florida 32611, USA
N. Smythe, Smithsonian Tropical Research Institute, Box 2072, Balboa, Panama
O. Sousa, Department of Microbiology, Ciudad Universitaria "Dr Octavio Méndez Pereira," El Cangregio, Apartado Estafeta Universitaria, Panama City, Panama
A.B. Taber, Animal Behaviour Research Group, Department of Zoology, South Parks Road, Oxford 0X1 3PS, UK
E. Wing, Florida State Museum, University of Florida, Gainesville, Florida 32611, USA
C.A. Woods, Florida State Museum, University of Florida, Gainesville, Florida 32611, USA

## Capybara

C.J.R. Alho, Departamento de Biologia Animal, Universidade de Brasilia, 70910 Brasilia, DF, Brazil
K.P. Bland, Department of Veterinary Physiology, Royal (Dick) School of Veterinary Studies, Summerhall, Edinburgh EH9 1QH, Scotland, UK
H. Bonilla Encizo, ICA Tibaitata, Apartado Aereo 151123, Bogotá, Colombia
P.R. Cheeke, Rabbit Research Center, Department of Animal Science, Oregon State University, Corvallis, Oregon 97331, USA
T.J. Cunha, P.O. Box R, LaVerne, California 91750, USA
J.F. Eisenberg, Florida State Museum, University of Florida, Gainesville, Florida 32611, USA
F. Golley, Institute of Ecology, University of Georgia, Athens, Georgia 30601, USA
E. Gonzalez-Jimenez, Facultad de Agronomía, Universidad Central de Venezuela, Maracay, Venezuela
H. Hemmer, Institute of Zoology, Johannes Gutenberg University, Anemonenweg 18, D-6500 Mainz-Ebersheim, Germany
J.A. Howarth, School of Veterinary Medicine, University of California, Davis, California 95616, USA
S. Ingrand, I.N.R.A., Campus agronomique, B.P. 709, 97387 Kourou Cedex, French Guiana
N. Konnerup, 74 North Sunrise Boulevard, Camano Island, Washington 98292, USA
J.E. Jackson, Terraza Turrado 80, 5730 Villa Mercedes, San Luis, Argentina
D. Kleiman, National Zoological Park, Washington, D.C. 20008, USA
A. Lavorenti, Animal Sciences Department, ESALQ Universidade de São Paulo, C.P. 9, Piracicaba, São Paulo 13.400, Brazil

R.D. Lord, Biology Department, Indiana University of Pennsylvania, Indiana, Pennsylvania 15705, USA
D.W. MacDonald, Animal Behaviour Research Group, Department of Zoology, South Parks Road, Oxford OX1 3PS, UK
J.R. de Alencare Moreira, Agricultural Research Center of the Humid Tropics, EMBRAPA-CPATU, Caixa Postal 48, Belém, Para 66000, Brazil
B. Muller-Haye, Food and Agricultural Organization of the United Nations, Room F-524, Via delle Terme di Caracalla, 00100 Rome, Italy
J. Ojasti, Institute of Tropical Zoology, Apartado 47058, Universidad Central de Venezuela, Caracas 1041A, Venezuela
N.L. Rondon, Departamento de Engenharia Florestal, Centro de Ciências Agrárias, Universidada Federal de Mato Grosso, Cuiabá-MT, 78000, Brazil
N. Smith, Department of Geography, University of Florida, Gainesville, Florida 32611, USA
A.E. Sollod, Section of International Veterinary Medicine, Tufts University, 200 Westboro Road, North Grafton, Massachusetts 01536, USA
F. Sunquist, Route 1, Box 1434, Melrose, Florida 32666, USA
A.B. Taber, Animal Behaviour Research Group, Department of Zoology, South Parks Road, Oxford OX1 3PS, UK
K.D. Thelen, FAO Regional Office for Latin America and the Caribbean, Casilla 10095, Santiago, Chile
D. Wharton, New York Zoological Society, Bronx, New York 10460, USA
E. Wing, Florida State Museum, University of Florida, Gainesville, Florida 32611, USA

## Coypu

A.C. Carmichael, The Museum, Michigan State University, East Lansing, Michigan 48824-1045, USA
G. Chapman, Utah State University, Logan, Utah 84322, USA
L. Contreras, Universidad Católica de Valpariso, Avenida Brasil 2950, Casilla 4059, Valparaiso, Chile
D. Huss, P.O. Box 426, Menard, Texas 76859, USA
H.-G. Klos, Zoologischer Garten Berlin, Aktiengesellschaft, Hardenbergplatz, 1000 Berlin 30, Germany
P. Lutz, Rosenstiel School of Marine and Atmospheric Science, Biology and Living Resources, University of Miami, 4600 Rickenbacker Causeway, Miami, Florida 33149, USA
J.R. de Alencar Moreira, Agricultural Research Center of the Humid Tropics, Caixa Postal 48, Belém, Pará, 66000, Brazil
J. Szumiec, Experimental Fish Culture Station, Polish Academy of Sciences, Golysz 43-422 Chybie, Poland
C.A. Woods, Florida State Museum, University of Florida, Gainesville, Florida 32611, USA

## Giant Rat

Animal Research Institute, Council for Scientific and Industrial Research (CSIR), P.O. Box 20, Achimota, Ghana
J. Baranga, Department of Zoology, Makerere University, P.O. Box 7062, Kampala, Uganda

K.P. Bland, Department of Veterinary Physiology, Royal (Dick) School of Veterinary Studies, Summerhall, Edinburgh EH9 1QH, Scotland, UK

P. Brinck, Ecology Building, Department of Animal Ecology, University of Lund, S-223 62 Lund, Sweden

G. Dubost, Laboratorie d'Ecologie, 4 avenue Petit Château, 91800 Brunoy, France

D. Fielding, Centre for Tropical Veterinary Medicine, Royal (Dick) School of Veterinary Studies, University of Edinburgh, Easter Bush, Roslin, Midlothian EH25 9RG, Scotland, UK

D.C.D. Happold, Department of Zoology, Australian National University, Canberra, Australian Capital Territory 2601, Australia

J.C. Heymans, BP 1910, Cotonou, Benin

J.A. Kamara, c/o Njala University College, Private Mail Bag, Freetown, Sierra Leone

M.H. Knight, Kalahari Gemsbok Park, Private Bag X5890, Upington 8800, South Africa

M. Malekani, Départment de Biologie, Faculté des Sciences, Université de Kinshasa, B.P. 190, Kinshasa XI, Zaire

R.W. Matthewman, Centre for Tropical Veterinary Medicine, Royal (Dick) School for Veterinary Studies, University of Edinburgh, Easter Bush, Roslin, Midlothian EH25 9RG, Scotland, UK

A.J. Smith, Centre for Tropical Veterinary Medicine, Royal (Dick) School of Veterinary Studies, University of Edinburgh, Easter Bush, Roslin, Midlothian EH25 9RG, Scotland, UK

## Grasscutter

A. Alexander, Department of Biology, Natal University, King George V Avenue, Durban 4001, South Africa

E.O.A. Asibey, Department of Game and Wildlife, Ministry Post Office, Accra, Ghana

E.S. Ayensu, African Development Bank, 01 B.P. No. 1837, Abidjan 01, Ivory Coast

R. Baptist, Institute for Animal Production in the Tropics and Subtropics, University of Hohenheim (480), P.O. Box 700562, D-7000 Stuttgart 70, Germany

S.Y. Bimpong-Buta, Council for Law Reporting, P.O. Box M165, Accra, Ghana

P. Brinck, Department of Animal Ecology, University of Lund, Ecology Building, S-223 62, Lund, Sweden

H. Dosso, Institut d'Ecologie Tropicale, Abidjan, Ivory Coast

D. Fielding, Centre for Tropical Veterinary Medicine, Royal (Dick) School of Veterinary Studies, University of Edinburgh, Easter Bush, Roslin, Midlothian EH25 9RG, Scotland, UK

D.C.D. Happold, Department of Zoology, Australian National University, Canberra, Australian Capital Territory 2601, Australia

J. Hardouin, Institut de Medicine Tropicale "Prince Leopold," Nationalestraat 155, B-2000 Antwerp, Belgium

J.C. Heymans, BP 1910, Cotonou, Benin

A. Hutton, Garaina Farms, Scrubby Creek Road, Gympie MS:115, Queensland 4570, Australia

J.A. Kamara, c/o Njala University College, Private Mail Bag, Freetown, Sierra Leone

Laboratoire Centrale de Nutrition Animale (LACENA), B.P. 353, Abidjan 06, Ivory Coast

S.D. Lukefahr, International Small Livestock Research Center, Department of Food Science and Animal Industries, Alabama A&M University, P.O. Box 264, Normal, Alabama 35762, USA

R.-P. Mack, Projet Bénino-Allemand Conseiller Technique auprés du MDRAC, BP 504, Cotonou, Benin

G.H.G. Martin, Department of Zoology, Kenyatta University College, P.O. Box 43844, Nairobi, Kenya
R.W. Matthewman, Centre for Tropical Veterinary Medicine, Royal (Dick) School of Veterinary Studies, University of Edinburgh, Easter Bush, Roslin, Midlothian EH25 9RG, Scotland, UK
G.A. Mensah, Responsable PBAA (Projet Bénino-Allemand d'Aulacodiculture), P.B. 2359, Cotonou, Benin
T.A. Omole, Department of Animal Science, University of Ife, Ile-Ife, Nigeria
K.J. Peters, International Livestock Centre for Africa (ILCA), P.O. Box 5689, Addis Ababa, Ethiopia
F. Petter, Mammifères et Oiseaux, Museum National d'Histoire Naturelle Zoologie, 55 rue de Buffon 75005, Paris, France
W. Schröder, Institute of Animal Breeding, Albrecht-Thaer-Weg 1, 3400 Göttingen, Germany
A.J. Smith, Centre for Tropical Veterinary Medicine, Royal (Dick) School of Veterinary Studies, University of Edinburgh, Easter Bush, Roslin, Midlothian EH25 9RG, Scotland, UK
D.W. Thomas, Department of Biology, University of Sherbrooke, Sherbrooke, Quebec J1K 2R1, Canada
St. von Korn, Institute of Animal Breeding, Albrecht-Thaer-Weg 1, 3400 Göttingen, Germany
J. Walder, Presbyterian Rural Training Centre, P.O. Box 72, Bamenda, NW Province, Cameroon

## Guinea Pig

G.P. Bilong, Department of Pimary Industry, Monogastric Research Centre, P.O. Box 73, Lae, Papua New Guinea
K.P. Bland, Department of Physiology, Royal (Dick) School of Veterinary Studies, Summerhall, Edinburgh EH9 1QH, Scotland, UK
R. Bolton, Department of Anthropology, Pomona College, Claremont, California 91711, USA
H. Bonilla Encizo, ICA Tibaitata, Apartado Aereo 151123, Bogotá, Colombia
J.K. Camoens, Asian Development Bank, P.O. Box 789, Manila 2800, Philippines
A. Del Carpio R., Facultad de Zootecnia, Universidad Nacional "Pedro Ruiz Gallo," Calle Atahualpa No. 179, Apartado 48, Lambayeque, Peru
P.R. Cheeke, Rabbit Research Center, Department of Animal Science, Oregon State University, Corvallis, Oregon 97331, USA
T.J. Cunha, P.O. Box R, LaVerne, California 91750, USA
J. Descailleux, Laboratorio Genética Humana, C.P. 11010, Lima 14, Peru
J.A. Ferguson, Overseas Development Administration, Eland House, Stag Place, London, UK
D. Fielding, Centre for Tropical Veterinary Medicine, Royal (Dick) School of Veterinary Studies, University of Edinburgh, Easter Bush, Roslin, Midlothian EH25 9RC, Scotland, UK
J.K. Frenkel, Department of Pathology, University of Kansas Medical Center, Kansas City, Kansas 66013, USA
J. Hardouin, Institut de Médecine Tropicale "Prince Leopold," Nationalestraat 155, B-2000 Antwerpen, Belgium
R.E. Honegger, Zurich Zoo, Zurichbergstrasse 221, CH-8044 Zurich, Switzerland

J. Howarth, School of Veterinary Medicine, University of California, Davis, California 95616, USA
D.L. Huss, P.O. Box 426, Menard, Texas 76859, USA
J.E. Jackson, Terraza Turrado 80, 5730 Villa Mercedes, San Luis, Argentina
C. Jenkins, PNG Institute Medical Research, P.O. Box 60, Goroka, Eastern Highlands Province, Papua New Guinea
R. de Jong, Department of Tropical Animal Production, Marykeweg 40, 6709 PG Wageningen, The Netherlands
J.A. Kamara, c/o Njala University College, Private Mail Bag Freetown, Sierra Leone
P.J. Kohun, Department of Agriculture, PNG University of Technology, Lae, Papua New Guinea
N.M. Konnerup, 74 North Sunrise Boulevard, Camano Island, Washington 98292, USA
T.E. Lacher, Jr., Huxley College of Environmental Studies, Western Washington University, Bellingham, Washington 98225, USA
S.D. Lukefahr, International Small Livestock Research Center, Department of Feed Science and Animal Industries, Alabama A&M University, P.O. Box 264, Normal, Alabama 35762, USA
R.C. Malik, Department of Agriculture, PNG University of Technology, Lae, Papua New Guinea
R.W. Matthewman, Centre for Tropical Veterinary Medicine, Royal (Dick) School of Veterinary Studies, University of Edinburgh, Easter Bush, Roslin, Midlothian EH25 9RG, Scotland, UK
A.R. Moreno, Departamento de Producción Animal, Universidad Nacional Agraria, Apartado 456, La Molina, Lima, Peru
B. Muller-Haye, Food and Agriculture Organization of the United Nations, Room F-524, Via delle Terme di Caracalla, 00100 Rome, Italy
H.L. Polley, Programme Développement Plateau Bateke, CBZO, B.P. 4728, Kinshasa 2, Zaire
V. Purizaga A., Alcalde del Concejo Provincial de Pacasmayo, Dos de Mayo 360, Peru
V.T. Quirante, Bureau of Animal Industry, Research Center, Alabang 3124, Metro Manila, Philippines
C.V. Reddy, Faculty of Veterinary Science, Andhra Pradesh Agricultural University, Rajendranagar, Hyderabad 500 030, India
O.W. Robinson, Department of Animal Science, North Carolina State University, Box 7621 Raleigh, North Carolina 27695-7621, USA
A.J. Smith, Centre for Tropical Veterinary Medicine, Royal (Dick) School of Veterinary Studies, University of Edinburgh, Easter Bush, Roslin, Midlothian EH25 9RG, Scotland, UK
D. Wharton, New York Zoological Society, Bronx, New York 10460, USA
E. Wing, Florida State Museum, University of Florida, Gainesville, Florida 32611, USA

## Hutia

A.C. Allen, Hope Zoo, Kingston, Jamacia
R.S. Canet, Instituto de Zoología, Academia Ciencias de Cuba, Calle 214, esq. a Avenida 19, No. 17A 09, Reparto Atabey, Havana 16, Cuba
D. Kleiman, National Zoological Park, Washington, D.C. 20008, USA
K. Jordan, Department of Zoology, University of Florida, Gainesville, Florida 32611, USA

W.L.R. Oliver, Jersey Wildlife Preservation Trust, Les Augrès Manor, Jersey, Channel Islands, UK
J.A. Ottenwalder, Parque Zoológico Nacional (ZOODOM), Apartado 2449, Santo Domingo, Dominican Republic
E. Wing, Florida State Museum, University of Florida, Gainesville, Florida 32611, USA
C. Woods, Florida State Museum, University of Florida, Gainesville, Florida 32611, USA

## Mara

L.T. Blankenship, Animal Resources Center, School of Medicine, East Carolina University, Greenville, North Carolina 27858-4354, USA
G. Dubost, Laboratoire d'Ecologie, 4 avenue Petit Château, 91800 Brunoy, France
J.E. Jackson, Terraza Turrado 80, 5730 Villa Mercedes, San Luis, Agentina
D. Kleiman, National Zoological Park, Washington, D.C. 20008, USA
D.W. MacDonald, Animal Behaviour Research Group, Department of Zoology, South Parks Road, Oxford OX1 3PS, UK
W.H. Pryor, Jr., School of Medicine, East Carolina University, Greenville, North Carolina 27858-4354, USA
A.B. Taber, Animal Behaviour Research Group, Department of Zoology, South Parks Road, Oxford OX1 3PS, UK

## Paca

J.R. de Alencar Moreira, Agricultural Research Center of the Humid Tropics, EMBRAPA-CPATU, Caixa Postal 48, Belém, Pará 66000, Brazil
C.J.R. Alho, Departamento de Biologia Animal, Universidade de Brasilia, 70910 Brasilia, DF, Brazil
R.H. Baker, 302 North Strickland Street, Eagle Lake, Texas 77434-1841, USA
A.D. Cuarón O., Instituto de Historia Natural, Apartado Postal No. 6, Tuxtla Gutierrez, Chiapas 29000, Mexico
G. Dubost, Laboratoire d'Ecologie, 4 avenue Petit Château, 91800 Brunoy, France
L. de Escalante, Zoológico Nacional, Dirección del Patrimonio Natural y Cultural, San Salvador, El Salvador
J.K. Frenkel, Department of Pathology, University of Kansas Medical Center, Kansas City, Kansas 66013, USA
W. Hallwachs, Ecology and Systematics, Corson Hall, Cornell University, Ithaca, New York 14853, USA
G. Hislop, Wildlife Section, Forest Division, Ministry of Food Production, Port-of-Spain, Trinidad and Tobago
D.H. Janzen, Department of Biology, University of Pennsylvania, Philadelphia, Pennsylvania 19104, USA
D. Kleiman, National Zoological Park, Washington, D.C. 20008, USA
A. Lavorenti, Animal Sciences Department, ESALQ Universidade de São Paulo, C.P. 9, Piracicaba, São Paulo 13.400, Brazil
Y. Matamoros, Escuela de Medicina Veterinaria, Universidad Nacional, Apartado 86, Heredia, Costa Rica
D. Meritt, Jr., 2200 North Cannon Drive, Chicago, Illinois 60614, USA
C.Ma. Rojas G., CATIE, 7170 Turrialba, Costa Rica

O. Rosado, Ministry of Industry and Natural Resources, Belmopan, Belize
N. Smith, Department of Geography, University of Florida, Gainesville, Florida 32611, USA
N. Smythe, Smithsonian Tropical Research Institute, Box 2072, Balboa, Panama
V. Solís, Unión Internacional para la Conservación de la Naturaleza y de los Recursos Naturales, Oficina Regional para Centro América, Apartado Postal 91-1009-FECOSA, San José, Costa Rica
S. Vaidés, Centro Universitario del Norte Carrera Zootécnica, Cobán, A.V., Ciudad Guatemala 16001, Guatemala
E. Wing, Florida State Museum, University of Florida, Gainesville, Florida 32611, USA

## Vizcacha

H. Bonilla Encizo, ICA Tibaitata, Apartado Aereo 151123, Bogotá, Colombia
J.E. Jackson, Terraza Turrado 80, 5730 Villa Mercedes, San Luis, Argentina
E.L. Marmillon, Estancias del Conlara S.A., Casilla de Correo 451, 5800 Rio Cuarto, Argentina
K.D. Thelen, FAO Regional Office for Latin America and the Caribbean, Casilla 10095, Santiago, Chile
E. Wing, Florida State Museum, University of Florida, Gainesville, Florida 32611, USA

## Other Rodents

T.M. Butynski, Impenetrable Forest Conservation Project, Zoology Department, Makerere University, P.O. Box 7062, Kampala, Uganda (springhare)
T. Flannery, Australian Museum, 6-8 College Street, Sydney, New South Wales 2000, Australia (Solomon Island rodents)
Inter-African Bureau for Animal Resources, Maendeleo House, Monrovia Street, POB 30786, Nairobi, Kenya (springhare)
N. Jacobsen, Nature Conservation Division, Private Bag X209, Pretoria 0001, South Africa (springhare)
C.P. Kofron, Museum of Zoology, Louisiana State University, Baton Rouge, Louisiana 70893, USA (springhare)
T. Leary, Ministry of Natural Resources, Honiara, Solomon Islands (Solomon Island rodents)
C.A. McLaughlin, San Diego Zoo, San Diego 92112, California, USA (springhare)
R.P. Millar, Department of Chemical Pathology, Medical School, Observatory 7925, South Africa (springhare)
J.D. Skinner, Mammal Research Institute, University of Pretoria, Pretoria 0002, South Africa (springhare)
Southern African Wildlife Management Association, POB 413, Pretoria, South Africa (springhare)
L. Talbot, 6656 Chitton Court, McLean, Virginia 22101, USA (springhare)
M. Van der Merwe, University of Pretoria, Mammal Research Institute, Pretoria 0002, South Africa (springhare)
Wildlife Society of Zimbabwe, POB 3497, Harare, Zimbabwe (springhare)

# DEER AND ANTELOPE

## Mouse Deer

J.O. Caldecott, World Wildlife Fund Malaysia, 7 Jalan Ridgeway, 93200 Kuching, Sarawak, Malaysia
R. Chaplin, Burnhouse, Fountainhall, Galashiels, Selkirk TD1 2RX, UK
G.W.H. Davison, Zoology Department, Universiti Kebangsaan Malaysia, 43600 Bangi, Selangor, Malaysia
G. Dubost, Laboratoire d'Ecologie Générale, 4 avenue Petit Château, 91800 Brunoy, France
P. Grubb, 35 Downhills Park Road, London N17 6PE, UK
Liu Ruiqing, Kunming Institute of Zoology, Academia Sinica, Kunming, Yunnan Province, People's Republic of China
M. MacNamara, Fauna Research, Inc., 11 Park Avenue, Ardsley, New York 10502, USA
I. Muul, Integrated Conservation Research, P.O. Box 920, Harpers Ferry, West Virginia 25425, USA
M. Nordin, Bureau of Research and Consultancy, University Kebangsaan Malaysia, Bangi, Selangor, Malaysia
S. Pathak, Department of Genetics, M.D. Anderson Hospital and Tumor Institute, Houston, Texas 77030, USA
C.T. Robbins, Wildlife Biology Program, Washington State University, Pullman, Washington 99164-4220, USA
K. Robin, Tierpark Dählhölzli Bern, Dalmaziquai 149, CH-3005 Bern, Switzerland
C.R. Schmidt, Zoological Garden, CH-8044 Zurich, Switzerland
R.A. Sirimanne, Singapore Zooological Gardens, 80 Mandai Lake Road, Singapore 2572
L. Talbot, 6656 Chilton Court, McLean, Virginia 22101, USA
M.K. Vidyadaran, Faculty of Veterinary Science, Department of Animal Science, Universiti Pertanian Malaysia, Serdang 43400, Selangor, Malaysia
Wang Yingxiang, Kunming Institute of Zoology, Academia Sinica, Kunming, Yunan Province, People's Republic of China
D. Wharton, New York Zoological Society, Bronx, New York 10460, USA
G.C. Whittow, Department of Physiology, John A. Burns School of Medicine, University of Hawaii, 1960 East West Road, Honolulu, Hawaii 96822, USA

## Muntjac

C. Barrette, Département de Biologie, Faculté des Sciences et de Génie, Université Laval, Québec, G1K 7P4, Canada
K. Benirschke, 8457 Prestwick Drive, La Jolla, California 92037, USA
The British Deer Society, Church Farm, Lower Basildon, Reading, Berkshire RG8 9NH, UK
J.O. Caldecott, World Wildlife Fund Malaysia, 7 Jalan Ridgeway, 933200 Kuching, Sarawak, Malaysia
N. Chapman, Larkmead, Barton Mills, Bury St. Edmunds, Suffolk IP28 6AA, UK
M. Coe, Department of Zoology, University of Oxford, South Parks Road, Oxford OX1 3PS, UK

O. Dansie, The Chapel, 2 Hobbs Hall, Welwyn, Hertfordshire, UK
S.K. Dhungal, Department of Parks and Wildlife, P.O. Box 860, Kathmandu, Nepal
G. Dubost, Laboratoire d'Ecologie Générale, 4 avenue Petit Château, 91800 Brunoy, France
R.J. Goss, Division of Biology and Medicine, Brown University, Providence, Rhode Island 02912, USA
C.P. Groves, Department of Prehistory and Anthropology, Australian National University, P.O. Box 4, Canberra, Australian Capital Territory 2600, Australia
P. Grubb, 35 Downhills Park Road, London N17 6PE, UK
S.P. Harding, Animal Ecology Research Group, Zoology Department, South Parks Road, Oxford OX1 3PS, UK
S. Harris, Department of Zoology, University of Bristol, Woodland Road, Bristol BS8 1UG, UK
H. Heck, Catskill Game Farm, Inc., RD #1, Box 133, Catskill, New York 12414, USA
H.A. Jacobson, Fish and Wildlife Department, Mississippi State University, Mississippi State, Mississippi 39762, USA
Liu Ruiqing, Kunming Institute of Zoology, Academia Sinica, Kunming, Yunnan Province, People's Republic of China
Lu Houji, Department of Biology, East China Normal University, Shanghai 200062, People's Republic of China
M. MacNamara, Fauna Research, Inc., 11 Park Avenue, Ardsley, New York 10502, USA
V.J.A. Manton, Zoological Society of London, Whipsnade Park, Dunstable, Bedfordshire LU6 2LF, UK
S. Miura, Department of Biology, Hyogo College of Medicine, Mukogawa-cho 1-1, Nishinomiya, Hyogo-ken 663, Japan
I. Muul, Integrated Conservation Research, P.O. Box 920, Harpers Ferry, West Virginia 25425, USA
S. Pathak, Department of Genetics, M.D. Anderson Hospital and Tumor Institute, Houston, Texas 77030, USA
C.T. Robbins, Wildlife Biology Program, Washington State University, Pullman, Washington 99164, USA
B. Seidel, Tierpark Berlin, Tierklinik, DDR-1136 Berlin, Friedrichsfelde, Germany
Shi Liming, Kunming Institute of Zoology, Academia Sinica, Kunming, Yunnan Province, People's Republic of China
H. Soma, Department of Obstetrics and Gynecology, Tokyo Medical College Hospital, 1-7, Nishishinjuku 6, Shinjuku-ku, Tokyo 160, Japan
S. Stadler, Fakultät für Biologie, Universität Bielefeld, 4800 Bielefeld, Universitätsstrasse 25, Germany
L. Talbot, 6656 Chilton Court, McLean, Virginia 22101, USA
P.F. Taylor, Taylor Energy Company, The 2-3-4 Loyola Building, New Orleans, Louisiana 70112, USA
Wang Yingziang, Kunming Institute of Zoology, Academia Sinica, Kunming, Yunnan Province, People's Republic of China
C. Wemmer, Conservation and Research Center, National Zoological Park, Front Royal, Virginia 22630, USA
D. Wharton, New York Zoological Society, Bronx, New York 10460, USA
D.M. Wurster-Hill, Department of Pathology, Dartmouth Medical School, Hanover, New Hampshire 03755, USA
R.H. Yahner, 320 Forest Resources Laboratory, Pennsylvania State University, University Park, Pennsylvania 16802, USA

## Musk Deer

R. Chaplin, Burnhouse, Fountainhall, Galashiels, Selkirk TD1 2RX, UK
S.K. Dhungel, Department of Parks and Wildlife, P.O. Box 860, Kathmandu, Nepal
A.J. Gaston, Canadian Wildlife Service, Ottawa K1A 0E7, Canada
C.P. Groves, Department of Prehistory and Anthropology, Australian National University, P.O. Box 4, Canberra, Australian Capital Territory 2600, Australia
M.J.B. Green, IUCN Conservation Monitoring Centre, 219C Huntingdon Road, Cambridge CB3 0DL, UK
P. Grubb, 35 Downhills Park Road, London N17 6PE, UK
M.L. Hunter, Jr., Wildlife Department, University of Maine, Orono, Maine 04469, USA
H.A. Jacobson, Fish and Wildlife Department, Mississippi State University, Mississippi State, Mississippi 39762, USA
B. Kattel, P.O. Box 3070, Kathmandu, Nepal
Liu Riqing, Kunming Institute of Zoology, Academia Sinica, Kunming Yunnan Province, People's Republic of China
Lu Houji, Department of Biology, East China Normal University, Shanghai 200062, People's Republic of China
J. McNeely, Program and Policy Division, International Union for Conservation of Nature and Natural Resources, Avenue du Mont-Blanc, CH1196 Gland, Switzerland
B.W. O'Gara, Montana Cooperative Wildlife Research Unit, University of Montana, Missoula, Montana 59812, USA
C.T. Robbins, Wildlife Biology Program, Washington State University, Pullman, Washington 99164-4220, USA
B. Seidel, Tierpark Berlin, Tierklinik, DDR-1136 Berlin, Friedrichsfelde, Germany
Shi Liming, Kunming Institute of Zoology, Academia Sinica, Kunming, Yunnan Province, People's Republic of China
P.F. Taylor, Taylor Energy Co., The 2-3-4 Loyola Building, New Orleans, Louisiana 70112, USA
Wang Yingxiang, Kunming Institute of Zoology, Academia Sinica, Kunming, Yunnan Province, People's Republic of China

## Water Deer

A. Cooke, Nature Conservancy Council, Northminster House, Peterborough PE1 1UA, UK
L. Farrell, Nature Conservancy Council, Northminster House, Peterborough PE1 1UA, UK
F. Feer, Laboratoire d'Ecologie Générale, 4 avenue Petit Château, 91800 Brunoy, France
P. Grubb, 35 Downhills Park Road, London N17 6PE, UK
Liu Riuqing, Kunming Institute of Zoology, Academia Sinica, Kunming Yunnan Province, People's Republic of China
Lu Houji, Department of Biology, East China Normal University, Shanghai 200062, People's Republic of China
V.J.A. Manton, Zoological Society of London, Whipsnade Park, Dunstable, Bedfordshire LU6 2LF, UK
B. Seidel, Tierpark Berlin, Tierklinik DDR-1136, Berlin, Friedrichsfelde, Germany

R. Smith, Department of Zoology, Whiteknights Park, University of Reading, Reading, Berkshire RG6 2AH, UK
S. Stadler, Fakultät für Biologie, Universität Bielefeld, 4800 Bielefeld, Universitätsstrasse 25, Germany
P.F. Taylor, Taylor Energy Company, The 2-3-4 Loyola Building, New Orleans, Louisiana 70112, USA
C. Wemmer, Conservation and Research Center, National Zoological Park, Front Royal, VA 22630, USA

## South American Microdeer

M. Sampath, Trinidad and Tobago Wildlife Breeders and Farmers Association, c/o 5 ¾ Mile Mark, Penal Rock Road, Penal, Trinidad, West Indies (red brocket)
M. MacNamara, Fauna Research, Inc., 11 Park Avenue, Ardsley, New York 10502, USA (pudu)

## Duikers

J. Baranga, Department of Zoology, Makerere University, P.O. Box 7062, Kampala, Uganda
B. Chardonnet, Projet Petits Ruminants, B.P. 65, Atakpame, Togo
R.L. Cowan, 324 Henning Building, Pennsylvania State University, University Park, Pennsylvania 16802, USA
G. Dubost, Laboratoire d'Ecologie Générale, 4 avenue Petit Château, 91800 Brunoy, France
R.I.M. Dunbar, Department of Zoology, University of Liverpool, P.O. Box 147, Liverpool L69 3BX, UK
F. Feer, Laboratoire d'Ecologie Générale, 4 avenue Petit Château, 91800 Brunoy, France
K.R. Kranz, Zoological Society of Philadelphia, 34th Street and Girard Avenue, Philadelphia, Pennsylvania 19104, USA
S. Lumpkin, Department of Zoological Research, National Zoological Park, Washington, D.C. 20008, USA
I. Player, Wilderness Leadership School, P.O. Box 15036, Bellair 4006, Natal, South Africa
K. Pond, Department of Animal Science, North Carolina State University, Box 7621, Raleigh, North Carolina 27695, USA
S. Schweers, Bahnhofstrasse 29, D-6744 Kandee, Germany
J.D. Skinner, Mammal Research Institute, University of Pretoria, Pretoria 0002, South Africa
A.E. Sollod, International Veterinary Medicine, Tufts University, 200 Westboro Road, North Grafton, Massachusetts 01536, USA
L. Talbot, 6656 Chilton Court, McLean, Virginia 22101, USA
C.C. Udell, 14934 Valley Vista Blouvard, Sherman Oaks, California 91403, USA
G.A. Varga, Department of Dairy and Animal Science, 225B Borland Laboratory, Pennsylvania State University, University Park, Pennsylvania 16802, USA
V.J. Wilson, Duiker Research and Breeding Centre, Chipangali Wildlife Trust, P.O. Box 1057, Bulawayo, Zimbabwe

## Other Small Antelope

E. Asibey, Department of Game and Wildlife, Ministry Post Office, Accra, Ghana
J. Baranga, Department of Zoology, Makerere University, P.O. Box 7062, Kampala, Uganda
J.I. Boshe, College of African Wildlife Management, Mweka, P.O. Box 3031, Moshi, Tanzania (dikdik)
B. Chardonnet, Project Petits Ruminants, BP 65, Atakpame, Togo (oribi)
S.K. Dhungel, Department of National Parks and Wildife, Conservation, Babar Mahal, Kathmandu, Nepal (four-horned antelope)
R.I.M. Dunbar, Department of Zoology, University of Liverpool, P.O. Box 147, Liverpool L69 3BX, UK (dikdik, klipspringer)
F. Feer, Laboratoire d'Ecologie Générale, 4 avenue Petit Château, 91800 Brunoy, France
A.M. Goebel, Department of Biology, University of Texas at Arlington, Arlington, Texas 76019, USA
P.P. Hoppe, Am Hauenstein 13, 6706 Wachenheim, Germany (dikdik, suni)
K.R. Kranz, Zoological Society of Philadelphia, 34th Street and Girard Avenue, Philadelphia, Pennsylvania 19104, USA (royal antelope, beira, dikdik, klipspringer, suni, four-horned antelope)
I. Player, Wilderness Leadership School, P.O. Box 15036, Bellair 4006, Natal, South Africa
E.N.W. Oppong, UNDP/FGN/FAO Livestock Project, P.O. Box 6603, Jos, Plateau State, Nigeria
W.R. Pritchard, School of Veterinary Medicine, University of California, Davis, California 95616, USA
J. Skinner, Mammal Research Institute, University of Pretoria, Pretoria 0002, South Africa (oribi, steinbok)
A.E. Sollod, International Veterinary Medicine, Tufts University, 200 Westboro Road, North Grafton, Massachusetts 01536, USA (dikdik, klipspringer)
L. Talbot, 6656 Chilton Court, McLean, Virginia 22101, USA (dikdik, klipspringer, four-horned antelope)

# LIZARDS

## Green Iguana

K. Benirschke, 8457 Prestwick Drive, La Jolla, California 92037, USA
G.M. Burghardt, Department of Psychology, University of Tennessee, Knoxville, Tennessee 37996, USA
A. Ferreira, Parque Zoológico Nacional, Apartado 2449, Santo Domingo, Dominican Republic
H.S. Fitch, Route #3, Box 142, Lawrence, Kansas 66044, USA
G. Hemley, World Wildlife Fund, 1255 23rd Street, N.W., Washington, D.C. 20037, USA (for trade information)
R.W. Henderson, Milwaukee Public Museum, 800 West Wells Street, Milwaukee, Wisconsin 53233 USA
J. Higginson, Voluntaria del Cuerpo de Paz, #156 6a Avenida 1-46, Zona 2 Guatemala, Guatemala

J.B. Iverson, Department of Biology, Earlham College, Richmond, Indiana 47374, USA
S.R. Kellert, School of Forestry and Environmental Studies, Yale University, 205 Prospect Street, New Haven, Connecticut 06511, USA
F.W. King, Florida State Museum, University of Florida, Gainesville, Florida 32601, USA
G. Lardé, Ministerio de Agricultura y Ganadería Planes de Renderos, Km. 10 Col. Los Angeles 13, CP 01196 San Salvador, El Salvador
D. Marcellini, National Zoological Park, Washington, D.C. 20008, USA
J.M. Mora, 200 m sur Pulperia la Unión, Cinco Esquinas, Carrieal, Alajuela, Costa Rica
J.A. Ottenwalder, Parque Zoológico Nacional, Apartado 2449, Santo Domingo, Dominican Republic
G.C. Packard, Department of Zoology, Colorado State University, Fort Collins, Colorado 80523, USA
J.J. Pérez, Condominio Su. Antonio 7, C. San Antonio Abad, San Salvador, El Salvador
J. Phillips, Research Department, San Diego Zoo, Box 551, San Diego, California 92112, USA
D. Werner, Fundación Pro Iguana Verde, Apartado 1501-3000, Heredia, Costa Rica

## Black Iguana

H.S. Fitch, Route #3, Box 142, Lawrence, Kansas 66044, USA
R.W. Henderson, Milwaukee Public Museum, 800 West Wells Street, Milwaukee, Wisconsin 53233, USA
J. Higginson, Voluntaria del Cuerpo de Paz, #156 6a Avenida 1-46, Zona 2 Guatemala, Guatemala
D.M. Hillis, Department of Biology, University of Miami, Coral Gables, Florida 33124, USA
G. Lardé, Ministerio de Agricultura y Ganadería, Planes de Renderos Km. 10 Col. Los Angeles 13, CP 01196, San Salvador, El Salvador
D. Marcellini, National Zoological Park, Washington, D.C. 20008, USA
J.M. Mora, 200 m sur Pulpería La Unión, Cinco Esquinas, Carrieal, Alajuela, Costa Rica
G.C. Packard, Department of Zoology, Colorado State University, Fort Collins, Colorado 80523, USA
J.J. Pérez, Condominio Su. Antonio 7, C. San Antonio Abad, San Salvador, El Salvador
J. Phillips, Research Department, San Diego Zoo, Box 551, San Diego, California 92112, USA
D. Werner, Fundación Pro Iguana Verde, Apartado 1501-3000, Heredia, Costa Rica

## Rock Iguana

W. Auffenberg, Florida State Museum, University of Florida, Gainesville, Florida 32611, USA
D. Auth, Florida State Museum, University of Florida, Gainesville, Florida 32611, USA
T.J. Cullen, Cullen Vivarium, 3401 South 16th Street, Milwaukee, Wisconsin 53215, USA
A. Ferreira, Parque Zoológico Nacional, Apartado 2449, Santo Domingo, Dominican Republic

R.W. Henderson, Milwaukee Public Museum, 800 West Wells Street, Milwaukee, Wisconsin 53233, USA
J.B. Iverson, Department of Biology, Earlham College, Richmond, Indiana 47374, USA
J.A. Ottenwalder, Parque Zoológico Nacional, Apartado 2449, Santo Domingo, Dominican Republic
T.A. Wiewandt, Wild Horizons, Inc., P.O. Box 5118, Tucson, Arizona 85703, USA

## Tegu

C.A. Blanco, Dirección Nacional de Fauna Silvestre, Avenida Paseo Colon 922 2°piso, (1063) Buenos Aires, Argentina
O.E. Donadio, Museo Argentino de Ciencias Naturales, CC 220, Avenida Angel Gallardo 470, Suc. 5, 1405 Buenos Aires, Argentina
J.M. Gallardo, Museo Argentino de Ciencias Naturales, CC 220, Avenida Angel Gallardo 470, Suc. 5, 1405 Buenos Aires, Argentina
G. Hemley, World Wildlife Fund, 1255 23rd Street, N.W., Washington, D.C. 20037, USA
D. Marcellini, National Zoological Park, Washington, D.C. 20008, USA
E.L. Marmillon, Casilla de Correo 451, 5800 Rio Cuarto, Argentina
G.C. Packard, Department of Zoology, Colorado State University, Fort Collins, Colorado 80523, USA
D. Werner, Fundación Pro Iguana Verde, Apartado 1501-3000, Heredia, Costa Rica

# BEES

Local Agriculture Extension Services and beekeepers are often the best source of information for beginning beekeeping. In addition, the following organizations can provide information: International Bee Research Association, 16/18 North Road, Cardiff CF1 3DY, UK; International Agency for Apiculture Development, 6N 909 Roosevelt, St. Charles, Illinois 60174, USA

I. Abt, Centre for International Agricultural Development Cooperation, Tel Aviv, Israel
M. Adey, International Bee Research Association, 16/18 North Road, Cardiff CF1 3DY, UK
S.O. Adjare, Technology Consultancy Centre, University of Science and Technology, Kumasi, Ghana
All India Beekeepers' Association, 817 Sadashiv Peth, Poono 411 030, India
S.K.N. Atuahene, Forest Product Research Institute (Ghana Forestry Commission), University P.O. Box 63, Kumasi, Ghana
H.S. Brar, Department of Entomology, Punjab Agricultural University, Ludhiana 141 004, India
N. Bradbear, International Bee Research Association, North Road, Cardiff CF1 3DY, UK
D.M. Caron, Department of Entomology and Applied Ecology, University of Delaware, Newark, Delaware 19717, USA
B.S. Chahal, Department of Entomology, Punjab Agricultural University, Ludhiana 141 004, India
B. Clauss, Beekeeping Survey, Kabompo, Zambia

M.P. Coleman, International Agency for Apiculture Development, 6N 909 Roosevelt, St. Charles, Illinois 60174, USA
Director, or Resident Bee Specialist, Apiary Studies, Zoological Institute, Leningrad, USSR
D. DeJong, Departamento de Genética, Faculdade de Medicina de Ribeirao Preto, Universidade São Paulo, 14.049—Ribeirao Preto, São Paulo, Brazil
R.W. Dutton, Centre for Overseas Research and Development (CORD), University of Durham, South Road, Durham DH1 3LE, UK (*Apis florea*)
E.H. Erickson, Carl Hayden Bee Research Center, Agricultural Research Service, U.S. Department of Agriculture, 2000 East Allen Road, Tucson, Arizona 85719, USA
Faculdade de Medicina de Ribeirao Preto, Departamento de Genética, 14100—Ribeirao Preto, São Paulo, Brazil
R.D. Fell, Virginia Polytechnic Institute, Blacksburg, Virginia 24061, USA
K. Flottum, *Gleanings in Bee Culture*, 623 West Liberty Street, P.O. Box 706, Medina, Ohio 44258-0706, USA (editor)
J. Free, Rothamstead Experimental Station, Harpenden, Hertfordshire L5 2JQ, UK
D. Wolde-Semait Gebre-Wold, P.O. Box 7505, Addis Ababa, Ethiopia
C.B. Habarugira, AHO (Apiary), Mbara Stockfarm, Beekeeping Section, P.O. Box 4, Mbarara, Uganda
L.A.M. Hassan, Njiro Wildlifc (Beekeeping) Research Centre, P.O. Box 661, Arusha, Tanzania
R. Hoopingarner, Department of Entomology, University of Michigan, East Lansing, Michigan 48824-1115, USA
W.E. Kerr, Departamento de Biolgia, Universidade Federal de Uberlândia, 38400 Uberlândia, Brazil
P.G. Kevan, or Resident Bee Specialist, Department of Environmental Biology, University of Guelph, Guelph, Ontario N1G 2W1, Canada
K.K. Kigatiira, Beekeeping Branch, Ministry of Agriculture & Livestock Development, P.O. Box 30028, Nairobi, Kenya
H.C. Killins, 163 Trowbridge Avenue, London, Ontario N6J 3M2, Canada
H.M. Kiwuwa, AHL (Apiary), Department of Veterinary Services, Headquarters, Beekeeping Section, P.O. Box 7141, Kampala, Uganda
G. Lanarolle, Apiculture Project, Bindunuwewa, Bandarawela, Sri Lanka
Y. Lensky, Triwaks Bee Research Center, Hebrew University Faculty of Agriculture, Rehovot 76-100, Israel
M.D. Levin, Carl Hayden Bee Research Center, Agricultural Research Service, U.S. Department of Agriculture, 2000 East Allen Road, Tucson, Arizona 85719, USA
M. Singh Limbu, Godavari Ashram, GPO Box 50, Kathmandu, Nepal
I. Mann, 14 Millimani Road, Box 20360, Nairobi, Kenya
R.J. McGinley, Department of Entomology, NHB-105, Smithsonian Institution, Washington, D.C. 20560, USA (Apoidea, except *Apis*)
J. McKay, Peace Corps, American Embassy, Asunción, Paraguay
C.D. Michener, Snow Entomological Museum, University of Kansas, Lawrence, Kansas 66045, USA
R.A. Morse, Department of Entomology, Cornell University, Ithaca, New York 14853, USA
J. Morton, Morton Collectanea, University of Miami, Box 248204, Coral Gables, Florida 37724, USA (bee plants)
D.G. Mugisa, c/o Insure Consult, P.O. Box 30002, Nakiuubo, Uganda
M. Bin Muid, Department of Plant Protection, Faculty of Agriculture, Universiti Pertanian Malaysia, Serdang 43400, Selangor, Malaysia
H.C. Mulzac, International Agency for Apiculture Development, 393 Decatur Street, Brooklyn, New York 11233, USA

K.S. Najar, Faculty of Agriculture, University of Aleppo, Aleppo, Syria

B.E. Nightingale, Box 23, Njoro, Kenya

G. Patty, Foreign Agriculture Service, U.S. Department of Agriculture, 6095 South Building, Washington, D.C. 20250, USA

L.M. Phiri, Office of Chief Beekeeping Officer, P/BAG Mwekera, Kitwe, Zambia

R. Pickard, Zoology Department, University College, P.O. Box 78, Cardiff CF1 1XL, South Wales, UK

G.W. Robins, GPO Box 50, Kathmandu, Nepal

D. Roubik, Stop 105, Museum of Natural History, Smithsonian Institution, Washington, D.C. 20560, USA (stingless bees)

D. Sammataro, 7011 Spieth Road, Medina, Ohio 44256, USA (editor, *Cornucopia*)

G. Schechtel, Cnel. Bogado/Itapua, Paraguay

H. Shimanuki, Beneficial Insects Laboratory, Agricultural Research Service, U.S. Department of Agriculture, B-476 BARC-East, Beltsville, Maryland 20705, USA

O.R. Taylor, Department of Entomology, University of Kansas, Lawrence, Kansas 66045, USA

J.E. Tew, Agricultural Technical Institute, Ohio State University, Wooster, Ohio 44691, USA

M. Tomasko, Pennsylvania State University, 106 Patterson Building, University Park, Pennsylvania 16802, USA

P. Torchio, Agricultural Research Service, U.S. Department of Agriculture, Bee Laboratory, Utah State University, Logan, Utah 84322-5310, USA

R. Wadlow, 8381 San Mateo, Ft. Myers, Florida 33907, USA

D.W. Whitehead, International Agency for Apiculture Development, c/o 1000 Connecticut Avenue NW, #707, Washington, D.C. 20036, USA

Siriwat Wongsiri, Department of Biology, Chulalongkorn University, Bangkok, Thailand (*Apis cerana*)

Wu Yan-ru, Department of Insect Taxonomy and Faunistics, Institute of Zoology, Academia Sinica, 7 Zhongguancun Lu, Haitien, Beijing, People's Republic of China

# APPENDIX C

## BIOGRAPHICAL SKETCHES OF PANEL MEMBERS

RALPH W. PHILLIPS retired in 1982 from the post of deputy director general of the Food and Agriculture Organization of the United Nations (FAO), Rome, Italy, a post he held for four years. Among his earlier posts were that of professor and head, Animal Husbandry Department, Utah State University; senior animal husbandman in charge, Genetic Investigations, United States Department of Agriculture (USDA); chief, Animal Production Branch and deputy director, Agriculture Division, FAO; and executive director, International Organization Affairs, USDA. Among his special assignments were: serving as consultant on animal breeding to the governments of China and India for the U.S. Department of State in 1943–44; and as scientific secretary for agriculture of the United Nations Conference on Science and Technology for the Benefit of Developing Countries, in Geneva, Switzerland, 1962–63. Dr. Phillips holds a B.S. degree in agriculture from Berea College (1930), M.A. (1931) and Ph.D. (1934) degrees from the University of Missouri, and Honorary D.Sc. degrees from Berea College and West Virginia University. He has been awarded the Berea College Distinguished Alumnus Award and the USDA's Distinguished Service Award. He is author or coauthor of some 240 scientific papers, review papers, chapters in books, and books on various aspects of physiology of reproduction, genetics, livestock production, and international agriculture. In his research, writings, and international activities, Dr. Phillips has given particular attention to breeding in relation to the environment and to the identification and conservation of valuable animal genetic resources. He is also the author of a definitive history of FAO entitled *FAO: Its Origins, Formation and Evolution, 1945–1981* and an autobiography, *The World Was My Barnyard*.

EDWARD S. AYENSU is currently senior advisor to the president of the African Development Bank. He is also president of ESA Associates, Washington, D.C., and former director of the Office of Biological Conservation, Smithsonian Institution, Washington, D.C. A citizen of Ghana, he received his B.A. in 1961 from Miami University in Ohio, M.Sc. from The George Washington University in 1963, and his Ph.D. in 1966 from the University of London. His research interests cover many areas of tropical biology. An internationally recognized expert on topics relating to science, technology, and development, especially in developing countries, he has also published extensively on tropical plants. Dr. Ayensu chairs and serves as a member of many international bodies.

BONNIE V. BEAVER, professor of small animal medicine and surgery, Texas A&M University, College Station, is a specialist in animal behavior and problem behaviors, especially in domestic and laboratory animals. She received her B.S. and D.V.M. from the University of Minnesota and her M.S. from Texas A&M. In addition to being a popular speaker at scientific meetings, she is the author of five books and numerous book chapters and articles.

KURT BENIRSCHKE, professor of pathology and reproductive medicine, University of California at San Diego, received his M.D. from the University of Hamburg, Germany, in 1948. He served on the faculties of Harvard and Dartmouth medical schools before coming to San Diego in 1970. At the San Diego Zoo he initiated a research department to advance knowledge in endangered species and now serves as a trustee of that organization. He has written on comparative mammalian cytogenetics, vanishing species, and human reproductive pathology.

ROY D. CRAWFORD is professor of animal and poultry genetics at the University of Saskatchewan, Saskatoon, Saskatchewan, Canada. He received his B.S.A. from the University of Saskatchewan in 1955, his M.S. in animal genetics from Cornell University in 1957, and his Ph.D. in poultry genetics from the University of Massachusetts in 1963. He was employed as a scientist with the Research Branch of Agriculture Canada from 1957 to 1964 in Prince Edward Island and Nova Scotia. He joined the faculty of the University of Saskatchewan in 1964. He was made a fellow of the Agricultural Institute of Canada in 1986 in recognition of his teaching and research work. His research interests include single gene genetics of poultry, and conservation of genetic resources in poultry and livestock. He has discovered and studied many mutants in chickens; some of them have biomedical importance, including one shown to be an animal genetic model of human grand mal epilepsy; some of them are potentially useful in food production, including an albinism mutant that is being developed for autosexing of commercial chicken broilers. He maintains a very large conservation collection of poultry genetic resources at the University of Saskatchewan and has prepared an inventory and assessment of Canada's poultry and livestock genetic resources. Dr. Crawford is a member of the Expert Panel on Animal Genetic Resources Conservation and Management, FAO/UNEP, Rome, and is a member of the Animal Resources Committee, Canadian Council on Animal Care, Ottawa. He serves on the International Scientific Committee for the French journal *Génétique, Sélection, Évolution*. He is a member of the Rare Breeds Survival Trust (UK) and is a board member of the American Minor Breeds Conservancy (USA).

TONY J. CUNHA, distinguished service professor emeritus, University of Florida, Gainesville, and dean emeritus, California State Polytechnic University, Pomona, received his Ph.D. at the University of Wisconsin in 1944. He has given more than 100 lectures in livestock feeding and nutrition in 40 foreign countries. He served as chairman of the Animal Nutrition Committee of the National Academy of Sciences and as a member of the NAS-NRC Board of Agriculture and Renewable Resources, Latin American Science Board, and as chairman of the livestock committee of two world food studies by NAS-NRC. He served as a member of the organizing committee for the first two World Conferences on Animal Production in Rome (1963) and Washington, D.C. (1968). He served as a member of the Title XII Board of International Food and Agricultural Joint Committee on Research and as chairman of its research priorities committee 1977–1981. He is author, editor, coeditor, or contributor to 30 books and author of more than 1,423 scientific and professional articles. He is a winner of 42 campus, state, national, and international honors and awards.

DAVID E. DEPPNER, director of Trees for the Future in Silver Spring, Maryland, has been a consultant for international development projects involving livestock and poultry management and marketing for the past 14 years. He has served in 17 countries of Asia, Africa, and Central America. He has written on processing livestock rations under tropical conditions and about the Madurese breed of cattle found in East Java, Indonesia, where he spent two years studying this ancient breed. He is currently providing technical assistance to projects in several countries for the development of improved forage production as an answer to destruction of natural resources of tropical uplands caused by overgrazing. He received his B.Sc. in animal science from Ohio State University in 1954 and M.Sc. in livestock economics from Araneta University, the Philippines, in 1977.

ELIZABETH L. HENSON is the director of the American Minor Breeds Conservancy, based in Pittsboro, North Carolina. She received an M.A. in zoology from Oxford University in 1980 and an M.Sc. in domestic animal breeding from Edinburgh University in 1981. Her primary interests are in the conservation of rare and endangered breeds and varieties of domestic livestock as a genetic resource for changing agricultural needs. She represented Britain at the first international conference on domestic animal conservation in Hungary in 1982, and was a member of the Office of Technology Assessment Committee on grassroots strategies to maintain genetic diversity in 1985. She is executive secretary for four British breed associations and is a member of the British Rare Breeds Survival Trust Technical Panel.

DONALD L. HUSS was regional animal production officer of the FAO Regional Office for Latin America and the Caribbean, Santiago, Chile, before his retirement. He received a B.S. in 1949, an M.A. in 1954, and a Ph.D. in 1959, all at Texas A&M University. He was assistant professor of range management at Texas A&M before joining the Food and Agricultural Organization of the United Nations (FAO) in 1967. He founded the FAO Regional Office's Small Animals for Small Farms programme in 1980. Professional assignments and travels in Latin America, the Caribbean, Near and Middle East, and Africa have contributed to his knowledge and experience in microlivestock development. Dr. Huss was recognized by Texas A&M University by being chosen as the recipient of the Memorial Student Center Appreciation and Distinguished Service Awards in 1958 and 1960, respectively, and Honour Professor in the College of Agriculture in 1966–67. He also received the Society for Range Management's Outstanding Service and Achievement Award in 1975 and its Fellow Award in 1978.

DAVID RICHARD LINCICOME has been a guest scientist with the Animal Parasitology Institute, United States Department of Agriculture Experiment Station, Beltsville, Maryland, since 1978, having retired as professor of parasitology, Howard University, Washington, D.C., in 1970. He received the B.S. and M.S. degrees cum laude simultaneously in 1937 from the University of Illinois and a Ph.D. in tropical medicine from the Tulane Medical School, New Orleans, Louisiana, in 1941. His principal research interests have centered around morphologic studies on Acanthocephala, molecular exchanges of dependent cells and their environments, and diagnosis of parasitic diseases. He has been a breeder of Nubian and American pygmy goats for the past 20 years. Dr. Lincicome is currently a member of the Board of Directors of the American Dairy Goat Association and is past president of the National Pygmy Goat Association. He was founder and long time trustee of the American Dairy Goat Association Research Foundation. He is also past president of the Helminthological Society and currently serves the society as archivist. He received the Helminthological Society's Anniversary Award in 1975. He was founder and, for 27 years, editor of the journal *Experimental Parasitology*, and is the author and editor of more than 180 scientific contributions.

THOMAS E. LOVEJOY is a tropical biologist and ornithologist. He is assistant secretary for external affairs, Smithsonian Institution, Washington, D.C. He was formerly executive vice president for the World Wildlife Fund-U.S., chairman of the Wildlife Preservation Trust International, and a member of two commissions of the International Union for the Conservation of Nature and Natural

Resources. At present, Dr. Lovejoy is a principal investigator of the world's largest controlled ecological experiment, which is attempting to determine the optimum size for parks and reserves. This project, conceived and designed by Dr. Lovejoy, is called "the Minimum Critical Size of Ecosystems" and is a joint program of World Wildlife Fund-U.S. and Brazil's National Institute for Amazon Research. Dr. Lovejoy is also the principal advisor for NATURE (WNET/THIRTEEN, New York), a series that he started in 1980. As principal advisor, he recommends program content and oversees the factual accuracy of the program scripts. A member of 11 scientific societies, Dr. Lovejoy has received grants from 16 foundations and institutions and written more than 100 articles for various national and International publications. He has published three books, *Key Environments*, Pergamon Press, Oxford; *Nearctic Avian Migrants in the Neotropics*, a Department of the Interior publication; and *Conservation of Tropical Forest Birds*, an ICPB publication. He is currently working on *The Magnificent Exception*, a book on people and the biosphere.

ARNE W. NORDSKOG is professor emeritus, Department of Animal Science, Iowa State University. He received his B.S., M.S., and Ph.D. degrees from the University of Minnesota between 1937 and 1943. His training and principal area of research has been in quantitative genetics, but by about 1960 his interests shifted to immunogenetics and more recently to molecular genetics. He has traveled widely, spending two years as an instructor in agriculture at the University of Alaska (1937–39), and has been an NSF Research Fellow at Cal Tech (1960), a visiting professor at the University of Minnesota (1966), an FAO lecturer at the Indian Veterinary Institute (1973), and an FAO-sponsored lecturer on poultry breeding in China (1979). He has acted as a consultant to a commercial breeder in Japan for 20 years. He has been the major professor at Iowa State for more than 60 M.S. and Ph.D. candidates and has published more than 150 scientific papers. Dr. Nordskog is an honorary member of the Norwegian Poultry Breeding Association, a fellow of the Poultry Science Association, and a fellow of the AAAS. In 1972, he was the recipient of the Poultry Science Distinguished Service Research Award.

LINDA M. PANEPINTO, swine research consultant, was director of the Colorado State University Swine Laboratory through 1988. She earned her B.S. in animal sciences at Colorado State University in 1972. In 1973 she joined the research team developing Yucatan miniature swine at Colorado State University, where she was given primary responsibility for colony management, protocol development, and genetic selection programs. In 1977, she designed an experimental research program for the development of a line of Yucatan pigs with a genetic propensity for exceptionally small size. She has continued her work in that area and has developed the Yucatan Micropig®, described elsewhere in this publication. Her other major professional area of interest has been the design of facilities and equipment for swine with emphasis on animal comfort and minimizing stress. Her invention, known as the Panepinto Sling, has been widely adopted as the primary restraint method for numerous medical schools and research facilities using swine in the laboratory. She has published extensively in the field.

KURT J. PETERS is professor of animal breeding and husbandry in the tropics and subtropics, University of Gottingen, and is currently director of research at the International Livestock Centre for Africa. He received his Dr. Agr. degree from the Technical University of Berlin in 1975. He has undertaken research in livestock production development in Southeast Asia and Africa. The major focus of his research has been small animals, with special attention given to the potential of unconventional animals. Early in 1985 he assumed his present position directing research at the International Livestock Centre for Africa.

JOHN A. PINO is a senior fellow of the National Research Council, Board on Agriculture, and is currently the project director of the study "Managing Global Genetic Resources: Agricultural Imperatives." He received his B.S. in agriculture in 1944–47 and Ph.D. in zoology in 1951 from Rutgers University. As an associate professor he taught and did research in the Department of Poultry Science at Rutgers until 1955 when he accepted a position with the Rockefeller Foundation as animal scientist with the Mexican Agricultural Program, becoming the associate director of that program in 1960. In 1965 he was transferred to the Rockefeller headquarters in New York and became director of the Agricultural Science Program in 1970. Most of his career has been in international agricultural development. He retired from the Foundation in 1983 and went to Washington to become agricultural science advisor at the Inter-American Development Bank until July 1986 when he accepted his present position. Dr. Pino has been a member of the Board on Agriculture since 1983 and previously from 1973 to 1977.

HUGH POPENOE is professor of soils, agronomy, botany, and geography, and director of the Center for Tropical Agriculture and International Programs (Agriculture) at the University of Florida. He received his B.S. from the University of California at Davis in 1951 and his Ph.D. in soils from the University of Florida in 1960. His principal research interest has been in the area of tropical agriculture and land use. His early work on shifting cultivation is one of the major contributions to this system. He has traveled and worked in most of the countries in the tropical areas of Latin America, Asia, and Africa. His current interests include improving indigenous agricultural systems of small landholders, particularly with the integration of livestock and crops. He was chairman of the Advisory Committee on Technology Innovation and a member of the Board on Science and Technology for International Development (under whose aegis this report is presented). He chaired the BOSTID report panels on water buffalo and little-known Asian animals. Currently, he is on the International Advisory Committee of the National Science Foundation and serves as U.S. Board Member for the International Foundation of Science.

MICHAEL HILL ROBINSON, director of the Smithsonian Institution's National Zoological Park, is an animal behaviorist and a tropical biologist. Before his appointment to the National Zoo, Dr. Robinson served as deputy director of the Smithsonian Tropical Research Institute in Panama, which he joined in 1966 as a tropical biologist. He received his Ph.D. from Oxford University after being awarded his B.S., summa cum laude, from the University of Wales. His scientific interests include predator-prey interactions, evolution of adaptations, tropical biology, courtship and mating behavior, phenology of arthropods, and freshwater biology. In the course of his studies, Dr. Robinson has done research in the United States and throughout the developing world. Recent publications include articles on predator-prey interactions, tropical forest conservation, reproductive behavior in spiders, and the function and purpose of zoos in relation to education and conservation. Dr. Robinson's favorite animals are cats, of all kinds.

KNUT SCHMIDT-NIELSON, J.B. Duke Professor of Physiology in the Department of Zoology at Duke University, has studied animal responses to extreme environmental conditions. His major emphasis has been on life in hot deserts, and he is widely recognized for his studies of camels and other desert animals. His research has involved field studies in North and South America, Africa, Asia, and Australia. Dr. Schmidt-Nielson has written several books, which have been translated into more than a dozen languages, and he has published several hundred research papers. He has been elected to the National Academy of Sciences, the Royal Society, the French Academie des Sciences, and several other academies.

ALBERT E. SOLLOD, is associate professor and head of the international veterinary medicine section at Tufts University School of Veterinary Medicine. He is currently stationed in Niger as chief of party of the integrated livestock project and policy advisor in the Ministry of Animal Resources. He has consulted in 15 countries in Africa and Asia, and his research interests include interdisciplinary systems analysis, indigenous pastoral technologies, monitoring change in agricultural production systems, and monitoring and assessing drought impact.

LEE M. TALBOT received his Ph.D. in geography and ecology from the University of California at Berkeley in 1963 and has worked on environmental and natural resource ecology and management in over 110 countries. At present he is visiting fellow at the World Resources Institute, Washington, D.C., and senior environmental consultant to the World Bank. He carried out pioneering research in Africa and elsewhere on the use of wild animals for food production. He has written more than 180 scientific and technical publications, including ten books and monographs.

CLAIR E. TERRILL, animal scientist, collaborator, Agricultural Research Service, United States Department of Agriculture (USDA), Beltsville, Maryland, received his Ph.D. from the University of Missouri in 1936, served briefly at the Georgia Agricultural Experiment Station, and joined the USDA at the U.S. Sheep Experiment Station, Dubois, Idaho, in the same year. His research concerned genetics and reproduction of sheep, leading to national and international responsibility regarding research and production of sheep, goats, and other animals, with primary emphasis on increasing efficiency of production of meat, wool, and other products.

CHRISTIAN M. WEMMER is assistant director for conservation and captive breeding programs at the National Zoological Park and is also in charge of the zoo's 3,100 acre Conservation and Research Center in Front Royal, Virginia. He is vice chairman of the IUCN Deer Specialist Group and a member of the Mustelid and Viverrid Specialist Group of the same organization. His interest in evolutionary and conservation biology has been motivated by frequent travel to South Asia and his role as scientific coordinator of the Smithsonian-Nepal Terai Ecology Project. For the past 12 years he has coordinated the development of facilities and programs at the Conservation and Research Center and with Dr. R. Rudran has promoted conservation training and wildlife research in developing nations. He has published over 50 papers on various aspects of mammalian biology and conservation and has co-edited with Benjamin Beck one book on Père David's deer. His edited volume "The Biology and Management of the Cervidae" was published by the Smithsonian Institution Press.

DANNY C. WHARTON is associate curator, Animal Departments, at New York Zoological Park, Bronx, New York. He received a B.S. from the College of Idaho and an M.Sc. in 1975 from the School for International Training in Brattleboro, Vermont. His Ph.D. in biology was earned at Fordham University. Dr. Wharton was a Peace Corps volunteer to Ecuador 1969–71 and a Fulbright scholar to Germany 1976–77. His research interests have been in the genetic and demographic management of small populations. He works on several committees including the IUCN/SSC Captive Breeding Specialist Group, Species Survival Plan Committee of the American Association of Zoological Parks and Aquariums, and is species chairman for the North American Propagation Group for the Snow Leopard.

CHARLES A. WOODS is curator of mammals at the Florida State Museum and a professor of zoology at the University of Florida. He received his B.S. in zoology from the University of Denver and his Ph.D. in zoology from the University of Massachusetts. He worked at the University of Vermont from 1970 to 1979 when he assumed his present position at the University of Florida. His principle research interests have been in the areas of mammalian ecology (Rodentia) and systematics and evolution and he is especially concerned with island biology. He has spent many years working in the West Indies on a variety of projects and has worked closely with the government of Haiti in establishing a plan for the National Parks of Haiti and in completing a biogeophysical inventory of the natural resources of Hispaniola. He is the ecological consultant for the Institut de Sauvegarde du Patrimoine National in Haiti. He is the author of a number of scientific articles on the fauna of the Antilles including a multivolume series on the fauna of the mountains of Haiti.

THOMAS M. YUILL is associate dean for research and graduate training of the School of Veterinary Medicine, assistant director of the Agricultural Experiment Station, and professor of pathobiology and of veterinary science at the University of Wisconsin-Madison. He received his B.S. in wildlife management from Utah State University in 1959 and his Ph.D. jointly in veterinary science and wildlife ecology in 1964 from the University of Wisconsin. His principal research interests are animal health and diseases of wildlife, including those transmissible to domestic animals and to man. He worked in Thailand for two years and has had active research programs in Colombia for 17 years, and Costa Rica for 5 years. He has recently become involved in animal health and production development in the Gambia, West Africa. Dr. Yuill is an executive committee member and immediate past president of the Organization for Tropical Studies and currently serves as president of the Wildlife Disease Association. He completed a five-year term as Chairman of the U.S. Virus Diseases Panel of the U.S.-Japan Cooperative Biomedical Sciences Program.

# INDEX OF SPECIES

## Board on Science and Technology for International Development

ALEXANDER SHAKOW, Director, Strategic Planning and Review, The World Bank, Washington, D.C., *Chairman*

### Members

PATRICIA BARNES-MCCONNELL, Director, Bean/Cowpea CRSP, East Lansing, Michigan

JORDAN J. BARUCH, President, Jordan Baruch Associates, Washington, D.C.

PETER D. BELL, President, The Edna McConnell Clark Foundation, New York, New York

BARRY BLOOM, Professor, Department of Microbiology, Albert Einstein College of Medicine, The Bronx, New York

JANE BORTNICK, Assistant Chief, Congressional Research Service, Library of Congress, Washington, D.C.

GEORGE T. CURLIN, The National Institute of Allergy and Infectious Diseases, The National Institutes of Health, Bethesda, Maryland

DIRK FRANKENBERG, Director, Marine Sciences Program, University of North Carolina, Chapel Hill, North Carolina

RALPH HARDY, Boyce-Thompson Institute for Plant Research, Cornell University, Ithaca, New York

FREDERICK HORNE, Dean, College of Sciences, Oregon State University, Corvallis, Oregon

ELLEN MESSER, Allan Shaw Feinstein World Hunger Program, Brown University, Providence, Rhode Island

CHARLES C. MUSCOPLAT, Executive Vice President, Molecular Genetics, Inc., Minneapolis, Minnesota

JAMES QUINN, Amos Tuck School of Business, Dartmouth College, Hanover, New Hampshire

VERNON RUTTAN, Regents Professor, Department of Agriculture and Applied Economics, Saint Paul, Minnesota

ANTHONY SAN PIETRO, Profcssor of Plant Biochemistry, Department of Biology, Bloomington, Indiana

ERNEST SMERDON, College of Engineering and Mines, University of Arizona, Tucson, Arizona

GERALD P. DINEEN, Foreign Secretary, National Academy of Engineering, National Research Council, Washington, D.C., *ex officio*

JAMES B. WYNGAARDEN, Foreign Secretary, National Academy of Sciences, National Research Council, Washington, D.C., *ex officio*

# Board on Science and Technology for International Development

Publications and Information Services (HA-476E)
Office of International Affairs
National Research Council
2101 Constitution Avenue, N.W.
Washington, D.C. 20418 USA

## How to Order BOSTID Reports

BOSTID manages programs with developing countries on behalf of the U.S. National Research Council. Reports published by BOSTID are sponsored in most instances by the U.S. Agency for International Development. They are intended for distribution to readers in developing countries who are affiliated with governmental, educational, or research institutions, and who have professional interest in the subject areas treated by the reports.

BOSTID books are available from selected international distributors. For more efficient and expedient service, please place your order with your local distributor. (See list on back page.) Requestors from areas not yet represented by a distributor should send their orders directly to BOSTID at the above address.

## Energy

33. **Alcohol Fuels: Options for Developing Countries.** 1983, 128pp. Examines the potential for the production and utilization of alcohol fuels in developing countries. Includes information on various tropical crops and their conversion to alcohols through both traditional and novel processes. ISBN 0-309-04160-0.

36. **Producer Gas: Another Fuel for Motor Transport.** 1983, 112pp. During World War II Europe and Asia used wood, charcoal, and coal to fuel over a million gasoline and diesel vehicles. However the technology has since been virtually forgotten. This report reviews producer gas and its modern potential. ISBN 0-309-04161-9.

56. **The Diffusion of Biomass Energy Technologies in Developing Countries.** 1984, 120pp. Examines economic, cultural, and political factors that affect the introduction of biomass-based energy technologies in developing countries. It includes information on the opportunities for these technologies as well as conclusions and recommendations for their application. ISBN 0-309-04253-4.

## Technology Options

14. **More Water for Arid Lands: Promising Technologies and Research Opportunities.** 1974, 153pp. Outlines little-known but promising technologies to supply and conserve water in arid areas.
ISBN 0-309-04151-1.

21. **Making Aquatic Weeds Useful: Some Perspectives for Developing Countries.** 1976, 175pp. Describes ways to exploit aquatic weeds for grazing, and by harvesting and processing for use as compost, animal feed, pulp, paper, and fuel. Also describes utilization for sewage and industrial wastewater. ISBN 0-309-04153-X.

34. **Priorities in Biotechnology Research for International Development: Proceedings of a Workshop.** 1982, 261pp. Report of a workshop organized to examine opportunities for biotechnology research in six areas: 1) vaccines, 2) animal production, 3) monoclonal antibodies, 4) energy, 5) biological nitrogen fixation, and 6) plant sell and tissue culture. ISBN 0-309-04256-9.

61. **Fisheries Technologies for Developing Countries.** 1987, 167pp. Identifies newer technologies in boat building, fishing gear and methods, coastal mariculture, artificial reefs and fish aggregating devices, and processing and preservation of the catch. The emphasis is on practices suitable for artisanal fisheries. ISBN 0-309-04260-7.

## Plants

25. **Tropical Legumes: Resources for the Future.** 1979, 331pp. Describes plants of the family Leguminosae, including root crops, pulses, fruits, forages, timber and wood products, ornamentals, and others.
ISBN 0-309-04154-6.

47. **Amaranth: Modern Prospects for an Ancient Crop.** 1983, 81pp. Before the time of Cortez, grain amaranths were staple foods of the Aztec and Inca. Today this nutritious food has a bright future. The report discusses vegetable amaranths also. ISBN 0-309-04171-6.

53. **Jojoba: New Crop for Arid Lands.** 1985, 102pp. In the last 10 years, the domestication of jojoba, a little-known North American desert shrub, has been all but completed. This report describes the plant and its promise to provide a unique vegetable oil and many likely industrial uses. ISBN 0-309-04251-8.

63. **Quality-Protein Maize.** 1988, 130pp. Identifies the promise of a nutritious new form of the planet's third largest food crop. Includes chapters on the importance of maize, malnutrition and protein quality, experiences with quality-protein maize (QPM), QPM's potential uses in feed and food, nutritional qualities, genetics, research needs, and limitations. ISBN 0-309-04262-3.

64. **Triticale: A Promising Addition to the World's Cereal Grains.** 1988, 105pp. Outlines the recent transformation of triticale, a hybrid between wheat and rye, into a food crop with much potential for many marginal lands. Includes chapters on triticale's history, nutritional quality, breeding, agronomy, food and feed uses, research needs, and limitations. ISBN 0-309-04263-1.

67. **Lost Crops of the Incas.** 1989. 415pp. The Andes is one of the seven major centers of plant domestication but the world is largely unfamiliar with its native food crops. When the conquistadores brought the potato to Europe, they ignored the other domesticated Andean crops—fruits, legumes, tubers, and grains that had been cultivated for centuries by the Incas. This book focuses on 30 of the "forgotten" Incan crops that show promise not only for the Andes but for warm-temperate, subtropical, and upland tropical regions in many parts of the world. ISBN 0-309-04264-X.

70. **Saline Agriculture: Salt-Tolerant Plants for Developing Countries.** 1989, approx 150pp. The purpose of this report is to create greater awareness of salt-tolerant plants and the special needs they may fill in developing countries. Examples of the production of food, fodder, fuel, and other products are included. Salt-tolerant plants can use land and water unsuitable for conventional crops and can harness saline resources that are generally neglected or considered as impediments to rather than opportunities for development. ISBN 0-309-04266-6.

## Innovations in Tropical Forestry

35. **Sowing Forests from the Air.** 1981, 64pp. Describes experiences with establishing forests by sowing tree seed from aircraft. Suggests testing and development of the techniques for possible use where forest destruction now outpaces reforestation. ISBN 0-309-04257-7.

40. **Firewood Crops: Shrub and Tree Species for Energy Production.** Volume II, 1983, 92pp. Examines the selection of species of woody plants that seem suitable candidates for fuelwood plantations in developing countries. ISBN 0-309-04164-3.

41. **Mangium and Other Fast-Growing Acacias for the Humid Tropics.** 1983, 63pp. Highlights 10 acacia species that are native to the tropical rain forest of Australasia. That they could become valuable forestry resources elsewhere is suggested by the exceptional performance of Acacia mangium in Malaysia. ISBN 0-309-04165-1.

42. **Calliandra: A Versatile Small Tree for the Humid Tropics.** 1983, 56pp. This Latin American shrub is being widely planted by the villagers and government agencies in Indonesia to provide firewood, prevent erosion, provide honey, and feed livestock.
ISBN 0-309-04166-X.

43. **Casuarinas: Nitrogen-Fixing Trees for Adverse Sites.** 1983, 118pp. These robust, nitrogen-fixing, Australasian trees could become valuable resources for planting on harsh eroding land to provide fuel and other products. Eighteen species for tropical lowlands and highlands, temperate zones, and semiarid regions are highlighted.
ISBN 0-309-04167-8.

52. **Leucaena: Promising Forage and Tree Crop for the Tropics.** 1984, 2nd edition, 100pp. Describes a multi-purpose tree crop of potential value for much of the humid lowland tropics. Leucaena is one of the fastest growing and most useful trees for the tropics.
ISBN 0-309-04250-X.

## Managing Tropical Animal Resources

32. **The Water Buffalo: New Prospects for an Underutilized Animal.** 1981, 188pp. The water buffalo is performing notably well in recent trials in such unexpected places as the United States, Australia, and

Brazil. Report discusses the animals's promise, particularly emphasizing its potential for use outside Asia. ISBN 0-309-04159-7.

44. **Butterfly Farming in Papua New Guinea.** 1983, 36pp. Indigenous butterflies are being reared in Papua New Guinea villages in a formal government program that both provides a cash income in remote rural areas and contributes to the conservation of wildlife and tropical forests. ISBN 0-309-04168-6

45. **Crocodiles as a Resource for the Tropics.** 1983, 60pp. In most parts of the tropics, crocodilian populations are being decimated but programs in Papua New Guinea and a few other countries demonstrate that, with care, the animals can be raised for profit while protecting the wild populations. ISBN 0-309-04169-4.

46. **Little-Known Asian Animals with a Promising Economic Future.** 1983, 133pp. Describes banteng, madura, mithan, yak, kouprey, babirusa, javan warty pig, and other obscure but possibly globally useful wild and domesticated animals that are indigenous to Asia.
ISBN 0-309-04170-8.

68. **Microlivestock: Little-Known Small Animals with a Promising Economic Future.** 1990, 460pp. Discusses the promise of small breeds and species of livestock for Third World villages. Identifies more than 40 species, including miniature breeds of cattle, sheep, goats, and pigs; eight types of poultry; rabbits; guinea pigs and other rodents; dwarf deer and antelope; iguanas; and bees.
ISBN 0-309-04265-8.

## Health

49. **Opportunities for the Control of Dracunculiasis.** 1983, 65pp. Dracunculiasis is a parasitic disease that temporarily disables many people in remote, rural areas in Africa, India, and the Middle East. Contains the findings and recommendations of distinguished scientists who were brought together to discuss dracunculiasis as an international health problem. ISBN 0-309-04172-4.

55. **Manpower Needs and Career Opportunities in the Field Aspects of Vector Biology.** 1983, 53pp. Recommends ways to develop and train the manpower necessary to ensure that experts will be available in the future to understand the complex ecological relationships of vectors with human hosts and pathogens that cause such diseases as malaria, dengue fever, filariasis, and schistosomiasis. ISBN 0-309-04252-6.

60. **U.S. Capacity to Address Tropical Infectious Diseases.** 1987, 225pp. Addresses U.S. manpower and institutional capabilities in both the public and private sectors to address tropical infectious disease problems. ISBN 0-309-04259-3.

## Resource Management

50. **Environmental Change in the West African Sahel.** 1984, 96pp. Identifies measures to help restore critical ecological processes and thereby increase sustainable production in dryland farming, irrigated agriculture, forestry and fuelwood, and animal husbandry. Provides baseline information for the formulation of environmentally sound projects. ISBN 0-309-04173-2.

51. **Agroforestry in the West African Sahel.** 1984, 86pp. Provides development planners with information regarding traditional agroforestry systems — their relevance to the modern Sahel, their design, social and institutional considerations, problems encountered in the practice of agroforestry, and criteria for the selection of appropriate plant species to be used. ISBN 0-309-04174-0.

69. **The Improvement of Tropical and Subtropical Rangelands.** 1989. This report characterizes tropical and subtropical rangelands, describes social adaptation to these rangelands, discusses the impact of socioeconomic and political change upon the management of range resources, and explores culturally and ecologically sound approaches to rangeland rehabilitation. Selected case studies are included.
ISBN 0-309-04261-5.

## General

65. **Science and Technology for Development: Prospects Entering the Twenty-First Century.** 1988. 79 pp. This report commemorates the twenty-fifth anniversary of the U.S. Agency for International Development. The symposium on which this report is based provided an excellent opportunity to describe and assess the contribution of science and technology to the development of Third World countries and to focus attention on what science and technology are likely to accomplish in the decade to come.

## Forthcoming Books from BOSTID

Applications of Biotechnology to Tradtional Fermented Foods (1991)
Neem (1991)
Vetiver: The Hedge Against Erosion (1991)
Forestry Research in the Tropics (1991)

## BOSTID Publication Distributors

### United States:

*Agribookstore*
1611 N. Kent Street
Arlington, VA 22209

*agAccess*
PO Box 2008
Davis, CA 95617

### Europe:

*I.T. Publications*
103–105 Southhampton Row
London WC1B 4HH
United Kingdom

S. Toeche-Mittler
*TRIOPS* Department
Hindenburgstr. 33
6100 Darmstadt
Germany

*T.O.O.L. Publications*
Sarphatistraat 650
1018 AV Amsterdam
Netherlands

### Asia:

*Asian Institute of Technology*
Library & Regional
Documentation Center
PO Box 2754
Bangkok 10501
Thailand

*National Bookstore*
Sales Manager
PO Box 1934
Manila
Philippines

*University of Malaya Coop. Bookshop Ltd.*
Universiti of Malaya
Main Library Building
59200 Kuala Lumpur
Malaysia

*Researchco Periodicals*
1865 Street No. 139
Tri Nagar
Delhi 110 035
India

*China Natl Publications Import & Export Corp.*
PO Box 88F
Beijing
China

### South America:

*Enlace Ltda.*
Carrera 6a. No. 51–21
Apartado Aero 34270
Bogotá, D.E.
Colombia

### Africa:

*TAECON*
c/o Agricultural Engineering Dept.
P.O. Box 170 U S T
Kumasi
Ghana

### Australasia:

*Tree Crops Centre*
P.O. Box 27
Subiaco, WA 6008
Australia

## For More Information

To receive more information about BOSTID reports and programs, please fill in the attached coupon and mail it to:

Board on Science and Technology for International Development
Publications and Information Services (HA-476E)
Office of International Affairs
National Research Council
2101 Constitution Avenue, N.W.
Washington, D.C. 20418 USA

Your comments about the value of these reports are also welcome.

---

Name ______________________________
Title ______________________________
Institution ______________________________
Street Address ______________________________
______________________________
City ______________________________
Country ______________ Postal Code ______________

68

---

Name ______________________________
Title ______________________________
Institution ______________________________
Street Address ______________________________
______________________________
City ______________________________
Country ______________ Postal Code ______________

68